**A Cultural History of Heredity**

# A Cultural History of Heredity

## Staffan Müller-Wille and Hans-Jörg Rheinberger

The University of Chicago Press :: Chicago and London

Staffan Müller-Wille is a senior lecturer and research associate with the
ESRC Centre for Genomics in Society and the Centre for Medical History,
both at the University of Exeter.
Hans-Jörg Rheinberger is director at the Max Planck Institute for the
History of Science in Berlin. They are the editors of *Heredity Produced:
At the Crossroad of Biology, Politics, and Culture, 1500–1870.*

The University of Chicago Press, Chicago 60637
The University of Chicago Press, Ltd., London
© 2012 by The University of Chicago
All rights reserved. Published 2012.
Printed in the United States of America
21 20 19 18 17 16 15 14 13          2 3 4 5

ISBN-13: 978-0-226-54570-7 (cloth)
ISBN-10: 0-226-54570-9 (cloth)

Originally published as *Vererbung: Geschichte und Kultur eines
biologischen Konzepts*
© Fischer Taschenbuchverlag, Germany, 2009

Library of Congress Cataloging-in-Publication Data
Müller-Wille, Staffan, 1964–
    [Vererbung. English]
    A cultural history of heredity / Staffan Müller-Wille and Hans-Jörg
Rheinberger.
        pages ; cm
    Translation of: Vererbung: Geschichte und Kultur eines biologischen
Konzepts.
    Includes bibliographical references and index.
    ISBN-13: 978-0-226-54570-7 (cloth : alkaline paper)
    ISBN-10: 0-226-54570-9 (cloth : alkaline paper)   1. Heredity—
History. 2. Genetics—History.   I. Rheinberger, Hans-Jörg. II. Title.
    QH438.5.M8513 2012
    576.5—dc23                                        2011050826

♾ This paper meets the requirements of ANSI/NISO Z39.48-1992
(Permanence of Paper).

At the first glance, the only "law" under which the greater mass of the facts the author has brought together can be grouped seems to be that of Caprice,— caprice in inheriting, caprice in transmitting, caprice everywhere, in turn.

WILLIAM JAMES, review of Charles Darwin's *The Variation of Animals and Plants under Domestication* (1868)

# Contents

Preface     ix

1    Heredity: Knowledge and Power     1
2    Generation, Reproduction, Evolution     15
3    Heredity in Separate Domains     41
4    First Syntheses     71
5    Heredity, Race, and Eugenics     95
6    Disciplining Heredity     127
7    Heredity and Molecular Biology     161
8    Gene Technology, Genomics, Postgenomics:
      Attempt at an Outlook     187

Notes     219
Bibliography     263
Index     303

## Preface

This book is about the genesis and historical evolution of a biological concept: the concept of heredity and the phenomena it refers to. Heredity is the subject of a whole array of scientific disciplines, most notably genetics. Yet we want this book to be more than, and different from, the history of a scientific discipline and its fundamental ideas. A number of excellent books on this subject already exist, most of them written by geneticists who approached the history of their discipline in a manner that must be recognized as having been far from naïve.[1] François Jacob's *The Logic of Life* is likely to remain an unsurpassed example of this genre. In it, the molecular biologist and Nobel laureate identifies the "stages" (*étapes*) that knowledge of heredity and reproduction passed through from the sixteenth to the twentieth centuries. Inspired by the French school of historical epistemology, Jacob did not consider these stages as stations on "the royal road of ideas, retracing the confident march of progress towards what now appears to be a solution." Yet even Jacob's "history of heredity" (this is the subtitle of his book) possesses an unmistakable vanishing point, aiming, as it were, at a concept that was established only by modern molecular genetics: the genetic program. Taking this notion for granted, Jacob analyzes not only life itself, but also the history of its conceptualization as a succession of ever higher, or rather, ever more deeply differentiated, structures, as a "series of organizations fitted into one another like nests of boxes or Russian dolls": visible structures, organs, cells, chromosomes, genes, and molecules, to mention only the most important ones.[2]

We will go beyond such diachronic history by describing the synchronic political, medical, and technological contexts in which knowledge of heredity was both generated and applied. What we will study is therefore less the history of a particular science or research program than the history of a "knowledge regime," which had its starting points in separate and heterogeneous contexts, coalesced along different paths, and then influenced other cultural domains in multiple ways.[3] Along these lines, we do not see science as being different from, or even opposed to, culture. Rather, we regard it as one cultural domain alongside others, such as the arts, technology, or politics, all of which interact in the reproduction and transformation of social values. This does not amount to a denial of the very special nature of science. Our discussion will remain focused on the production of scientific knowledge as a cultural domain sui generis. Scientific values—"truth values," as philosophers of science would call them—possess their own peculiar nature, and their production tends to follow its own historical trajectories. To lose sight of the special character of science would risk losing sight of what makes it distinct and underlies its cultural power: the authority it claims to possess with regard to questions about nature, human nature in particular. We therefore support Henning Schmidgen in his reading of Georges Canguilhem's philosophy of science: "The meaning of a scientific concept does not simply reveal itself horizontally, with respect to other words and texts. It can be grasped only by recourse to the forces that now or in the past have taken hold of an 'object' (like health, heredity, or the mind)."[4]

This focus on cultural forces has consequences for our presentation of the history of heredity. In the first chapter, we outline our analytic framework by exploring some central metaphors that Francis Galton used in the latter half of the nineteenth century to defend his view that heredity was a subject in need of scientific scrutiny. The second chapter then takes a step back and tackles the question of why it was so late that heredity became a subject of biological speculation. At the same time, this chapter traces changes in the ways in which the generation and reproduction of organic beings was conceptualized in the seventeenth, eighteenth, and early nineteenth centuries, thus providing an important contextual background for the account of the history of heredity that we unfold in the subsequent six chapters.

This account will take the shape of an hourglass.[5] In the early modern period, as we explain in chapter 3, thinking in terms of heredity started to take hold in a broad *epistemic space* that incorporated elements from different contexts, such as medicine, breeding, and trade. This development created a set of coordinates that made it possible to conceive of reproduction no longer only as personalized and individual generation of offspring, but also as the transmission and redistribution of a more or less atomized biological substance. Heredity thus not only moved into the center of the life sciences in the course of the nineteenth century, as we will show in chapter 4; it also assumed a powerful biopolitical dimension that we will explore in chapter 5. This conjunction created a research dynamic in the late nineteenth and early twentieth centuries by which the idea of heredity was condensed into *epistemic things*—most prominently, genes—that could be traced and manipulated experimentally. Chapters 6 and 7 survey the research technologies, as well as their political and economic contexts, associated with classical and molecular genetics, respectively—the two disciplines that have defined the twentieth century as the "century of the gene."[6]

Finally, in chapter 8, we discuss how the reification of heredity is beginning to be reversed again in the current conception of the genome and in the strategies developed to study its interactions with the body. A spatialization of the concept seems to take place once more, this time, however, not in relation to disparate domains of practice brought together, but rather from within. Nowadays the epistemic space of heredity manifests itself in the increasingly complex internal structures and architectures of the cellular machinery serving the transmission and expression of traits. Both heredity and the gene may very well have seen their day as research objects at the forefront of bioscientific inquiry in this context, receding into the background of unquestioned assumptions. Under opposite signs, heredity has become the malleable matter of politics, familial affairs, and the marketplace once more. What used to be the bedrock and very substance of cultural traditions was for a while reduced to universal laws of nature, only to emerge as a repository for shaping the future of humanity.

In writing the history outlined in the preceding paragraphs, we have taken care to distinguish between representational regimes on the one hand and associated epistemic objects on the other. Time and again, we will use the notion of an "epistemic space"—that is, the broader realm in which a scientific concept takes shape—or refer to "epistemic things"—that is, the objects of a particular research technology or experimental system. Although such objects are again and again represented in texts, images, collections, and experimental systems, they nevertheless elude full and final representation. If this were not the case, they would not be able to stimulate research, an activity that is always oriented toward what is not yet known.[7] We believe that it is important to distinguish between epistemic objects and their discursive and institutional representation in what we called "knowledge regimes" above. It is, after all, from the gap between the two—the fact that an epistemic object can never be unambiguously defined and represented without receding into the background of universally recognized assumptions, standards, and technologies, and thus in fact ceasing to be scientifically interesting—that the history of a concept like heredity gains its fascinating dynamics.

Like every book, this one has its history. It presents our synthesis of the results of a long-term research project carried out at the Max Planck Institute for History of Science in Berlin and the ESRC Research Centre for Genomics in Society at the University of Exeter. This collaborative project was dedicated to the "cultural history of heredity" and was additionally supported by the Government of Liechtenstein (Karl Schaedler Grant), the German Academic Exchange Service, the British Council, the British Academy, the Arts and Humanities Research Council, and the Wellcome Trust. Five international conferences and a number of smaller meetings allowed us to meet with and draw on the expertise of colleagues from all over the world and from a great variety of disciplinary backgrounds.[8]

A large number of colleagues and friends have had their input in the making of this book as a consequence. Special thanks go to Peter McLaughlin and Wolfgang Lefèvre for their role in the initial shaping of the project. For inspiration, discussion, and advice we would further like to thank, in no particular order: Jean Gayon, Djatou Medard, Carlos López Beltrán, Edna Suarez, Ana Barahona, Soraya de Chadarevian, Nick Hopwood, Henning Schmidgen, Dan Kevles, Evelyn Fox Keller, Jonathan Hodge, Pascal Germann, Andrew Mendelsohn, Philipp Sarasin, Mary Terrall, Shirley Roe, Bernardino Fantini, Michel Morange, Renato Mazzolini, John Dupré, Barry Barnes, Diane Paul, Paul Weindling, Luc Berlivet, Fred Churchill, Gar Allen, Eva Jablonka, Bernd Gausemeier, Paul Griffiths, Christina Brandt, John Waller, John van Wyhe, Gianna Pomata, Vincent Ramillon, Phil Wilson, Rafi Falk, Jean-Paul Gaudillière, Ilana Loewy, Edmund Ramsden, Volker Roelcke, Helmut Müller-Sievers, François Duchesneau, Marc Ratcliff, Peter Beurton, Marsha Richmond, Ken Waters, Lenny Moss, Roger Wood, Jenny Bangham, Marianne Sommer, Adam Bostanci, Stefan Borchers, Jennifer Marie, Vítězslav Orel, Stephen Pemberton, Brad D. Hume, Uwe Hoßfeld,

Michael Hagner, Michal Šimůnek, Ohad Parnes, Pietro Corsi, Theodore Porter, Francesco Cassata, Yoshio Nukaga, Sabina Leonelli, Laure Cartron, Judith Friedman, Bob Olby, Nathaniel Comfort, Didier Debaise, David Sabean, Ida Stamhuis, Bengt Olle Bengtsson, Ulrike Vedder, Jean-Christophe Coffin, Susan Lindee, Miguel García-Sancho, Christophe Bonneuil, Anne Cottebrune, Alexander von Schwerin, Christine Hauskeller, Marcel Weber, Jonathan Harwood, Paul White, Sigrid Weigel, Nils Roll-Hansen, Jeffrey Schwarz, Manfred Laubichler, Massimo Mazzotti, María Jesús Santesmases, Helga Satzinger, Snait Gissis, Stefan Willer, Stephen Snelders, Volker Hess, Peter Harper, Jan Witkowski, Jim Secord, Igal Dotan, Maureen O'Malley, Gregory Radick, Bert Theunissen, Maria Kronfeldner, Keith Benson, Maaike van der Lugt, Dick Burian, Charles de Miramon, Judy Jones Schloegel, Angela Creager, Norton Wise, Veronika Lipphardt, Luis Campos, Jane Calvert, and Yossi Ziegler. It remains to thank those who have helped materially in bringing about this book. The English text is based on a German version published with S. Fischer Verlag, and we would like to thank this publisher for generously waiving copyright for translations into other languages. We are also grateful to Karen Merikangas Darling, who supported its publication with unusual enthusiasm at the University of Chicago Press and saw the project through its various stages with diligence and foresight; to Adam Bostanci for the rigor with which he transformed a rather Germanic first version of our manuscript into readable English; to Norma Sims Roche for careful copyediting; and to Daniela Kelm and Hannes Bernhagen for checking and tracing countless quotations and references. Very special thanks go to Antje Radeck and Birgitta von Mallinckrodt.

# 1

## Heredity: Knowledge and Power

Genealogy is the oldest kind of logic. Humans have always made use of kinship and descent in mythology, in philosophy, and in the sciences to describe the constitution of the world. Key concepts of ancient logic such as genus and species possess genealogical connotations, and the relationships among these concepts were modeled on relations of parentage. According to Porphyry's *Isagoge*, an introduction to Aristotle's *Categories* from late antiquity, a species relates to its genus in the same way "as Agamemnon is an Atreid [i.e., one of the sons of Atreus] and a Pelopid and a Tantalid and, finally, of Zeus."[1] Lines of descent reflect a degree of continuity, so that something can be learned about the properties of a present-day individual by following the chain of his or her ancestors. Descent forms a common fate, which acts sometimes like a blessing and sometimes like a curse under which—as the philosopher of religion Klaus Heinrich put it—"nobody *can* but repeat the configurations (because they give him substance) that already were the configurations of the actions of his parents."[2]

This way of thinking—this "mental disposition toward origin myths" (*ursprungsmythische Geisteslage*), as Heinrich characterized it[3]—is alive and well today. In 2002, the publishing house of the Swedish daily newspaper *Dagens Nyheter* issued a biography of Karl Vilhelm

Ossian Dahlgren, professor of botany at the University of Uppsala in the 1920s and 1930s. Dahlgren was a prominent supporter of eugenic and race-biological policies. He even directed his convictions against his own children, who took after his small and black-haired wife. In all seriousness, he advised them to refrain from having children. His advice was obviously ignored, for it was his own granddaughter, the journalist Eva F. Dahlgren, who became his biographer. About the reasons that motivated her to tell the story of his life, she said in an interview that the thought of her grandfather having been a race biologist had caused her to feel ashamed, even if she did not share his convictions. Yet she added, "I also have my grandfather's irregular teeth, and who knows what else went into the heritage?"[4] Such a remark may appear anachronistic from a strictly genetic point of view, but in a certain sense Dahlgren is right. In her book, she recalls how much her grandfather impressed her when she was a child. To describe family influences as "heritage," and to allow for connotations that derive from modern genetics, is perfectly legitimate in everyday and literary contexts. Genetic images lend themselves to fabricating life histories of varying degrees of sophistication.

Faced with the ubiquity of such figures of thought, many will be surprised to learn that a strictly *naturalistic* concept of heredity—according to which that which is transmitted from generation to generation through the germ line counts as the "first," strictly biological, nature of a human being—is a relatively recent product of science. Until well into the nineteenth century, a medical tradition dating back to antiquity recognized six "non-natural things" (*res non naturales*) as well as seven "natural things" (*res naturales*) that together determined the constitution and character of individuals. The non-natural things included things that could be influenced, like nutrition, environment, rest, exercise, bodily excretions and psychological states, whereas the natural things encompassed components of the body (like spirits, elements, humours, and organs) as well as aspects resulting from the overall organization of these components (like functions, capacities, and temperament).[5] The "natural things," moreover, were not linked to a hereditary substance, but were rather thought to result from the direct, physical creation of the embryo by both parents, or as some believed, by one of its parents only. Until the end of the eighteenth century, naturalists and physicians did not talk about biological "inheritance," a substance transmitted independently from the processes and circumstances of conception and development.[6] And it took until the middle of the nineteenth century for "heredity" in this sense to emerge as one of the central problems of biology, especially in the works of Charles Darwin and of his cousin Francis

Galton. With genetics and molecular biology in the twentieth century, heredity became a clearly delineated phenomenon that lent itself to experimental and mathematical treatment. Its existence was now beyond dispute, even if questions remained about its underlying mechanisms and how these could be exploited technologically to further analyze the phenomena of heredity.

Relatively precise dates can thus be given for the historical origins of the concept of heredity. In addition, the concept went through an initial phase of thorough problematization, followed by a phase of solidification into a measurable and manipulable research object. The roots of the knowledge regime from which the concept of heredity emerged extend, as we will see, to the early modern period, and in some instances even back to the early fourteenth century. This was the period in which Europeans began to conquer the globe, discovering at the same time that they were not its sole inhabitants; the period in which Europeans realized that progress and innovation could lead to the destruction of traditions, institutions, and even of whole cultures; the period in which Europeans gradually came to the conclusion not only that nature determines history, but also that nature itself has a history. As a consequence, it appeared possible to make history by dominating nature. These conclusions were reached in the course of interactions with other peoples and natural environments, in the context of a movement that was not simply an expansion into empty space, but which presupposed and initiated massive translocations and transplantations, both of humans and of the plants and animals they cultivated.[7] The knowledge regime of heredity—this is the central thesis of our book—started to unfold as people, goods, and the relationships they mediated began to move and change on a global scale.[8]

Thinking about heredity as originating from a knowledge regime that was oriented toward the mobilization, rather than the fixation, of the conditions of life has two implications. First, we cannot assume that heredity was a concept that had a definite meaning from the beginning. We are all familiar with ideas of inheritance or heritage that, above all, aim to capture a persistence of forms over time and are often employed to justify political authority and cultural traditions. In this sense, heritage becomes something worth upholding for its own sake. Ideological reifications of this kind are, of course, questionable.[9] But in scrutinizing them, one should not overlook the fact that the political dimensions of concepts of heredity can be and have been the result of very different classificatory and causal assumptions and that such concepts were used for widely different ideological purposes. For example, it makes a difference whether the inheritance of individual properties and the inheritance of species-typical properties are thought of as distinct processes or whether both are analyzed in a common framework. It similarly matters whether heredity is considered to be a reversible process to some degree or whether it is thought of as an irreversible series of distinct events. We will see that opinions with respect to these questions varied widely over time. The contours of consensus and of a scientific discipline of heredity emerged only gradually with certain conceptual and methodological suppositions, and have never been beyond contestation, as we experience now in the so-called postgenomic era.[10]

Second, our approach implies that these variations and changes in thinking about heredity are best understood in the context of the institutions and discursive practices that constituted the knowledge regime of heredity. The distinction that eighteenth-century naturalists began to draw between "constant" characters and characters that vary with climate and other environmental conditions, for example, was connected with attempts to acclimatize exotic plants and animals to the temperate regions of Europe. When naturalists sought to distinguish between these types of characters, especially in institutions such as botanical gardens, they soon discovered "constant varieties"; that is, varieties that always transmitted their often trifling differences, even under changing environmental conditions. In the history of investigations into hereditary phenomena, we encounter this kind of development repeatedly. Typically, when a generally accepted dichotomy such as constant versus varying character was subject to scrutiny, phenomena that called for alternative descriptions and explanations were brought to the fore. Once the taxonomies and causalities subsumed under the concept of heredity were put into practice and institutionalized, they furnished the very conditions for the recognition of evidence to the contrary, which in turn called for the resolution of ensuing conceptual challenges.

Before we begin to unravel this complicated history, we would like to expand these two points to sketch out our view of heredity for the benefit of the reader. The work of Francis Galton—whom some historians consider the modern founder of research into heredity[11]—provides a suitable starting point. Galton was indeed one of the first biologists who placed heredity at the center of their theorizing, and he employed a particularly rich arsenal of analogies and figures of thought in his attempts to make sense of heredity. As we will see in the following pages, some of these analogies and figures of thought are well suited for describing the dynamic character of the knowledge regime of heredity.

## Galton's Post Office: Heredity as an Epistemic Space

When we mentioned earlier that people did not speak about "heredity" in a biological sense until the end of the eighteenth century, we were in the first place making an observation about the use of language. Phenomena that we would nowadays explain in terms of heredity as a matter of course—for example, the fact that certain physical or psychological peculiarities run in families, or the curious phenomenon that children sometimes show similarities to their grandparents rather than their parents—were already recognized in antiquity.[12] The now dominant biological sense of the term "heredity," however, resulted from a metaphorical transfer of a juridical concept to a description of the generation and propagation of living beings. Generally speaking, such a transfer simply did not take place before the eighteenth century. Such historical facts about terminological use evoke the curiosity of historians of science. After all, analogies between juridical and biological inheritance seem to offer themselves quite readily. The inheritance of property is often connected to kinship, and thus—from today's perspective at least—to a biological relationship. Why then should people not have talked about the transmission of physical traits in terms of "inheritance" in the past?

In the original juridical sense, heredity, just like inheritance, refers to the distribution of status, property, and other goods in temporal succession and according to a system of rules and distinctions that regulate how these goods are passed on to other persons upon the death of a proprietor. The distinctions usually discriminate between groups of various degrees of kinship, and the rules determine under which conditions these groups can claim a certain portion of the inheritance. Thus in many regions of early modern Europe, inheritance rules were laid down according to which property would always be passed on to the firstborn—an institution known as majorat or primogeniture. Today many countries possess laws that guarantee at least a portion of an inheritance to all close kin, independently of gender or birth order. The historical and cultural diversity of such regulations is enormous. The rather trivial but essential point we want to make here is that they always involve classifications of kin and rules of distribution.

This basic consideration already puts us in a position to identify an important precondition that had to be met before a transfer of ideas of inheritance from the sphere of law to that of biological reproduction could take place. The reproduction of organisms first had to appear as a process in which something was transmitted and redistributed in a regular fashion. And it is precisely this perspective that is missing in the works of great premodern and early modern natural philosophers such as Aristotle, William Harvey, or René Descartes. As we will see in the next chapter, their observations and speculations focused instead on the procreative act of generating an individual. To mark out the contours of a system to which the metaphor of inheritance, so rich in connotations, could be applied in a productive and useful way, one had to go beyond this point of view; it was necessary to establish connections between individual acts of procreation. Understanding propagation as a kind of inheritance, we can already say at this point, presupposed an identification of factors at work in propagation that persisted beyond the transitory moment of creating an individual being.

Francis Galton made this point repeatedly in his 1876 essay "A Theory of Heredity." Heredity, he claimed, could be explained only if one presupposed an organic system that endured across, and thus supported, successive generations. Galton called this system the *stirp* (from the Latin *stirps*, "descendant" or "stock," also used in the sense of "family branch," "son," and "heir"). According to Galton, the *stirp* was "the sum-total of the germs, gemmules, or whatever they may be called, which are to be found, according to every theory of organic units, in the newly fertilised ovum."[13] To illustrate the structure of the *stirp*, Galton resorted to analogies that appear curious, even awkward. Among other things, he compared heredity to a post office:

> Ova and their contents are, to biologists looking at them through microscopes, much what mail-bags and the heaps of letters poured out of them are to those who gaze through the glass windows of a post office. Such persons may draw various valuable conclusions as to the postal communications generally, but they cannot read a single word of what the letters contain. All that we may learn concerning the constituents of the stirp must be through inference, and not by direct observation; we are therefore forced to theorise.[14]

Galton stressed that such comparisons were not meant to be "idle metaphors, but strict analogies," analogies worth "being pursued, as they give a much-needed clearness to views on heredity."[15] A closer look at his comparison to a post office reveals what Galton had in mind when he called for concrete models of heredity. The comparison operates on two levels. On a first level, by referring to a communication technology that was developing rapidly at the time, Galton described heredity as a complex arrangement of spatial elements. The mail bags stand for the cyto-logical space of the fertilized egg, "a space not exceeding the size of the head of a pin";[16] the letters for the elements of the *stirp*, the "germs and gemmules" that fill this cytological space but at the same time inhabit the much wider space of "postal communications" among ancestors and descendants. Put differently, heredity is defined by a structure, the *stirp*, that comprises both genealogical and cytological relationships inasmuch as the former must somehow be represented in the latter. Galton ex-pressed this idea in the following terms: "everything that reached [a per-son] from his ancestors must have been packed in his own stirp."[17]

On a second level, Galton's comparison to a post office illustrates how heredity becomes a subject for scientific research. The post office building, with its windows, represents the various tools and technolo-gies that scientists employ to make sense of the phenomena of heredity. Microscopes alone, as Galton admitted, were certainly not enough. The description of heredity as a complex system of postal communications observed through the windows of a post office suggests that an equally complex configuration of technologies needs to be in place before one can discern the contours of heredity. To mention only the most obvious of these technologies, reliable means for recording genealogical data are necessary to observe the genealogical dimension of heredity. The com-parison of this assemblage of technologies to a post office is significant in two further respects. First, it contains heredity within its own phe-nomenal space, separating it from other biological phenomena such as development or nutrition. Second, it makes the relationships among the elements of heredity observable rather than demanding direct character-ization of these elements themselves. Owing to the low resolution power of the instruments at his disposal, the observer may not be able to read the "letters," but he or she still remains in a position to say something about "postal communications in general."

As characterized by Galton, heredity can therefore be seen as an "epistemic space." This term underscores what is peculiar about hered-ity in comparison with other subjects of biological research, which are investigated as "epistemic objects" within particular experimental set-tings.[18] Heredity, in Galton's time, constituted a domain of research to be mapped out by means of taxonomies and regularities, rather than an object of research that was to be identified by determining its proper-ties and functions. It is our aim in this book to reconstruct the various historical developments and cultural domains that contributed to the emergence, transformation, and eventual compaction of this epistemic space. Following Michel Foucault's approach in *The Order of Things*,

François Jacob described the shift from thinking in terms of generation to thinking in terms of heredity as a succession of different paradigms separated by sharp epistemological breaks. Rather than emphasizing their differences, we will focus on the transitions between these ways of thinking. As we will see, the formation of the discourse of heredity depended on a whole assemblage of locally circumscribed and highly disparate cultural contexts. It was only in the long run that more globally connected configurations of technologies and institutions, such as botanical gardens, hospitals, genealogical and statistical archives, breeding stations and associations, and chemical and physiological laboratories, began to form.

It was not right from the start, with the clear aim of elucidating heredity in mind, that connections were established among the aforementioned institutions through exchange of materials and experiences. Ultimately, the many historical connections that formed the knowledge regime of heredity have to be seen against the background of the formation of nation-states, with their centralized bureaucracies, capitalist economies, and imperialist aspirations. But neither was heredity simply a construction invented after the fact for the ideological justification of a new socioeconomic order. Rather, heredity appeared as a new subject for scientific inquiry *within* this emergent world order. As we probably are only beginning to realize today—in times when genetic screening, testing, and patenting pervade all sectors of social and economic life and with the synthetic powers of genomics on the horizon—the epistemic space that heredity eventually came to constitute was itself intimately entwined with the reconfiguration of social life that went along with long-term historical processes like industrialization, urbanization, and globalization.[19]

## Galton's Parliament: Heredity as a Biopolitical Notion

Galton was not only one of the first biologists who tried to lay the foundations of a general theory of heredity; he was also one of the first to suggest that this theory could and should be applied to the management of human affairs. In his *Inquiries into Human Faculty and Its Development* (1883), he proposed to erect a "science of improving stock" or "of the cultivation of race" on the basis of heredity theory, and he suggested that this new science be called *eugenics* (earlier, as he reported in a footnote, he had considered the term *viriculture*, literally the "cultivation of men").[20] Eugenic doctrines proliferated in many countries in the late nineteenth and early twentieth centuries, accompanied by a flood of journals and popular pamphlets, by the establishment of national and international associations calling for the application of these doctrines, and finally by the creation of legal frameworks by which individual states, especially in northern Europe and North America, endorsed these doctrines through measures like enforced sterilization. The violent culmination of eugenics, especially in the euthanasic and genocidal policies of Nazi Germany, has sometimes distracted historians' attention from the fact that eugenics was an extraordinarily broad movement. Eugenic arguments also featured in working-class movements, in calls for the creation of public health care systems, in the fight for the emancipation of women, and in the struggle of national and ethnic minorities for greater autonomy.

Galton's essay "A Theory of Heredity" also offers a revealing metaphor for the political dimensions of the concept of heredity. Galton's *stirp* was not a static structure, but supposed to consist of elements— the "germs and gemmules" containing hereditary predispositions—that were engaged in unsettled dynamic relationships. Galton thought that this permanent struggle at the microscopic level of life was best compared to "the events of political life, such as those connected with the struggle for place and power, with election, and with representation." "We know," he continued,

> that the primary cells divide and subdivide, and we may justly compare each successive segmentation to the division of a political assemblage into parties, having, thenceforward, different attributes. We may compare the stirp to a nation and those among its germs that achieve development, to the foremost men of that nation who succeed in becoming its representatives; lastly we may compare the characteristics of the person whose bodily structure consists of the developed germs, to those of the house of representatives of the nation.[21]

From this comparison to political events, heredity emerges as a fundamentally contingent process. Individual germs, Galton conceded, are propagated "true to their kind," but which of the many germs contained in the *stirp* get a chance to develop depends entirely on their position in the overall framework of the *stirp* and on their ability to prevail over other germs in the resulting struggle of conflicting forces within this framework. These ideas allowed Galton to explain some of the more bewildering facets of heredity; for example, that siblings sometimes differ quite considerably from one another, even though one had to assume that they developed from the same common *stirp*. "A strict analogy and explanation of all this," Galton maintained,

> is afforded by the well-known conditions and uncertainties of political elections. We have abundant experience that when a constituency is very varied, trifling circumstances are sufficient to change the balance of parties, and therefore, although there may be little real variation in the electoral body, the change in the character of its political choice at successive elections may be abrupt. A uniform constituency will always elect representatives of a uniform type; and this result precisely corresponds to what is found to occur in animals of pure breed, whose stirp contains only one or a very few varieties of each species of germ, and whose offspring always resemble their parents and one another. The more mongrel the breed, the greater is the variety of the offspring.[22]

At the time of his writing, the political analogy chosen by Galton was highly charged. England had gone through a series of political crises, each of which resulted in franchise reform, first extending voting rights to broader sections of the population, then making the sizes of voting districts more comparable, and finally introducing secret ballots (First Reform Act 1832, Second Reform Act 1867, Ballot Act 1872). In terms of Galton's analogy, the extension of franchise implied that the political sovereign increasingly became a "mongrel." As far as we know, Galton himself voted for the Conservatives throughout his life, and he was explicitly skeptical about extending franchise, fearing that this would result in political instability and cultural degeneration.[23] At the same time he opposed aristocratic privileges and supported a liberal individualism according to which only personal talents and accomplishments ought to determine social status.[24] His naturalistic theory of inheritance, the mathematical algorithms he developed for the statistical analysis of heredity, and his proposals for eugenic reform all reflected a political standpoint consistent with the interests of the professional middle class of the Victorian age.[25]

The analogy Galton drew between processes of political representation and the transmission and development of germs reveals more than his personal political standpoint. Like his comparison of heredity to a post office, this analogy operated on two levels. On the one hand, there was the microscopic space of the *stirp*, filled with elementary units of life struggling for the realization of their potentials. On the other hand,

there were macroscopic populations, falling apart into competing parties, nations, or other political factions. And again, these two realms of
life were related. Talents and qualities that came to the fore in the political struggle within populations were somehow prefigured in the *stirp*,
and any changes in the balance of powers in political life would result in
the differential reproduction of parts of the population and thus influence the future composition of the *stirp*. This was the core of Galton's
program of eugenics, which pragmatically aimed to promote the preferential propagation of the most talented parts of a population.

Galton's understanding of heredity thus had little to do with ideas
of an unavoidable fate incorporated, as it were, in flesh and blood. Just
as political relationships could be actively formed—as demonstrated by
the contemporary franchise reforms that Galton opposed—heredity likewise appeared susceptible to active design if one abstracted from individual acts of procreation and focused on the structures and dynamic relationships that governed successive generations. "I conclude that each
generation has enormous power over the natural gifts of those that follow," Galton claimed in his book *Hereditary Genius*, and he added that
"it is a duty we owe to humanity to investigate the range of that power,
and to exercise it in a way that, without being unwise towards ourselves,
shall be most advantageous to future inhabitants of the earth."[26] In his
opinion, aristocratic privileges, celibacy, and philanthropy were institutions that inevitably led to the degeneration of populations, while a
strictly meritocratic and selectionist regime, which encouraged the reproduction of the talented and inhibited the propagation of the weak
and feeble-minded, could only result in the progressive improvement of
populations.

For Galton and his contemporaries, contemplating heredity thus involved little reflection about the past. "The question of man's past was
not what had worried Galton," as Cyrill Darlington, one of the most ob-
stinate twentieth-century epigones of Galtonian eugenics, emphasized in
his introduction to a reprint of *Hereditary Genius* that was published in
1962.[27] This apodictic statement finds illuminating confirmation in Galton's curious reinterpretation of the Christian idea of original sin: "The
sense of original sin would show, according to my theory, not that man
was fallen from a high estate, but that he was rising in moral culture with
more rapidity than the nature of his race could follow."[28] Having fallen
behind their own moral development with respect to their heritable dispositions, humans become embroiled in an open-ended struggle between
different nations or races. The same struggle for space and power reigns
among the germs, which, though hidden in the *stirp*, strive for their

development—a struggle that, according to Galton, can be controlled and given direction only through forms of targeted selection.

Galton's political analogies are noteworthy because they illustrate that questions about biological inheritance were not neutral questions that were deflected, if they were deflected at all, into the realm of politics by the interests of those who tried to answer them. As it was articulated, the concept of heredity itself had a political dimension. The speculations and proposals that the concept gave rise to therefore serve as a paradigmatic example of a complex set of discursive practices that Michel Foucault called a "biopolitical dispositive." In *The Will to Knowledge*, Foucault proposed a historical periodization of forms of governance according to which a new form of power grounded in a "deployment of sexuality" emerged toward the end of the eighteenth century and complemented older forms of governance based on the "deployment of alliance." Foucault characterized the latter as the "system of marriage, of fixation and development of kinship ties, of transmission of names and possessions," which is proper to every society and whose primary function it is "to reproduce the interplay of relations and maintain the law that governs them." The deployment of sexuality, on the other hand, does not obtain its "reason for being . . . in reproducing itself, but in proliferating, innovating, annexing, creating, and penetrating bodies in an increasingly detailed way, and in controlling populations in an increasingly comprehensive way."[29] It forms the basis of a new form of political power, "bio-power" as Foucault called it, a power that operates "at the level of life, the species, the race, and the large-scale phenomena of population."[30] With this new form of governance, a "bio-politics of the population" became feasible, which "brought life and its mechanism into the realm of explicit calculations and made knowledge-power an agent of transformation of human life."[31]

The knowledge regime of heredity is an important part of the biopolitical dispositive because the processes that were subjected to it—as is evident from Galton's analogies of a post office and a parliament that we have analyzed in this chapter—reached beyond the generation of individual beings and included the relationships and forces that affect whole populations and shape their development in decisive ways. We will consider these ideas in more detail in the next chapter. Galton formulated this thought at the end of *Hereditary Genius* in the following way:

> Nature teems with latent life, which man has large powers of
> evoking under the forms and to the extent which he desires. We
> must not permit ourselves to consider each human . . . as some-

thing supernaturally added to the stock of nature, but rather as a
segregation of what already existed, under a new shape, and as a
regular consequence of previous conditions.[32]

Knowledge of heredity confers formative power, Galton implies, and it
does so not simply by lending itself to occasional interventions, but due
to its very structure. This is a fundamental point that we will keep in
mind throughout this book. Distinctions between valuable and dispos-
able forms of life, on which eugenicists and racial theorists based their
proposals for societal reform, have often been described—certainly for
good reason—as ideologically motivated attempts to reify and perpetu-
ate social distinctions. After all, it was usually privileged elites of society
that sought to advance policies of eugenics and racial purification. But
at the same time, knowledge of heredity could also be, and often was,
employed in movements that aimed at breaking up what were perceived
as fossilized social structures and opening up new horizons for political
action. For a critical approach to heredity, it is therefore not enough to
simply ignore the options that this concept implies. Today the question
is no longer whether it is possible to take practical measures based on genetic knowledge; the question is whether we want a society that employs
such measures, and if so, how far we want them to go.

## 2 Generation, Reproduction, Evolution

In contemporary biology, "heredity" refers to the transmission of genetic information from generation to generation. The importance of this process for the reproduction of organic beings seems so obvious that it may be difficult to believe that heredity did not always occupy a central place in reflections on and speculations about the life of organisms. Yet until the mid-eighteenth century, the term "heredity" was absent from theoretical writings on the generation and propagation of living beings. Much the same is true for the expression "reproduction." The French naturalist Georges-Louis Leclerc, Comte de Buffon introduced this term into the life sciences in 1748 when he inquired after the "secret cause" by which nature "enables beings to propagate their kinds."[1] Buffon, director of the Jardin du Roi in Paris and author of a natural history in 36 volumes, borrowed the expression from contemporary theology, in which it referred to the resurrection of the dead on doomsday. In this context, reproduction literally meant the re-creation of bodies that had fallen apart into their components.[2]

What contemporary biology identifies as an essential feature of all living systems—namely, their ability to reproduce themselves more or less identically through nutrition, growth, and propagation—thus seems to have escaped the attention of natural philosophers and physicians prior to the eighteenth century; and this despite the fact that plants and animals, and the human body in particular, have always been privileged objects of scientific curiosity and inquiry. How can we make sense of this apparent neglect? François Jacob put forward a convincing answer in his *Logic of Life*. Before the eighteenth century, he writes,

> living beings did not reproduce. . . . The generation of every plant and every animal was, to some degree, a unique, isolated event, independent of any other creation, rather like the production of a work of art by man.[3]

According to Jacob's interpretation, premodern theories of propagation understood generation essentially as an act of creation. Such creative acts did not have to presuppose a necessarily supernatural *creatio ex nihilo*—one only has to think of parental organisms literally making their offspring. Yet whatever way one wants to understand creation, it is a concept that simply does not leave room for a distinction between hereditary transmission and individual development. Most significantly, the propagation of organisms was not separated from the series of circumstances that accompanied copulation, conception, pregnancy, birth, and even weaning in mammals, and undeniably had a formative influence on the developing child. Development and inheritance were not seen as two distinct, autonomous processes that had to be kept separate in studying the reproduction of organisms. This distinction, so decisive for the rise of modern biology, began to be made only in the nineteenth century or, according to some historians, as late as the beginning of the twentieth century, when genetics established itself as a discipline.[4] And it is indeed not a trivial distinction, because phenomena of hereditary transmission become manifest only once offspring do in fact develop. Heredity, one can say in hindsight, is not something that exists apart from generation and development, but consists in a process that extends through generation and development.

There is something else that we need to add to this account: Contrary to popular belief, the premodern world was not a world of unchanging species, at least not in the sense that like *always* begets like.[5] On the contrary, that world was full of transmutations, monstrous births, and bizarre unions. Aristotle, for instance, maintained that each being is normally generated by a "thing bearing the same name"; a man, for example, is generated by a man, a horse by a horse.[6] But he also conceded the possibility that "two animals different in species produce offspring which differs in species; for instance, a dog differs in species from a lion,

FIGURE 2.1 The so-called Scythian lamb. Woodcut from Claude Duret, *Histoire Admirable des Plantes* (Paris: Nicolas Buon, 1605), p. 330.

and the offspring of a male dog and a female lion is different in species."[7] He even used this principle to explain the origin of exotic species. Like many ancient naturalists, he believed that animals of different species met at water-holes in the African deserts, copulated with each other, and generated new beings.[8] Pliny reported similar stories in his natural history, and the textual traditions of the Arab-Islamic and Latin-Christian Middle Ages added accounts of many hybrid species, such as between camel and panther (the giraffe, hence called *camelopardalis*), between eel and viper, and yes, even between plants and animals, as in the case of the so-called Scythian lamb (fig. 2.1).[9]

Such reports sometimes arose as a consequence of mistranslations, but that was certainly not always the case. They are still found in the encyclopedic natural histories of the Renaissance, such as Conrad Gesner's four-volume *Historia animalium* (1551–1558). And in the eighteenth century, René-Antoine Ferchault de Réaumur created a sensation among learned Europeans when he reported that he had succeeded in crossing a rabbit with a chicken, although he kept silent about the result of this alleged hybridization.[10] Nature's promiscuous joy in combining species seemed to know no bounds. John Locke reported in his *Essay Concerning Human Understanding* (1690) that he had seen with his own eyes "a

Creature, that was the Issue of a Cat and a Rat, and had the plain Marks of both about it."[11] According to Locke, such observations discredited attempts to define species as isolated units of propagation that differed from each other in constant characters—a definition that had just been proposed for the first time by the botanist John Ray in his three-volume *Historia plantarum* (1683), although even Ray admitted to being aware of cases of transmutation that seemed to cause problems for his definition.[12]

It was not only miscegenations that were seen to cause transmutations. Every more or less profound morphological change in the development of an organism—in insects, for example, the metamorphosis from larva to pupa to imago—was understood as an instance of species transmutation. Scholars in the early modern period tended to explain these cases of "equivocal generation" (*generatio aequivoca*), as they were called, via a variety of natural processes, extraordinary contingencies, and supernatural events. Naturally occurring processes of fermentation and putrefaction provided an explanation for the "spontaneous generation" (*generatio spontanea*) of living beings low on the scale of nature from dead matter. Rotting meat, for example, obviously produced maggots.[13] Among the contingent factors that profoundly influenced the development of embryos, the imaginative power of the mother enjoyed particular popularity.[14] Similarities between a putative father and a child were therefore considered by the courts to be rather unreliable evidence in paternity disputes, because it was generally assumed that such similarities could arise from chance events. For example, the mother might have thought about another man while conceiving the child of her rightful husband.[15] Reports of monstrous births, which were interpreted either as products of extraordinary or "preternatural" circumstances or as signs of divine intervention, are legion in the scientific and popular literature of the sixteenth and seventeenth centuries. They attest to an interest in the extraordinary and miraculous that disappeared from the sciences only in the course of the eighteenth century.[16] And in most cases, these monstrosities were regarded not only as deviations from the norm of a species, but as being of an entirely different species. While the range of species remained unchanged—it was generally believed that the number and essence of specific forms that matter could assume was limited and determinate[17]—individual beings were thought to have the capacity to undergo transformations during their lifetime or to generate beings of another kind.

As fanciful as such reports about spontaneous, equivocal, and hybrid generation may appear today, the idea that nature was creative and promiscuous rested neither on naïve credulity nor on an uncontrolled impulse to fabulate. It rather reflected the analogical style of reasoning that governed and propelled knowledge production in the premodern world. The reason for similarities between offspring and their procreators was not sought in a stable and self-sufficient substance that was passed on from generation to generation and determined the properties of individual beings. It was rather to be found in the fact that the generation and development of individual beings usually occurred under similar circumstances and thus usually involved similar materials and causes. Conversely, this view implied that any deviation from the ordinary course of events—such as the *mesalliances* described by Aristotle and Locke—would produce equally deviant results. "All things are governed by law" is the conventional translation of the opening sentence of the Hippocratic tract *De genitura* ("On seed") from the fifth or fourth century BC, which is perhaps the earliest theoretical text on generation.[18] It is instructive, however, to compare this with the Renaissance Latin translation: "Law indeed strengthens everything" (*lex quidem omnia corroborat*), one reads there, and the translator, Girolamo Mercuriale (1530–1606), notes carefully that "law" (νόμος) meant "customs, pasture, region, tribe" (*instituta, pascua, regionem, classem*) in the Greek dialect used in *De genitura*.[19] In *De genitura*, "law" did not refer to universal laws of nature but, more mundanely, to the persistence of local traditions and environments.

In this premodern perspective—which prevailed, as we will see, well into the seventeenth century—specific and individual similarities between ancestors and descendants resulted from similarities in the particular constellations of climatic, economic, political, and social factors, from the persistence of a "fabric in which the warp and the woof correspond respectively to localities and to lineages," as the anthropologist Claude Lévi-Strauss once formulated this idea.[20] The phenomenon of procreation—that like tends to beget like—was thus as trivial as it was precarious. It was trivial insofar as the production of like offspring was stabilized and reinforced by municipal rules and regularities, such as the regular reoccurrence of certain weather conditions with the seasons; and it was precarious insofar as transgressions and disturbances—miscegenations, for example—or unusual meteorological events could always result in the generation of unlike offspring. The view that reproduction was governed by its own laws did not fit into this broader perspective. For this reason Aristotle often modified his saying that "man generates man" and instead adopted the formula that "man and the sun . . . generate man."[21]

Thus conceived, generation remained a singular, local event. Apart from discussing individual instances of procreation, natural philosophers had little to say about phenomena of organic reproduction. Parents literally made their offspring. Premodern texts about the generation and propagation of living beings therefore abounded with metaphors taken from the crafts as well as from the mechanical and alchemical arts. René Descartes, for example, described the formation of the embryo as a procedure in which the seed "is fermented and well concocted by maternal heat, so that its parts are mixed together all the more thoroughly."[22] He explicitly compared this to the brewing of beer, referring to both male and female "seminal materials" as substances "which act on each other as a kind of yeast," and he proffered one consideration only that vaguely pointed toward inheritance; namely, that "the scum formed on beer is able to serve as yeast for another brew."[23]

The lasting influence of thinking in terms of generation or "making offspring" is evinced by a well-known literary parody that dates from the second half of the eighteenth century. Laurence Sterne's novel *The Life and Opinions of Tristram Shandy, Gentleman* (1760) begins by recounting the conception of the unhappy hero of the tale. Tristram Shandy's father had the habit of fulfilling his marital obligations on the eve of the first Sunday each month, always after having wound up the big clock in the hallway. On one of these occasions, as Tristram was just about to be conceived, his wife interrupted the usual routine by exclaiming, "Pray, my dear, have you not forgot to wind up the clock." The bewilderment and distraction this careless question provoked in her husband was the unfortunate start for the miserable life of Tristram Shandy. As he retrospectively reasons at the beginning of the novel,

> I wish either my father or my mother, or indeed both of them, as they were in duty both equally bound to it, had minded what they were about when they begot me; had they duly consider'd how much depended upon what they were then doing;—that not only the production of a rational Being was concern'd in it, but that possibly the happy formation and temperature of his body, perhaps his genius and the very cast of his mind;—and, for aught they knew to the contrary, even the fortunes of his whole house might take their turn from the humours and dispositions which were then uppermost.[24]

In this chapter we are going to outline the doctrines of generation that preceded the discourse of heredity and provided much of the context for understanding heredity in the nineteenth century. We will not be able to cover all debates and figures of thought that have surrounded the concept of generation since antiquity. This has been accomplished in great detail by Erna Lesky for ancient and medieval times and by Jacques Roger for the Renaissance and the Enlightenment.[25] Instead, we focus on some genealogical phrases and expressions that were used by generation theorists to describe the relationship between ancestors and descendants. These descriptions, as we will see, underwent an interesting evolution from the seventeenth to the nineteenth century that already reveals much about the rise of the discourse of heredity.[26] While early modern generation theorists emphasized diachronic relations of direct, linear descent, emphasis shifted toward the synchronic dimension of a common pool of dispositions transmitted from generation to generation with the nineteenth century.

## Generation, Patriarchy, and Autonomy

William Harvey's explanation of blood circulation, as set out in his *Anatomical Disputation Concerning the Movement of the Heart and Blood in Living Creatures* of 1628, is often considered to be the starting point of modern physiology based on experiment and quantification. Much less historical influence is usually attributed to his *Anatomical Exercises on the Generation of Animals*, which appeared more than twenty years later (in 1651) and discussed the beginning and propagation of organic life, especially in animals.[27] This is surprising, given that the concept of "circulation" (Lat. *circulatio*), which at the time referred not only to the movement of blood in the body, but also to the rotation of the skies, to alchemical procedures of distillation, and to generational cycles of organisms, unites both works to such an extent that Harvey has been called a "lifelong thinker" about questions of circulation.[28] With Jacques Roger, one might even contend that Harvey's *Exercises on the Generation of Animals* are of greater historical interest precisely because they failed in explaining the problems raised to the satisfaction of the author.[29]

Like his contemporaries, Harvey did not distinguish between the development of an individual organism and the transmission of properties or dispositions from one generation to the next. For him, explaining generation involved explaining the procreation of individual beings. Questions of inheritance were therefore not what occupied him in his *Exercises on the Generation of Animals*. The embryo was the "work" (*opus*) of its procreators, a product of materials, instruments, and potencies that entered into play from both parental sides during conception.

Accordingly, similarities trivially resulted from similar constellations of causal factors that were involved in the conception, growth, and development of individual organisms. As Harvey explained in the *Exercises*:

> Now, I maintain that the offspring is of a mixed nature, inasmuch as a mixture of both parents appears plainly in it, in the form and lineaments, and each particular part of its body, in its color, mother-marks, disposition to diseases, and other accidents. In mental constitution, also, and its manifestations, such as manners, docility, voice, and gait, a similar temperament is discoverable. For as we say of a certain mixture, that it is composed of elements, because their qualities or virtues, such as heat, cold, dryness, and moisture, are there discovered associated in a certain similar compound body, so, in like manner, the work of the father and mother is to be discerned both in the body and mental character of the offspring, and in all else that follows or accompanies temperament.[30]

Against the backdrop of this concept of procreation as a "mixture" of parental substances and properties, a host of questions arose, all of which are alien to modern biology. First of all, explaining similarities between parents and offspring was perceived to be a multilevel problem. Why offspring share the same specific form with their parents, which causes determine the sex of children, and how similarities with respect to individual features such as hair or eye color originate—these different questions were to be answered on the basis of different principles. Another problem concerned the roles of the sexes in the production of offspring. Did both sexes contribute equally, so that one could speak of male and female seed? Or was it the male, as Aristotle had proposed, that contributed form and efficient cause through seed, while the female provided only passive matter through menstrual blood? Finally, queries arose regarding the temporal relationship between the instant of conception and the differentiation of the embryo. Most generation theories assumed that the embryo was generated during or shortly after the moment of conception. The product of the mixture of male and female generative substances was regarded as a homogeneous mass from which the structure of the embryo surfaced organ by organ, a process referred to as "epigenesis." This assumption harmonized well with a doctrine known as "creatianism," the idea that the soul of each individual being was directly created by God on the occasion of each individual act of procreation (not to be confused with the more general idea of creationism, i.e., that God created and designed the world). Other theories, in contrast, started with the assumption that the embryo was somehow preformed in the male or female seed. This could either be a consequence of the co-agulation of material particles derived from the various organs of the parental body, thus representing the future body *en miniature* (a doctrine known as pangenesis), or a consequence of the budding off of a part of the ensouled parental body or of the parental soul alone—a position known as "traducianism." According to such preformationist theories, the embryo was in a sense already alive before conception and needed only a triggering impulse and an adequate environment for full development. In viviparous animals, for instance, the uterus provided this environment. The dispute between creatianists and traducianists was of great theological significance because traducianists tended to view generation as an event that, to some extent at least, was not in need of divine intervention (even if most traducianists of course believed that God had created the world).[31]

The set of questions just described intersected in a profound problem that had occupied natural philosophers since antiquity, and that provides a good starting point for discussing Harvey's contribution to the debates raging over generation in the seventeenth century. The problem was raised not only in relation to generation, but also in relation to projectile movement and the formation of economic surplus value.[32] In a nutshell, it consisted in the following: In generation—at least in the generation of higher animals—two beings are involved, the male, "which generates in another," and the female, "which generates in itself," to use Aristotle's formulations.[33] However, this implied that something (the male seed in the case of generation) remained active in a new place (the female body) although it was displaced from its original source of motion, just as a projectile continued to move after its detachment from the throwing hand, and just as capital accrued additional value through its mere circulation. To understand why this constituted a profound problem, one has to recall that Harvey and many of his contemporaries, including Descartes, remained committed to an Aristotelian understanding of causation in one respect: they all rejected the idea of action at a distance and were convinced that cause and effect were simultaneous and contiguous events. Causes did not precede their effects, but shared a point of contact, so to speak, with their effects. As Harvey formulated the problem, "The knot therefore remains untied . . . , namely: how the semen of . . . the cock forms a pullet from an egg . . . especially when it is neither present in, nor in contact with, nor added to the egg."[34]

Harvey tried to solve this problem, as Aristotle had before him, by restricting the role of the parents to the formation of the first rudiments of the embryo. The subsequent growth and differentiation of the latter could then be explained by assuming that the embryo, once it had been procreated, was endowed with the capacity to nourish and fashion itself out of the materials provided by the maternal body. In explaining this capacity, Harvey employed a genealogical metaphor that is of great significance for understanding premodern concepts of generation. "What Aristotle says of the generation of the more perfect animals," he maintained,

> is confirmed and made manifest by all that passes in the [chicken] egg, viz.: that all the parts are not formed at once and together, but in succession, one after another; and that there first exists a particular genital particle, in virtue of which, as from a beginning, all the other parts proceed. . . . And this particle is like a son emancipated [*filius emancipatus*], placed independently, a principle existing of itself, from whence the series of members [*membrorum ordo*] is subsequently thrown out, and to which belongs all that is to conduce to the perfection of the future animal.[35]

Harvey returned to this comparison frequently in the *Exercises*. It shows unmistakably that he believed that the formation of the embryo was initially dominated by the direct action of its progenitor, just as a father forms his son by educating him. According to Aristotle—and Harvey was willing to follow him in this—the male seed had no material part in the formation of the embryo. The semen rather formed the embryo out of matter contributed by the female, in a manner analogous to a smith who forms a sword through the skillful application of cold and heat.[36] In this sense, the embryo was preformed during conception by the animated paternal seed, which somehow "knew" how to fashion it out of the passive matter provided by the mother. Once having been thus formed, however, the embryo continued its life independently. It was now endowed with the capacity for self-perfection and produced the different parts of its own body successively, by epigenesis, just as a son provides for himself upon release from paternal authority.[37]

In their speculations on generation, both Harvey and Aristotle exploited the genealogical figure of patrilineal succession, according to which each individual is determined by the immediate relationship to its male progenitor. In this model, the role of the female is reduced to providing the arsenal of materials and instruments employed by the embryo for its own perfection, albeit Harvey, in contrast to Aristotle and in accordance with Galenic tradition, sometimes granted a more active role to the female in generation.[38] The maternal body was a household, so to speak, that became animated only through the growing embryo. This becomes particularly evident when Harvey compares the embryo to an ancestral household deity, the *lar familiaris* of ancient Rome, "whence life proceeds to the body in general, and to each of its parts in particular."[39] The embryo, after conception, enjoys autonomy and leads a life of its own. At one point, Harvey even went so far as to deny any material connection between the developing embryo and the maternal body, maintaining that, in chickens, "the egg . . . is free and unconnected . . . , rolling round within the cavity of the uterus and perfecting itself."[40]

Aristotle held that the starting point for the independent life of the embryo was an organic body—namely, the heart—that resulted from the action of male seed on female menstrual blood. Harvey's embryological studies can be seen as an attempt to characterize this starting point more closely. Through the anatomical study of chicken eggs and pregnant deer, he sought to empirically retrace the "processes nature brings into play" in generation, from the fully developed fetus all the way back to its first beginnings, the immediate product of conception.[41] Rather unsurprisingly, Harvey located this starting point further back than Aristotle. It turned out that it was not a little heart, but what Harvey called the "pulsating point" (*punctum saliens*) from which the independent life of the embryo proceeded: a minuscule pulsating droplet of blood, hardly visible and as yet without any discernable internal structure, but endowed with the potential to produce structure. In its inconspicuousness it occupied the threshold between life and death—"a leaping point, of the colour of blood, so small that at one moment, when it contracts, it almost entirely escapes the eye, and again, when it dilates, it shows like the smallest spark of fire."[42]

Harvey thus radicalized Aristotle's solution of the problem of generation. The embryo had an independent life of its own from its very first, still undeveloped, unstructured, and ultimately invisible beginnings. Ironically, this solution only exacerbated the original problem. It became impossible for Harvey to establish an immediate, physical connection between the embryo and any concrete product of the sexual act, be it the mixture of male and female seed of the Hippocratic-Galenic tradition or be it, in accordance with Aristotle, the female menstruum animated by the semen. In Dietlinde Goltz's words, Harvey faced the "paradox of the empty uterus," which threatened contemporary ideas of a patrilineal chain and of the patriarchal power that the father had

over his wife and their children.[43] As the embryo appeared to lead an independent life from its very beginning, there was nothing that pointed to
an immediate causal and material involvement of both male and female
parent in the production or making of offspring.

To ensure that some kind of effect on the embryo could be granted
to parents, Harvey therefore had to invoke action at a distance after all.
The effects of generative substances, and especially the male seed, he
proposed, could be compared to the effects of a "contagion" in transmitting disease, of a magnet in attracting or repulsing other magnets,
of the stars in influencing sublunar events, and of the imagination in
representing distant objects.[44] Such models, however, denied the immediate, bodily participation of father and mother in the creation of offspring. The male seed became the carrier of a universal "plastic force,"
and the female seed or "egg" an equally universal but material carrier
of a potential for life, growth, and development. Whether generation
was viviparous, oviparous, equivocal (in the case of insects, for instance,
which underwent a metamorphosis), or spontaneous, "all animals,"
Harvey claimed, "are engendered from an oviform primordium; I say
oviform, not as meaning that it has the precise configuration of an egg,
but the nature and constitution of one; this being common in generation, that the vegetal primordium whence the foetus is produced . . . pre-
exists."[45] Such a "primary something," moreover, was not simply without structure, but preformed. It consisted of "a moisture inclosed in some
membrane or shell; a similar body, in fact, having life within itself either
actually or potentially."[46] "Many animals, especially insects," Harvey
believed, "arise and are propagated from elements and seeds so small as
to be invisible, (like atoms flying in the air), scattered and dispersed here
and there by the winds."[47]

In assuming that the world is not only populated by visible beings,
but is also full of invisible, or only microscopically visible, individual
beings—a worldview that has been characterized as "microsubstantialism"[48]—Harvey moved furthest away from the position of Aristotle and
came closest to the atomistic and preformationist natural philosophies
of his time. This raised profound questions regarding the (patrilineal)
chain in which like was supposed to beget like. If both formative powers
and their material substrates—"eggs" encapsulating potential life—were
universal, how should there be any limit to their interaction? Would not
everything be able to "generate into" everything else, resulting in a tur-
moil of transmutations? After all, Harvey was prepared to acknowledge
the reality of equivocal generation, in which living beings were procreated from beings of a different species, or even from nonliving matter.[49]

In order to avoid the aporias that plastic forces and preformed germs implied, Harvey recurred to an age-old solution. He subordinated generation to a cosmological order that was dominated by cyclical movements. The movement of the sun, planets, and stars, the seasonal cycle, the body of the state (with the king as its "heart" and "sun"), the chemical processes of distillation—all these analogous cyclical movements were coordinated to ensure the regular generation of like from like.[50] Parental bodies thus emerged as subordinate instances of a universal cosmic order that generated, sustained, and perpetuated life through cyclical movements, from the macrocosm of the solar system all the way down to the microcosm of the human body. As Harvey explained:

> The male and female, therefore, will come to be regarded as merely the efficient instruments, subservient in all respects to the Supreme Creator, or father of all things. In this sense, consequently, it is well said that the sun and [man] engender man; because, with the advent and secession of the sun, come spring and autumn, seasons which mostly correspond with the generation and decay of animated beings. So that the great leader in philosophy says: "The first motion [of the fixed stars] is not the cause of generation and destruction; it is the motion of the ecliptic [i.e., the apparent annual movement of the sun] that is so, this being both continuous and having two movements; for, if future generation and corruption are to be eternal, it is necessary that something likewise move eternally, that interchanges do not fail, that of the two actions one only do not occur.[51]

In this overarching image of cosmic order, each individual being is defined by its position in relation to its procreators and the particular circumstances that accompanied its procreation, especially its position with respect to celestial bodies like the sun—the climate in its original, astronomical meaning—and thus its position within the seasonal cycle. Each individual being results from a particular constellation of causal factors (which include its parents) that is reinstalled again and again through the movement of the skies to allow for the re-creation of individual beings. Harvey refers to the resulting cycles of generation and corruption as "interchanges" (*mutationes* in the Latin original).[52] Indeed, they can be seen as exchanges, albeit fundamentally asymmetrical ones in which each generated being takes the place of its particular progenitor. Generation and corruption are not processes that cancel each other out stochastically, but each generation must be followed in time by decay and subsequent regeneration—just as the sun moves up and down in the annual cycle.[53] Harvey reinstates the figure of a patrilineal chain, which was threatened by the autonomy he ultimately had to ascribe to the individual being, by projecting patriarchal relations onto nature at large with the help of ancient cosmological models of a closed universe in which each being occupies its "natural place."[54] While this solution accounted for an orderly succession of beings, it simply left no room for the transmission of a universal, hereditary substance.

Harvey, court physician to both James I and Charles I, wrote the largest part of his *Exercises on the Generation of Animals* during the time when Charles I was already imprisoned by Parliamentary forces, and he completed the manuscript in 1649, the year in which Charles I was executed. He was a close friend of Thomas Hobbes and sympathized with the latter's theory of an absolute royal power over individuals that are equal by nature. This is not to say that the *Exercises* were written to legitimate Charles's pretensions to the throne. But Harvey's conception of generation touched contemporary political debates about the autonomy and equality of individuals and the universality of creative forces and substances that radically questioned traditional, essentially patriarchal power relationships.[55] His characteristic ambivalence with regard to positions that a history of ideas would portray as contradictory—preformation versus epigenesis; the maternal body as a matrix of all life versus the paternal body as the indispensable and exclusive source of life; the embryo as the "work" of its parents versus the embryo as an autonomous being with a life of its own; causality of direct action versus action at a distance—was a response to the political turmoils of his time. Although this response was not unambiguous, it came to define a conceptual framework within which the problem of generation continued to be discussed in the eighteenth and nineteenth centuries.

## Laws of Generation and the Reproduction of Species

Late-seventeenth-century and early-eighteenth-century speculation about generation was increasingly dominated by theories according to which living beings preexisted in their microscopic germs. In order to understand these theories, it is important to distinguish carefully between preexistence and preformation. Preformationist theories maintained only that a germ capable of development is not an undifferentiated mass, as implied by epigenesis, but already possesses a certain structure that in one way or another preconditions the form of the future, developed

body. This idea did not necessarily imply that the adult body is represented in the germ in all its detail on a smaller scale, and it was fully compatible, as seen in the last section, with the idea that the preformed germ
was created in one of the parental bodies or during conception. The first
early modern preformation theories—which were proposed by medical
writers like Fortunio Liceti, Giuseppe degli Aromatari, and Emilio Parisano, who taught at northern Italian universities around the time when
Harvey studied there—thus went only as far as to proclaim that the embryo was produced and began its life prior to conception in the male
testes.[56]

Preexistence theories went a step further. In their most explicit versions, they claimed that the germs of all future living beings were created directly by God at the beginning of time and that each part of these
beings was already preformed in the germs. The germ thus contained
a miniature representation of all the parts of the organism that would
develop from it, including the germs it contained of all its future descendants. All members of the genealogical chain were thus already encapsulated in a series of germs within germs in the first living beings—a
relationship referred to as *emboîtement*. Successive generations resulted
from the evolution—in the sense of unfolding—of these germs.[57] Development was thus reduced to mere growth of preexistent organic structures. The only unresolved question was which sex contained the series
of germs, the male or the female, thus defining the alternative positions
of animalculism and ovism. Animalculist theories received a considerable boost from a widely discussed letter by Antony van Leeuwenhoek,
which was published in the Royal Society's *Philosophical Transactions*
in 1679. In this letter, Leeuwenhoek reported that he had observed "ani-
malcules" in male semen under the microscope. "And when I saw them,"
Leeuwenhoek declared, "I felt convinced that, in no full-grown human
body, are there any vessels which may not be found likewise in sound
semen."[58]

Few naturalists defended such extreme versions of preexistence. A
broad spectrum of opinions existed with respect to questions such as
how exactly future organisms were represented in the germs, where
and when they were generated, and how they developed. Nicolas Malebranche, one of the most prominent supporters of preexistence in the
late seventeenth century, for instance, clearly rejected the idea that germs
contained miniature representations of the adult organism. "All I'm saying," he corrected those who had accused him of holding such a view, "is
that all the organic parts of [creatures] are formed in their [germs], and
fit so well with the laws of motion that [they] can grow and take their

shape . . . just through their own construction [*par leur propre construction*] and the efficacy of the laws of motion, without God's providing finishing touches through extraordinary providence."[59] Preexistence theories shared only the assumption that all germs of all future organisms had been created in the beginning and that there was no need for divine intervention in generative processes subsequently. According to preexistence theories, one might say, there was no need for any "generation" in the full sense of that word at all. And there was likewise no need for anything like hereditary transmission.

Preexistence theories were thus not simply opposed to epigenetic theories such as those proposed by Aristotle or Harvey. They rather separated and radicalized two aspects of those theories. On the one hand, they maintained that offspring were largely independent from their parental procreators. On the other hand, all offspring were seen to be dependent on a "divine watchmaker." Precisely because germs were wholly preformed by divine creation at the outset, they were able to nourish and to develop themselves solely by mechanical means, without need for an external agent initiating, sustaining, or guiding these processes. Preexistence theories thus abstracted from individual acts of physical procreation and reduced genealogical descent to a formal relationship of mere succession. Parents were not so much creators but rather containers of their offspring, and the latter developed independently and successively, subject to universal laws of motion only.[60] The world of preexistence theories was a world of autonomous individuals and devoid of patriarchal powers over offspring, a world in which mechanical laws governed uncreative matter. The dependence of creatures on a creator existed only at the hypothetical starting point, beyond which creation, in the sense of adding something qualitatively new, no longer played any role.

One radical version of this elimination of everything but formal genealogical relationships was proposed by a naturalist whose significance for eighteenth-century theorizing about generation has so far escaped the attention of historians. We refer to the Swedish botanist and physician Carl Linnaeus, or Carl von Linné, as he called himself after having been ennobled in 1761. His first publication in 1735, entitled *Systema Naturae*, is generally seen as the starting point of modern taxonomy, the science of classifying plants and animals. But Linnaeus was interested in topics other than the mere division and denomination of organisms. In 1746, he published a short essay on the "marriages of plants" (*Sponsalia plantarum*), which aimed to demonstrate that "plants live no less than animals."[61] By far the greatest part of his argument in this respect consisted in establishing that plants, just like animals, reproduce sexually. In discussing this topic, he rejected all common generation theories of his time, regardless of whether they relied on assumptions about specific interactions between the generative substances—"effervescences and precipitations of the ancients," as he put it—or whether they speculated about the seat and beginning of embryonic life. He was particularly critical of ovist and animalculist versions of preexistence theories, arguing that the implausibility of a complete *emboîtement* as well as the existence of hybrids, which combined parental characters, refuted these theories. Interestingly, Linnaeus himself remained agnostic about generation. "We do not know how generation occurs," he simply stated in conclusion.[62]

But if nothing is to be said about the generation of an individual being, how can anything be said about generation at all? The answer Linnaeus offered in *Sponsalia plantarum* consisted in drawing attention away from causal connections and instead focusing on what he called the "process of generation" (*processus generationis*); that is, on the regular series of changes and movements that plants undergo when they propagate.[63] Linnaeus's descriptions of these processes border on pornography. The reader learns that the plant world knows of "voluptuously gaping vulvae" and of "males ejaculating their prolific substance."[64] It is no surprise, then, that Caspar Friedrich Wolff—who some twenty years later revived epigenetic accounts of generation and development in a book boldly entitled *Theorie von der Generation* (1764)—stung Linnaeus with the following charge: "He may well know how to copulate, but has no understanding of the theory of generation."[65]

Indeed, the detailed descriptions that Linnaeus compiled of the functions and movements of plant organs during fertilization remained devoid of content as far as the actual production of offspring was concerned. On the basis of these descriptions, however, Linnaeus derived a number of empirical generalizations about temporal relationships in reproductive processes: for example, "the flower always appears before the fruit"; "anthers and pollen always appear before the fruit"; "stigmas always flower at the same time as the anthers."[66] These generalizations added up to the famous Linnaean dictum that "all life proceeds from an egg" (*omne vivum ex ovo*), and they provided him with support for an original concept of generation, that of "continued generation" (*generatio continuata*). For Linnaeus, generation was not a singular event, but a process that continued throughout growth and development and eventually led to the generation of new individuals of the same species. "There is no fresh creation [*nova creatio*]," he maintained apodictically in his *Philosophia botanica* (1751), "but continued generation [*generatio continuata*]."[67] Linnaeus's theory of generation, in other words, was a theory of the reproduction of species.[68]

That Linnaeus's theory of generation was in fact a theory of the reproduction of species becomes especially apparent in his rather idiosyncratic theory of divine creation. Two simple generalizations served him as a starting point for arguing that God had initially created only a single adult individual in the case of hermaphroditic organisms, and only a single pair of individuals in the case of organisms with two distinct sexes (this was creationism, albeit of a peculiar variety). "Each egg," he observed, "produces offspring highly similar to its parents," and with each generation "the number of individuals is multiplied," so that "there is at this time a greater number of individuals in each species than there was originally." By simply "counting backwards," as he put it, one would thus inevitably come to the conclusion that the "series" of descendants constituting a particular species issued from a single progenitor, who in turn could only have been the work of the hands of God.[69] As Linnaeus explained in an essay entitled "On the Increase of the Habitable Earth" (1744):

> If we trace this train back in the opposite order, and consider the ascending series, we shall find the number of individuals less and less at every step; so that many owe their origin to few, these to still fewer, and so on till the decreasing progression terminates in an individual pair; and the first link in this chain of secondary causes must be referred to an act of creation in Deity.[70]

The argument Linnaeus makes here is based on a peculiar inversion of a common representation of ancestry. In so-called ancestor tables, the ancestry of an individual is depicted by retracing the lines of descent that connect a given individual with its ancestors. As the number of ancestors grows exponentially with each generation, it is easy to see that one would rapidly arrive at an ancestral generation that encompasses the entire human species, and thus constitutes no particular ancestry at all. The only way to avoid this result is by stopping the procedure after a certain number of generations or by singling out a particular line of descent (in most cases that of male ancestors was chosen). Both procedures are arbitrary from a formal point of view, but they succeed in representing a particular individual through a unique set of ancestors.

In contrast, Linnaeus's analysis started with the undifferentiated mass of individuals belonging to a particular species, including the human species, in the present and reduced this mass to one specific individual or pair of individuals. As a consequence, and unlike the construction of ancestor tables, Linnaeus's procedure thus represents each and every individual member of a species as essentially the same, and as essentially different from members of other species, no matter what particular line of ancestry gave rise to that individual. The only thing that changes with descent, despite the myriad of causes that can play a role in the generation of individual beings, is the number of individuals. Generation is reduced to a strictly formal relation of descent, constituting the species as a set of individuals determined in essentially the same uniform way by a "procreative unit" instituted by God at the beginning of time.[71]

With this inversion, Linnaeus split apart two aspects of genealogy. Genealogical relationships usually imply causal as well as classificatory relationships: the causal relationship of producing another individual and the classificatory relationship of identity by virtue of common descent. The separation of these two aspects found full expression in Linnaeus's influential definitions of the taxonomic concepts of species and variety. According to these definitions, varieties are distinguished by differences owed to particular, local circumstances that accompany the production of offspring—"place or accident," as Linnaeus put it. Species, on the other hand, possess properties—"specific differences" (*differentia specifica*) in the technical language of the time—that are passed on from generation to generation under all circumstances and obey universal "laws of generation" (*leges generationis*), which again were instituted at the time of the Creation. Like preexistence theorists before him, Linnaeus saw no necessity for further creative interference once God had set into play the generative forces at the beginning of time.[72]

Linnaeus's conviction that there are "laws of generation" that govern the succession of beings came to the fore in another interesting consideration of his. Harvey, as we have seen in the previous section, used the economic metaphor of "interchanges" to express the idea that each generated being steps into the place of its procreator. Linnaeus expanded this metaphor to encompass synchronic relationships among organisms as well. As all individuals of a given species are essentially the same by virtue of their presumed uniform relationship to an original "procreative unit," they can also replace one another in what Linnaeus called the "economy of nature" (*oeconomia naturae*); that is, the network of nutritional relationships that extends throughout animate and inanimate nature. "All natural things," as Linnaeus wrote in an essay dedicated to the economy of nature, "should contribute and lend a helping hand to preserve every species; [so] that the death and destruction of one thing should always be subservient to the restitution of another."[73]

Members of different species offer one another products and services in the form of their own prolific bodies, and the economy of nature is thus characterized by symmetrical relationships of mutual benevolence and indulgence. In a powerful image, Linnaeus compared this system to a "weekly market," where "at first one only sees how a great mass of people spreads out in this or that direction, while nevertheless each of them has his home [*domicilium*], from where he approached and to which he will proceed."[74] Harvey's cosmological cycle that guaranteed the orderly succession of beings had been replaced by inexorable natural laws that governed the reproduction of species from within.

*Germs and the Evolution of Species*

According to one popular account, it was an age-old, religiously and metaphysically motivated tradition that led Linnaeus to propose that species remain "constant."[75] This account fails to recognize that Linnaeus's notion of laws of generation actually marked a radical break with tradition. As we have seen in the previous section, Linnaeus was not interested in the processes and contingencies that resulted in the generation of individual beings, but rather in a general theory of reproduction that described the multiplication of individuals within species as a regular phenomenon. This interest clearly sets Linnaeus apart from generation theorists of the seventeenth century such as Harvey. With the assumption of laws of generation, transmutations as well as spontaneous and equivocal generations were excluded from naturalists' considerations as a matter of principle; there was no room any more for occasional violations of these laws. The "life of the species"—as the German naturalist Carl Friedrich Kielmeyer would call it toward the end of the eighteenth century[76]—that is, the vital processes that connect a multiplicity of beings, rather than constituting the life of an individual, became the focus of a science that received the name of "biology" around 1800.

Both Michel Foucault and François Jacob have identified "organization" as the key concept of this "new" science of biology.[77] We do not want to challenge the significance of this conceptual innovation. Nevertheless, for a fuller understanding of the nature of the science of biology, it seems pertinent to recognize the interindividual and intraspecific dimensions that many of its concepts, including "organization," gained around 1800.[78] Instead of referring, as in earlier times, to individual bodies, organic functions like generation, growth, development, nutrition, and sensation were increasingly perceived as reproductive functions physically constituting the reality and unity of species.[79] As Buffon put it programmatically in his influential *Discourse on the Manner of Studying and Expounding Natural History* (1749), "The history [. . . of the species] ought to treat only relations, which the things of nature have among themselves and with us. The history of an animal ought to be not only the history of the individual, but that of the entire species."[80] The evolution of the concept of generation illustrates this change of perspective particularly well. Throughout the eighteenth century, its reference remained largely restricted to individual acts of procreation. It was only in the course of the early nineteenth century that "generation" acquired its secondary connotation of a group of individuals born around the same time.[81] In parallel, the expression "regeneration" was increasingly used to designate the ability of organisms to re-create damaged or lost parts of the body.

Far from being a paradoxical concomitant, the focus on organization that emerged around 1800 dovetails with the growing attention that naturalists paid to the "life of the species." As a matter of fact, reproduction was at the heart of Kant's influential concept of a "natural purpose" (*Naturzweck*), which he introduced to get a grasp on the kind of causality proper to organized beings. According to Kant, organisms were special not simply because they were organized, but rather because they organized *themselves.*[82] Applying this perspective to reproduction—that is, to the production of an organized being by another organized being like itself—directed attention to the internal organization of the germs. After all, the germs constituted the only physical link between the individuals that succeeded one another in reproduction. How a small part that was separated from an organism could reconstitute a similar organism in its entirety thus became the central problem of biology. That is why the description of the fantastic regenerative abilities of the freshwater polyp *Hydra* by Abraham Trembley created such a stir in 1744 (fig. 2.2). Each and every tiny piece of this organism, so it appeared, was able to reproduce the whole polyp. This discovery inspired Charles Bonnet, a contemporary naturalist and philosopher from Geneva, to proclaim that each animal "is a world inhabited by other animals."[83]

Two alternative, but not mutually exclusive, conceptions of the germ and its ability to bring forth beings of its own kind emerged in the eighteenth century and persisted throughout the nineteenth century alongside each other. The first, which was formulated paradigmatically in Johann Friedrich Blumenbach's concept of a "formative drive" (*Bildungstrieb*), understood reproduction as resulting from forces that acted at a distance, in analogy with Newtonian gravitation. In the second, which found its equally paradigmatic formulation in Buffon's notion

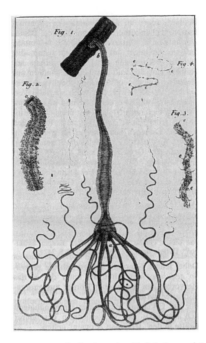

**FIGURE 2.2** Freshwater polyp (*Hydra*). Copperplate from Abraham Trembley, *Mémoires pour servir à l'histoire d'un genre de polypes d'eau douce, à bras en forme de cornes* (Leiden: Verbeek, 1744), pl. 5.

of "organic molecules," reproduction resided in living matter passed on from generation to generation and organized according to specific "inner moulds."[84] Preformation and epigenesis thus ceased to be opposing positions with respect to the mode and moment in which an individual being was generated by its divine or parental procreator. The two views became aspects of one and the same process that permeated the life of populations. When preformation and epigenesis clashed once more in a notorious debate between Albrecht von Haller and Caspar Friedrich Wolff, the question was no longer whether individual beings were "preformed" in the germs from which they arose. The dispute was now more general: Did the ability to reproduce always presuppose some kind of organic structure (Haller's preformationist position), or was it sufficient to assume some unstructured substance equipped with a specific vital force (Wolff's epigenetic position)?[85] Life increasingly became a playing field for organic forces and molecules that manifested themselves in the generation of individuals, but also enjoyed a life of their own that endured beyond the lives of individuals. Germs budded off from organisms, united in sexual propagation, and occasionally gave rise to new or-

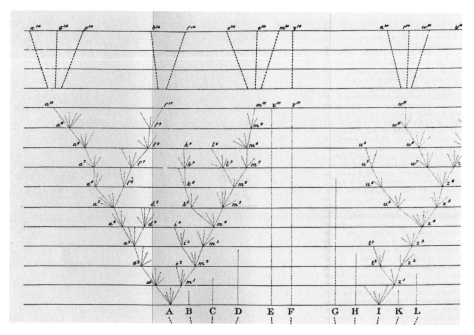

FIGURE 2.3 Diagram from Charles Darwin, *On the Origin of Species* (London: Murray, 1859).

ganisms. In comparison with the emphasis that generation theorists like Harvey placed on vertical relations of descent—even if moderated by assumptions about cosmological cycles—one can speak of a horizontalization of life at the end of the eighteenth century.

The single illustration in Charles Darwin's *On the Origin of Species* (fig. 2.3) helps to convey what we mean by "horizontalization." The illustration consists of a diagram that represents the relationships among species in a bifurcating structure, in accordance with Darwin's statement that the so-called natural system of organisms "is genealogical in its arrangement, like a pedigree."[86] A closer look at the diagram reveals some interesting details, however. As drawn, the pedigree does not possess a unique point of origin common to all species, nor does it unfold in a linear or otherwise regular manner. Lines of descent rather follow a "long and circuitous course," as Darwin put it,[87] which results from the competition among and eventual displacement of varieties and species. In the diagram, the only element of a regular order cuts across genealogical relations: horizontal lines demarcate time slices that, as Darwin explains, include "a thousand . . . , better . . . ten thousand generations" and at the same time represent "successive geological formations."[88]

These formations, defined by two unrelated time frames, generations of organisms and geological strata, provided the stage for the antagonistic forces that Darwin's theory of natural selection called on—the stage, in Darwin's words, for "an occasional scene, taken almost at hazard, in a slowly changing drama."[89] But more importantly for our theme, they also provided room for a concept of reproduction that went beyond the generation of individual beings and encompassed the circulation of germs and dispositions within populations. The biological phenomena that engaged Darwin's lifelong interest—the "associated life" in colonial organisms like bryozoans; the growth of coral reefs; the propagation of fruit trees by scions; the "life" and "death" of species on the scale of earth history—were defined in this epistemic space that connected the life of microscopic germs with episodes in earth history.[90] Darwin described this space in a harrowing series of analogies:

> The fertilized germ of one of the higher animals, subjected as it is to so vast a series of changes from the germinal cell to old age—incessantly agitated by what Quatrefages well calls the *tourbillon vital* [i.e., vital whirlpool]—is perhaps the most wonderful object in nature. It is probable that hardly a change of any kind affects either parent, without some mark being left on the germ. . . . we must believe that it is crowded with invisible characters, proper to both sexes, to both the right and left side of the body, and to a long line of male and female ancestors separated by hundreds or even thousands of generations from the present time: and these characters, like those written on paper with invisible ink, lie ready to be evolved whenever the organization is disturbed by certain known or unknown conditions.[91]

If we compare this passage with Harvey's theory of organic reproduction, two important differences emerge. First, the passage demonstrates that Darwin's theory of heredity abstracted considerably from the personal relations between ancestors and descendants. Although he conceded the possibility of an inheritance of acquired characters, he put forward the unmistakable idea that the true carriers of hereditary properties are not parents and their respective offspring, but submicroscopic entities that are distributed anew in each generation and therefore percolate through succeeding generations. Second, and even more fundamentally, the passage indicates a remarkable inversion in comparison with Harvey's conception of generation. In the latter, emphasis was placed on the diachronic dimension of linear descent in which parental organisms, in conjunction with the seasonal cycle, actually make their offspring. Darwin, in contrast, paints a picture in which the synchronic or "horizontal" dimension predominates. In his view, a common reservoir of dispositions, passed down from the sum total of ancestors and competing for their realization, is redistributed among individuals with each new generation. Life has thus assumed a form that is no longer dominated by direct relations among organisms, but rather by the interactions among circulating units of life—cells, germs, elementary dispositions, or organic molecules, as the case may be—that are transmitted and redistributed from generation to generation.[92]

Darwin had already used the term "heredity" to speak about the transmission of dispositions and characters, and what he had to say on the subject will be discussed in detail in chapter 4. What remains to be emphasized in this chapter is that the transition to the discourse of heredity did not simply eliminate earlier conceptions of generation. Instead, genealogical figures of thought persisted, and modern conceptions of heredity lodged themselves in this wider framework of ideas. Even in the current age of reproductive technologies, the discourse of generation retains its dominant position. Now as ever, parents feel that they make their children, and they seek counsel from physicians who advise them not only with genetic knowledge, but also with knowledge about how environmental factors can influence the health of their child. It seems as evident to us as it was to our predecessors that the behavior of its parents, and especially of its mother during pregnancy, is of great consequence for the development of a child.

This leads us to the important conclusion that knowledge of heredity was not simply dictated by common sense. Knowledge of heredity had to prevail *against* appearances, as illustrated by the fact that the germs and dispositions that Darwin and his seventeenth- and eighteenth-century predecessors speculated about remained largely invisible to them. Modern theories of heredity pinolicate a threshold in the life sciences that the French epistemologist Gaston Bachelard has diagnosed to be constitutive for modern science in general. They are largely counterintuitive, not directly relevant to individual lives, and derive from organized forms of knowledge production, or, as Bachelard put it, from "reality, as it is revealed in the scientific laboratory."[93] Perhaps the emergence of heredity forms a belated parallel to the break with the Aristotelian tradition that Alexandre Koyré discerned in the work of Galileo.[94] In the next chapter, we examine the specific contexts that produced the fissures that eventually grew into a break of this kind in the life sciences.

# 3   Heredity in Separate Domains

At the beginning of the previous chapter, we highlighted the fact that the concept of heredity began to be applied to biological phenomena only around 1800. Carlos López Beltrán has shown in detailed studies that French physicians in particular—and especially psychiatrists like Prosper Lucas with his *Traité philosophique et physiologique de l'hérédité naturelle* (1847–1850)—popularized a biological-medical or, in contemporary parlance, "philosophical and physiological" meaning of the term that soon made its way into all major European languages.[1] The terms *hérédité* and *heredity*—both derive from the Latin *hereditas*, meaning inheritance or succession—then quickly lost their original juridical meaning in French and English. The *Oxford English Dictionary* lists 1540 as the year of the earliest occurrence of "heredity" in its legal sense, marking it as "obsolete today,"[2] while the earliest references to "heredity" in the modern, biological sense date from 1863 (Herbert Spencer's *Principles of Biology*) and 1869 (Francis Galton's *Hereditary Genius*). *Black's Law Dictionary* similarly defines "heredity" as "archaic" for "hereditary succession, and inheritance."[3] The American *Corpus Juris Secundum* defines heredity only as "a universal law of organic life."[4] The original meaning appears to have been forgotten altogether. In German-speaking countries, the corresponding term *Vererbung* was first used metaphorically in a biological context by Immanuel Kant in his anthropological writings of the 1770s and 1780s, to which we will return later. In contrast to the French and English terms, however, *Vererbung* never completely lost its legal connotations. All three expressions finally acquired additional meanings in the course of the nineteenth century by being applied to processes of societal or cultural life.

Of course, natural philosophers and physicians had noticed long before the end of the eighteenth century that special talents, physical peculiarities, and diseases tended to recur in offspring, or that such properties sometimes skipped a generation and reappeared in grandchildren. The crucial point is that these phenomena were not conceptualized in terms of inheritance, with one remarkable exception. Since the late medieval period, physicians had classified diseases that were restricted to certain families as "hereditary diseases" (*morbi haereditarii*) and compared them to the goods that were transmitted along family lines.[5] However, such comparisons reflected the conviction that the generation of offspring was dominated by a constellation of causal factors that persisted over generations, rather than a view that abstract properties or dispositions were transmitted and redistributed with each generation. Thus Jean Fernel maintained in his *Medicina* (1554) that "children succeed their fathers, and are no less inheritors [*haeredes*] of their infirmities as of their properties [*possessionum*]."[6] That people in the sixteenth and seventeenth centuries associated landed rather than mobile property with such statements transpires in the translation of Fernel's proposition in Robert Burton's *The Anatomy of Melancholy* (1621): *possessionum* is here rendered as "lands."[7] A comparable perspective is evident in discussions about the causes of similarities between grandchildren and grandparents in the Aristotelian and Galenic traditions. The problem consisted in explaining not how properties had been transmitted, but rather how the same causal agents that had been involved in the generation of ancestors could apparently remain active in the generation of remote descendants.[8]

These historical observations about the use of the terminology of inheritance point to some fundamental matters that also affect the structure of this chapter. François Jacob characterized the transition from thinking in terms of generation to thinking in terms of inheritance around 1800 as a break or rupture that sharply separates two styles of thought, or epistemes. We are already in a position, however, to note that the supposed rupture was preceded by a long history of observations, practices, and considerations that, in hindsight, can be interpreted as focusing on phenomena of heredity, but did not form a coherent discourse. Instead, we find discussions of heredity in widely different genres of literature, in medical textbooks, encyclopedic natural histories, natural philosophical tracts, and religious and political pamphlets. In other words, inheritance was not a unified biological object in early modern thought, but was rather a subject distributed across different and often disparate domains.

Moreover, heredity surfaced as a subject for speculation only at the margins of these domains. We take it for granted today that each disease possesses a genetic and thus heritable component, even if such components influence the development of diseases in variable degrees. Physicians in the late medieval and early modern period, on the other hand, considered hereditary disease as a peculiar form of disease alongside other types of disease. Unsurprisingly, these family diseases did not attract a great deal of attention in times when contagious diseases, such as the plague, were still barely controllable. Hereditarily transmitted diseases were therefore seen as curiosities. Thus, Michel de Montaigne asked himself, when he began to suffer from gallstones at forty-five years of age just as his father had at the same age, what kind of "prodigy" (*monstre*) was hidden in the male seed, so that it transmitted "impressions" not only of the bodily conformation, but also of the "thoughts and inclinations of our fathers." After all, Montaigne also shared a pronounced antipathy toward physicians with his father. For Montaigne, heredity was not a generalizable natural phenomenon, but an example of the "miracles in obscurity" with which nature confronts humans on a daily basis. What was most intriguing about the gallstones was that they appeared at the same age in both father and son, although the elder had not yet developed this ailment when he generated his offspring. All other examples that Montaigne listed to illustrate heredity share the eccentric character of curiosities: members of the Roman family Lepidus were often born with one eye covered by cartilage; according to ancient tradition, a Theban tribe was distinguished by a lancelike birthmark; and, if one believed Aristotle, some of the Greek tribes practicing "women . . . in common" determined paternity by means of such bodily characters.[9]

Montaigne's assorted examples point to a further important attribute of early modern thinking about heredity. A transmission of characters was initially perceived only in cases that involved individual peculiarities and deviations. Normal characters were not conceived in the same manner. It is very revealing in this respect that it was familial *diseases* that first brought to the fore the metaphorical potential of concepts of inheritance in the early fourteenth century. In marked contrast to what one would perhaps expect from today's vantage point, naturalists and

physicians were thus not inspired to speak about heredity by the sem-
blance of permanence and necessity that is often associated with tradition. What initially inspired hereditarian figures of thought were instead
individual, contingent deviations that were consistently transmitted to
descendants once they had arisen.

A final peculiarity of the concept of heredity can be brought out by
considering an obscure technical meaning that the German term for heredity, *Vererbung*, once possessed alongside its more general sense, and
that it shared with the Latin *hereditas*, the root of the corresponding
English and French terms. "*Vererbung* . . . occurs," we read in volume
47 of Johann Heinrich Zedler's *Universal-Lexicon* of 1746, "if an inher-
itor dies before he can accept or reject an inheritance, and if that inheritance, neither really accepted nor rejected by him, falls and is transmitted
to those indicated as his inheritors by law."[10] In this sense, "heredity"
refers to the transmission of goods that may formally have become the
property of inheritors, but were never actually appropriated by them,
and therefore could also not become subject to their will. Technically,
heredity has nothing to do with claims to property that result from concrete familial relationships, even if courts in practice usually attend to
such relationships. In this narrower sense of "heredity," what is inherited—and by whom—is solely determined by law, which may, but certainly does not have to, pay heed to "family matters."[11]

If, in this chapter, we delve into the period before a general concept
of biological inheritance was established, we need to relinquish the tight
scientific definition that the term acquired in the nineteenth century and
take ourselves back to a time when the linguistic conventions and social
institutions that later supported the biological concept of heredity did
not yet exist. We will have to seek out the beginnings of the knowledge
regime of heredity in a multiplicity of social domains that were neither
discursively nor materially related to one another in a coherent way.
Below, we group these domains into three loosely defined complexes.
Initially, we consider the legal and political contexts in which heredity
played an important role. Long before heredity was applied to biological phenomena, legal regulation of inheritance and succession became
an issue of intense debates between families, territorial states, and the
church. In these debates, "blood relationships" increasingly gained in
significance. We then turn to medical contexts, which in the eighteenth
century provided one of the chief arenas for the conceptual clarification
and differentiation of heredity. The final section in this chapter will consider natural history, breeding, and anthropology, three areas in which

heritable and environmentally stable variation attracted attention toward the end of the eighteenth century.

## Law and Politics

If one considers the distribution of knowledge of heredity in early modern Europe, one is struck by the profound asymmetry mentioned above.
While the regulations that determined how properties were passed along
familial lines were the subject of long-standing political and legal debates, it was not until much later that inheritance became a focal point of
interest to the sciences concerned with living beings. According to a well-
known thesis advanced by the historical anthropologist Jack Goody, in
early medieval times the church began to interfere with strategies that
had been developed to hold together and secure family belongings—such
as certain forms of adoption, cousin marriage, concubinage, marriage
with affines (relatives by marriage, like a brother-in-law), and remarriage
of divorced persons. The church did so, if we follow Goody's account,
to weaken the social ties within and among families and to free properties from family claims. The properties would then, at least in principle,
be alienable and could be used for private donations that would allow
the church to retain its independence from worldly powers.[12] The elev-
enth- and twelfth-century Gregorian reforms achieved this through the
institution of celibacy, the abolition of lay investiture, and the extension
of incest prohibitions to relatives up to the seventh degree. Moreover,
for the purpose of these prohibitions, kinship was defined according to
the more inclusive Germanic system, which calculated degrees of kinship
among two relatives by counting the number of generations that separated them from their last common ancestor rather than the number of
generative acts that lay between them, as had been usual in the Roman
system (fig. 3.1).

Incest prohibitions were made even more restrictive by the inclusion
of both cognates (relatives by blood) and affines in such calculations.
The emerging territorial states, although they competed with the church
in many respects, adopted similar systems of bureaucracy and of property, and they supported these regulations.[13] As a consequence, any marriage, and thus any alliance between families, had effectively become
subject to special dispensations from the church or state authorities by
the end of the seventeenth century. Of course, the authorities did not always intervene in family affairs on these grounds, but in situations of political instability or conflict, such interventions could be expedient.[14]

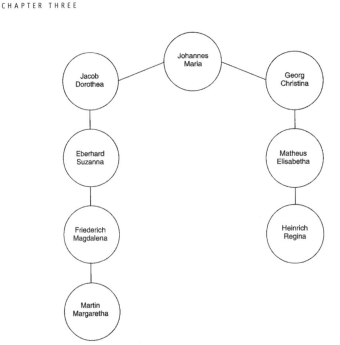

**FIGURE 3.1** Scheme illustrating the Germanic and Roman systems of determining degrees of kinship. According to the Roman system, Friederich and Heinrich are related by six degrees; according to the Germanic system, by three degrees only. From Staffan Müller-Wille & Hans-Jörg Rheinberger, eds., *Heredity Produced: At the Crossroads of Biology, Politics, and Culture, 1500–1870* (Cambridge, MA: MIT Press, 2007), p. 10, fig. 1.1.

These developments triggered a veritable explosion of genealogical activities that left their traces in a wealth of complex ancestry diagrams and heraldic representations (fig. 3.2).[15] An important innovation in this context was the introduction in the twelfth century of the *ipse* (Lat. "self") or *ego* (Lat. "me"). Rather than starting from a group of close relatives such as siblings—the traditional *truncus* or *stirps* (Lat. "stem" or "stock")—pedigrees and ancestor tables began to take an individual as the starting point for genealogical analysis.[16] On the other hand, noble families began to counteract the alienation and dispersion of family holdings by privileging agnatic—that is, exclusively male—lines of descent (so-called *lignages*) in the transmission of property. This practice found its most radical expression in the institution of the majorat or primogeniture, which laid down for all future generations that property had to pass to the firstborn son. Daughters, younger brothers, maternal relatives in general, and all male collateral relatives such as uncles and nephews were thus passed over. This did not necessarily mean that all

**FIGURE 3.2** Heraldic-genealogical representation of kin relations from a late medieval world chronicle. *Edward IV. Roll,* c. 1461. Courtesy Free Library of Philadelphia.

these relatives were excluded from participation in a family's property. David Sabean has shown that the particular form of such arrangements varied massively on a regional scale, sometimes privileging younger brothers, or even daughters. He argues that such lineage constructions did not serve to expropriate certain groups of close relatives, but rather to organize families around an inalienable property that was firmly held together.[17] Thus we can see why, in the middle of the thirteenth century, the legal scholar Henry Bracton compared the movement of inheritance to the straight line that is formed by a falling weight, rather than with the growth of a bifurcating, treelike structure, a comparison that would be more intuitive to us today.[18] Genealogical constructs were intimately linked with concerns about safeguarding family property against its dispersion.

This insight also allows us to explain the curious history of the notion of "noble blood." In the first half of the fourteenth century, conflicts between royal and aristocratic powers in France as well as in England led to a revival of ancient conceptions of "noble blood," but with a particular emphasis on genealogical descent along agnatic lines that had not existed in antiquity. Borrowing from contemporary literature on the breeding of dogs and falcons, supporters of this idea claimed that nobility was "in the blood" and that it was passed on from father to son. Such expressions remained metaphorical, however. For even in scholarly contexts—in legal opinions, for instance—one does not find direct connections between these expressions and the theories of generation that were prevalent at the time, which often postulated an intimate relation between blood and the generative substances, be it that menstrual blood was seen as the material from which the embryo grew or be it that the male seed was regarded as a perfected product of the blood contained in the spinal marrow.[19] Much the same can be said about theological and juridical arguments according to which marriage constituted a union of "flesh and blood," so that marriages between affines—for example, between a widow and her deceased husband's brother—had to be treated as incest and prohibited as well. Incest amounted to the "pollution of blood" in a moral sense and was illegal for this reason; a physiological justification beyond this argument was not really necessary.[20]

The lack of a closer relationship between metaphors of blood and theories of generation such as those we encountered in the last chapter can be explained by the fact that legal and political disputes about inheritance rarely aimed at establishing priority for blood relationships in general. On the contrary, depending on the particular situation, arguments were brought forward that supported either the exclusion of some close relatives, such as daughters, from inheritance or the inclusion of half-siblings, illegitimate or adopted children, close friends ("brothers in blood"), or even inhabitants of the same village, region, or nation in the circle of relatives that enjoyed participation in a common heritage or privilege. "Blood," writes the medieval scholar Charles de Miramon with respect to the revival of metaphors of blood in the fourteenth century, "was not something that diffused over widely extended populations. . . . The concept had . . . a geographic dimension. Blood had to manifest itself at a certain locality, on the battle field, at court, on an estate, in the intimate space of a family."[21] We will return to this relationship of "blood" and "soil" when we discuss animal breeding in the eighteenth century.

First, however, we would like to return to the strategy of managing inheritance by means of the majorat, or primogeniture. From a nineteenth-century perspective, Karl Marx conducted a detailed analysis of this institution in his *Critique of Hegel's Philosophy of Law* of 1843–1844, which led him to the following conclusion:

> When birth, as distinct from the other determinants, directly gives a position to a human being, his body makes of him *this particular* social functionary. *His body* is his *social* right. . . . It is therefore natural that the nobility should be proud of their blood, their descent, in short the *life-history of their bodies*; it is, of course, this *zoological* way of looking at things which has its corresponding science in *heraldry*. The secret of the nobility is *zoology*.[22]

This polemical diagnosis points out an underlying irony. As indicated above, the goal of inheritance regulations like the majorat was to defend family holdings against powers that sought to weaken family ties. Ironically, this goal was accomplished by placing even more emphasis on the position of an individual within a system of abstract genealogical relations that were mediated by generation and birth, ultimately to the detriment of the rights and claims of groups of relatives such as siblings.[23] This led Marx to describe the majorat as the "superlative of private property."[24] In the early modern period, this dialectic became all the more pronounced as mobile property gained in importance in relation to landed property.

The fact that inheritance law played a major role in the work of philosophers such as Georg Wilhelm Friedrich Hegel and Marx— another student of Hegel, Eduard Gans, wrote a four-volume book on

inheritance law and its history[25]—was an immediate consequence of the
French Revolution. The revolutionaries had criticized the majorat as a
"monstrous" institution that subordinated the interests of the living to
those of the long dead. On November 14, 1792, the so-called *substitu-
tions*—the French form of the majorat—were prohibited by decree. Less
than a year later, Article 28 of the constitution issued on June 24, 1793,
stated categorically but ambiguously, "One generation cannot subject
to its law the future generations." This statement was not meant to
establish absolute freedom of will with respect to inheritance. On the
contrary, the aim of this law was to suppress, or at least contain, the
elements of caprice that were contained in institutions like the majorat.
The Napoleonic Code, therefore, later furnished carefully crafted inheritance rules that were based on the complete divisibility of property and
on equal rights among all inheritors, including the wife. The code pre-
scribed that an inheritance should fall to "the children and descendants
of the deceased, to his ancestors and collateral relations." It was to be
divided according to a formula that resulted from an analysis of kinship
degrees based exclusively on the "number of generations," without "[re-
spect for] the nature nor the origin of property" and without "distinction
of sex or primogeniture." The only ones to be excluded from inheritance
were "natural," that is, illegitimate children.[26] Later, equal inheritance
came under attack once again, and primogeniture was reintroduced in
both France and Prussia. During the early decades of the nineteenth century, political discussions often revolved around patriarchal authority,
the rights of the firstborn, the status of wives and illegitimate children,
the value of landed property, and the abstract power of money. These
topics often formed the narrative of contemporary literature as well.[27] In
*Democracy in America* (1835–1840), the French state theorist Alexis de
Tocqueville even described equal inheritance as the fundamental precondition of democracy, as a "machine" that, once set in motion, "divides,
distributes, and disperses both property and power."[28]

Another salient development in this period of transition around 1800
needs to be mentioned in this context. From the mid-eighteenth century, marriage patterns in all social strata underwent substantial change,
which continued well into the nineteenth century. Marriages between
close kin, especially between first cousins, increasingly became the norm,
although prohibitions of such unions persisted in many places. Once
again, this can be understood as a strategy for holding family property
together by establishing mutual loyalties between family lines, a strategy
that was both reinforced and counteracted by the increasing capitaliza-
tion of society. The Darwin and Wedgewood families afford a nice ex-

ample of this strategy in Victorian England. They established multiple ties through cousin marriages over several generations, though this also fanned degeneration fears among several members of the family, particularly Charles Darwin, who had married his first cousin Emma Wedgewood.[29] David Sabean associates this widespread practice with a general shift "from vertical to horizontal relationships, from 'clan' to 'kindred,'" and he explains the preference for cousin marriage by highlighting that marriage was increasingly understood as a relationship "among people on the same cultural, class, and stylistic plane."[30] Close parentage could serve as a guarantee for such a relationship of equality. Bourgeois circles in the nineteenth century therefore engaged with equal, if not even more, fervor in the kinds of genealogical inquiries that had previously been the preoccupation of the noble classes. We will come back to this topic in chapter 5.

The three developments sketched out above—the increasing importance of agnatic lines of descent in the early modern period, their replacement with closely knit networks of kindred at the dawn of modernity, and the first attempts to establish inheritance regimes based on equal rights and full divisibility of property—have one thing in common. They reflect increasing attempts to ground social relationships in an analysis of kinship that prioritized individuals over groups of kin and blood relationships over other kinds of relationships (such as wardenship). However, we should emphasize again what we have already pointed out with respect to the medieval period: ideologies of "blood" were only loosely connected to generation theories during the entire early modern period, including the eighteenth century. In the latter part of the seventeenth century, for example, the Italian lawyer Paolo Zacchia provided a careful analysis of contemporary generation theories in his *Quaestiones Medico-Legales* (1661), one of the most influential works in early modern legal medicine. He argued that, in deciding paternity disputes, one should rely on similarities in "temperament," as indicated by similar habits, inclinations, and diseases. In practice, however, judges remained highly skeptical about similarity as a reliable indicator of genealogical descent, and this skepticism persisted well into the nineteenth century. After all, so the argument ran, similarities might be the result of accidental causes such as maternal imagination.[31]

Political and legal questions of inheritance were thus relatively independent from the problems that occupied physicians, natural philosophers, and naturalists when they discussed generation. Of course, unmistakable echoes of political themes can be found in the generation theories examined in chapter 2. The debate between traducianists and creatianists, for instance, turned on the question of how kinship was to be evaluated against the relationship with God, and thus the church, that all humans shared. If the soul of an individual, as traducianists held, was a continuation of the parental soul transmitted by generation, then secular relations of parentage carried some weight, whereas creatianists who believed that each soul was directly created by God had to come to the conclusion that all humans were "brothers" and "sisters" alike. A similar case can be made about the preexistence theories that were so popular in the seventeenth century. They deprived patriarchal authority, and its institutional counterpart, the majorat, of an important foundation by maintaining that the embryo developed in complete independence from its worldly procreators. Despite such occasional points of contact and reference, however, eugenic ideas were rarely put forward before the second half of the nineteenth century. The expertise of a physician at the late medieval court of Aragon, and a few eugenic treatises that appeared at the end of the eighteenth century, remain rather exceptional.[32] That one could organize society on the basis of a theory of heredity may have been on the horizon of the imaginable since Plato's breeding fantasies in *The Republic*, but in premodern times such a view never developed into a school of thought that guided political action.

### Medicine

As mentioned earlier, medicine was the first domain in which metaphors of inheritance were employed in a biological context, particularly to designate diseases that recurred within families. Up to the eighteenth century, however, this usage was sporadic, mostly consisting of more or less offhand remarks that some disease, such as gout or leprosy, tended to be heritable. Nevertheless, three contributions to this tradition stand out as systematic attempts to understand hereditary diseases. In the second decade of the fourteenth century, the northern Italian physician Dino del Garbo grappled with the question, "if a disease, that exists in the father, can be heritable in the son." In 1605, the Spanish physician Luis Mercado, who served as court physician to Philip II of Spain from 1592 to 1598, published an entire treatise dedicated to hereditary diseases. And at the end of the eighteenth century, the Parisian Société Royale de Médecine received several essays on the same topic written in response to two prize competitions, issued in 1788 and 1790, respectively. Although far from continuous and coherent, this body of medical work is significant for our discussion, for it resulted in several conceptual clarifications and distinctions that provide part of the background for the epistemic space of heredity that emerged in the first decades of the nineteenth century.

In scholastic medicine, as Maaike van der Lugt has emphasized, hereditary diseases played only a minor role. The dominant doctrine of disease was that of humoral pathology, which defined diseases as disturbances in the balance of the four body humors, blood, phlegm, yellow bile, and black bile.[33] Each of these humors was associated with a pair of qualities, and the bodily constitution or "temperament" of a person was determined by the mixture of the humors. Depending on which of the humors and associated qualities predominated, one thus distinguished between sanguine (hot and wet), phlegmatic (cold and wet), choleric (hot and dry), or melancholic (cold and dry) temperaments. What we would call "environmental factors," on the other hand, were accounted for, since Galen, under the heading of the "six non-naturals": light and air, nutrition, movement, sleep, excretions, and emotions. They were referred to as "non-naturals" because they could be influenced by the physician or patient. According to humoral pathology, diseases came about if deprivation or excess of one of these external factors moved the balance of humors beyond what was normal for a person with a given temperament. Treatment aimed to restore this balance by soliciting changes in lifestyle (for example, by keeping to a certain diet) or by forcing removal of surplus humors (for example, by bloodletting or prescribing vomitive agents). As we have seen, the phenomena of inheritance had little influence on medieval and early modern ideas of generation. Nor did they have a bearing on the predominant conception of disease. Diseases were brought on by incidental or periodically recurring factors, and they beset individual bodies; diseases were not thought of as entities that could be abstracted from individual bodies.[34]

Dino del Garbo's disputations, preserved in manuscripts that date from around 1320, are therefore a remarkable exception, for they deal explicitly and exclusively with hereditary diseases. Dino del Garbo taught theoretical and practical medicine at several universities in northern Italy (Bologna, Padua, Siena, and Florence), and he was one of the first European physicians who read and commented on Avicenna's *Canon medicinae*. At the beginning of this medical textbook, which remained in use throughout Europe well into the seventeenth century, Avicenna distinguished between contagious, local, and hereditary diseases, but offered no further discussion of this classification. Dino del Garbo, in contrast, carefully distinguished between truly hereditary diseases (*morbi ex hereditate*) and connate diseases (*morbi ex generatione*), and he did so on the basis of an analysis of the legal concept of inheritance. Connate diseases, according to Dino del Garbo, were caused by events that led to changes in the seed or embryo during conception and pregnancy. Hereditary diseases, in contrast, had to be present in one of the parents already. Otherwise, as Dino del Garbo argued, the analogy with the transmission of worldly goods would not be warranted. A similar distinction between hereditary and connate diseases was made slightly later by John of Gaddesden, a physician connected to the court of Edward II in England.[35]

Dino del Garbo's punctilious distinction was not widely adopted, even though the phrase "hereditary disease" subsequently appeared in European medical literature from time to time. Many authors applied the term to diseases that, according to Dino del Garbo, should have been considered connate, such as leprosy, which most authors, in accordance with popular opinion, believed to arise from intercourse during menstruation. The reason for this lack of interest in Dino del Garbo's distinction between hereditary and connate diseases was the continuing dominance of humoral pathology. Hereditary diseases, as Dino del Garbo clearly recognized, required a different explanatory framework. Referring back to Aristotle's theory of generation, he suggested that hereditary diseases should be explained not by a disturbance in the balance of humors, but by a permanent change in the "formative power" (*virtus formativa*) of the male seed.[36]

Luis Mercado's seminal treatise "On hereditary diseases" (*De morbis hereditariis*) appeared in 1605 in the second volume of his collected works. In it, Mercado—who had taught "medicine according to Avicenna" at the university of Valladolid since 1572 and later, as royal physician, had overseen medical affairs across the Spanish empire—undertook a systematic exploration and application of the idea of hereditary disease. Like Dino del Garbo, Mercado proceeded from the assumption that hereditary diseases resulted from a permanent change in the *vis formativa* of the seed, but in contrast to Dino del Garbo, and following Galen, he assumed that both the male and female parent produced seed that was mixed in generation. Hereditary diseases, as he put it, consisted in a changed "character" of the body that was "preternatural" (*praeter naturam*); that is, it deviated from what was the case under normal circumstances. This abnormal "character" and its continuation over generations, in turn, were attributed to the "power of the seed of the parents, grandparents, or great-grandparents." It was as if, Mercado explained, "by an instrument or some other cause nature so orders the bringing forth of individuals similar to oneself and deformed by the same defect."[37] The expression "character" was borrowed from the Greek and originally referred to instruments used for branding domestic animals or stamping money. Mercado also used the term *sigillatio*, derived from *sigillum* (Latin for "seal"), which in theology denoted the indelible, but invisible, nature of the sacraments.[38]

The appearance of Mercado's treatise coincides with an early-seventeenth-century divergence in conceptions of heredity into what, from the vantage point of the present, can be designated as "hard" and "soft" positions. Soft explanations identified some kind of infection or pollution of the seed by harmful humors as the cause of hereditary diseases. They appear "soft" in hindsight because the question of whether the harmful influence acted "horizontally" on the embryo—for example, during pregnancy—or whether it was passed on "vertically" to descendants through the seed was of minor significance. Both processes seemed equally possible. Within the framework of humoral pathology, such explanations tended to regard hereditary diseases as treatable. One merely had to neutralize the harmful influences. "Hard" explanations, in contrast, assumed the transmission of some feature of bodily constitution that was not easily influenced by external factors and predisposed developing individuals to the disease in question. This model allowed one to make a sharp distinction between hereditary diseases and other diseases, and in the late eighteenth century this model was accommodated in solidist conceptions of disease, which explained disease via lesions in the structure of the body.[39]

The rift between epigenetic and preformationist generation theories that appeared in the seventeenth century roughly coincides with this divergence between soft and hard explanations of hereditary disease. However, one has to be wary of assuming that these explanations are mutually exclusive. Just as epigenesis and preformation captured different aspects of generation, soft and hard conceptions of heredity were often employed in concert to account for different aspects of hereditary disease. Mercado, for example, often referred to hereditary diseases as "incurable," but according to his model of inheritance, the fate of a family stricken by such a disease was far from sealed. For one thing, adequate therapy could alleviate the symptoms of the disease or even suppress them completely, and such treatment could, if carried on for a number of generations, finally also penetrate to the hereditary "character" of the seed. In addition, Mercado believed that the diseased "character" of one partner could be compensated for in generation by the healthy character of the other partner. One could thus prevent the outbreak of a hereditary disease among one's children by choosing the right spouse, even if this did not put a stop to the transmission of the underlying disposition toward the disease to further generations. In this way, Mercado not only left room for medical advice and action with respect to hereditary diseases, but also provided an elegant explanation for the fact that hereditary diseases sometimes seemed to skip one or more generations.

Mercado's model of inheritance was thus flexible enough to leave room for conceptual innovations. In the course of the eighteenth century, diseases that were considered to be hereditary received more and more attention from physicians, as is evident from the two aforementioned prize competitions of the Société Royale de Médecine in Paris. Terms such as "character," "temperament," or "diathesis," which all referred to constitutional dispositions toward certain diseases, now became malleable concepts at the core of debates among physicians concerning the nature of hereditary diseases (*maladies héréditaires*).[40] The meaning of heredity became ever more nuanced as the result of a set of distinctions and criteria that were articulated and specified in these debates. For example, physicians began to distinguish between congenital, connate, and acquired diseases; observational criteria such as homochrony—the curious fact, already noticed by Montaigne, that hereditary diseases tend to afflict individuals at the same age—gained in significance; and a number of causal concepts like latency, as well as various versions of soft and hard mechanisms of heredity, were proposed to account for the complex and sometimes capricious patterns of hereditary transmission. Heredity thus gradually became a process of fundamental medical relevance.[41] Still, some voices—among those who sent essays to the prize competition of the Société Royale de Médecine, but also throughout the entire nineteenth century—continued to raise doubts about the idea that something like an inheritance of bodily characteristics and dispositions existed at all.

The reasons for the growing significance of heredity in medicine are complex, but two stand out, especially if one gives due consideration to the timing of the two prize competitions of the Société Royale de Médecine, published as they were in the two years on either side of the French Revolution of 1789. First, hereditary diseases were mobilized in political arenas. As Gianna Pomata observed, "a recurrent feature of the medical discourse on heredity in the eighteenth and early nineteenth century is the critique of the aristocratic family."[42] Accordingly, in the eighteenth century, physicians were most interested in so-called "noble" maladies like gout, which was believed to be softly inherited from ancestors who indulged in luxury and overconsumption. Charles Darwin's grandfather Erasmus shared this interest in hereditary diseases.[43] In the nineteenth century, after the aristocratic family model had buckled under sustained political pressure, the focus of medical attention shifted toward "degenerative" diseases like phthisis (tuberculosis) and madness. These diseases were typically ascribed to the rapidly growing class of landless and poor people migrating to urban centers, who were seen as a social threat. Concurrently, explanations tended to shift toward models of hard inheritance that left little room for betterment through changes in conditions of life.[44]

A second set of reasons for the changing medical views on hereditary diseases can be located in the new social role physicians acquired at the turn of the nineteenth century, especially in postrevolutionary France, but later on in the rest of Europe as well. The new scientific profile that physicians sought to establish for themselves made it advisable to acknowledge the limits of medicine, particularly in treating incurable, constitutional, and thus heritable diseases.[45] Additionally, the new responsibilities of physicians for public hygiene and health favored a definition of dangers such as heritable diseases, which lay hidden in the populace and which, consequently, only the expert could discern. This was the background of the two prize competitions of the Société Royale de Médecine, which in this respect was competing with the Académie Royale de Chirurgie, whose long-term secretary Antoine Louis had already published a treatise on hereditary disease in 1748.[46] Finally, the hospitalization of patients and the resulting attempts at medical statistics made possible unprecedented forms of representation, such as medical topographies and chronicles. Far beyond recording individual cases or family histories, it began to appear feasible to build up a profile of the population as a whole. The genealogical and epidemiological data that were consequently accumulated certainly constitute one of the most important prerequisites for disentangling complex patterns of familial diseases and for telling them apart from diseases caused by local influences or differences in lifestyle.[47]

If we now look back at the early history of hereditary disease, we can draw two conclusions. First, we find it striking that some of the physicians who first developed an interest in hereditary diseases, notably John of Gaddesden and Luis Mercado, were closely connected to royal courts and thus served not only as personal physicians but also as advisers on and overseers of health policies. Not much is known about the roots of scholastic thought about hereditary disease in the Arab-Islamic tradition, but it is certainly not a coincidence that Avicenna suggested that hereditary diseases form a special category at a time when the Arab-Islamic empire was at the height of its power. The work of Dino del Garbo highlights another context. He composed his disputations at a time when vehement and polemical debates raged at European universities about the nature and status of nobility. At the University of Paris, for example, a number of satirical arguments pursued the question of whether noblemen, like the races of some dogs, were characterized by long ears.[48] In general, it appears that hereditary diseases tended to become topical in contexts in which the "body politic" of whole populations, rather than the individual body, was at stake.

## Natural History, Breeding, and Anthropology

As we suggested at the beginning of this chapter, the history of the notion of hereditary disease leads to a surprising insight. Attempts at generalizing a medical concept of heredity did not, as one might expect, emerge from ever-growing attention to similarities between parents and offspring in general, from a fixation of the scientific mind on regularity at the expense of contingency and complexity. Considerations of and research into heredity were rather propelled by observations concerning the fluctuating patterns and processes—the appearance, distribution, and transmission of individual peculiarities, diseases, and monstrosities—that structured life at the subspecific level. Heredity, in other words, had more to do with variation than with the permanence of species.

Immanuel Kant's anthropological essays helpfully illustrate and expand this point. His essays, the first German writings that speak of inheritance in a biological sense, deal with a specific phenomenon, the existence of human varieties or "races," as Kant called them. Races differed in properties that mixed in "hybrid" unions but were otherwise "inevitably" transmitted to offspring, even under changed environmental conditions. For Kant, this observation undercut the distinction between specific forms and individual peculiarities. That the races could interbreed and produce viable and fertile offspring demonstrated, he argued, that human races belonged to one and the same species. By the same token, he held that the properties that distinguished the races had to be counted as individual peculiarities that characterized varieties, not species. But these same peculiarities, on the other hand, were infallibly reproduced in each generation even under new climatic conditions. To quote his favorite example, Portuguese living in Africa continued to produce white children and Africans who settled in Europe went on having black children. Race characteristics thus seemed to be subject to the same kind of regularity that otherwise governed specific characters. To take account of these facts, Kant proposed a concept of heredity (*Vererbung*) that brought together natural law and contingent family history. The dispositions, or *Anlagen*, for hereditary traits, such as those of the races, were present from the very beginning in the ancestral trunk of humanity. They were thus, in a sense, preformed and not acquired. But once they had been expressed as actual traits in reaction to a change in environment—that is, after having been acquired epigenetically—they were permanently and irrevocably transmitted from generation to generation.[49]

The way Kant set up the problem and the way he sought to solve it is a paradigmatic example of the emergence of heredity as a biological phenomenon. The problem did not manifest itself in the constancy of species, but rather in the patterns of variety that structure life at a sub-specific level. As long as such patterns coincided with locally circumscribed environments, they could be explained by the permanence of ties between living beings and their natural environments. Black and white skin color were simply reactions to specific climatic environments. If one were to apply the notion of inheritance to such situations at all, it would be the environment that "inherits" its inhabitants and impresses its character on them, rather than the other way round. The need to apply a complex metaphor like heredity to make sense of the proliferating phenomena of change and stability thus arose only when local ties were dissolved and the relationships between forms, places, and modes of transmission multiplied.

Within natural history, botanical gardens and menageries played a major role in creating such phenomena, a role comparable to that of hospitals in the study of hereditary disease toward the end of the eighteenth century. To be able to abstract heredity from environmental effects, organisms had to be transplanted from their native environments. And that is exactly what botanical gardens and menageries, mostly set up in the context of the mercantile and colonial expansion of Europe, were designed to do. Organisms from all over the world, and thus from vastly different climatic and ecological backgrounds, were assembled in one place and subjected to more or less controlled conditions, under which they subsequently reproduced. The exchange of specimens among institutions of this kind only enhanced the facility of making visible, observing, and describing even complex and obscure patterns of hereditary transmission, such as atavisms, segregation of characters, or mutations. It is therefore hardly surprising that botanical and zoological gardens, which were originally founded for the descriptive purposes of natural history, also provided the first sites at which questions of heredity were tackled experimentally.[50]

The first botanical gardens were established in the middle of the sixteenth century at the Universities of Padua and Pisa (exact dates, as well as priority, are debated).[51] In contrast to monastery gardens, which were cultivated mainly to provide local people with medicinal plants, university gardens exclusively served the purpose of representation. These gardens supported the education of medical students in pharmacology, served as reference collections for spice merchants and apothecaries, and sustained the research of naturalists. The significance of these gardens for the development of the life sciences lies in the fact that they operated as global institutions. Even the founding directors of the botanical gardens at Padua and Pisa were already traveling extensively to collect specimens, and they engaged in correspondence and exchange networks with other botanists across Europe with the aim of adding to their collections. In the course of the late sixteenth and seventeenth centuries, botanical gardens were founded in cities all over Europe (Bologna, 1567; Leiden, 1577; Heidelberg, 1593; Montpellier, 1593; Oxford, 1633). Alongside the universities, the courts, trading companies, and associations of apothecaries (linked with the Chelsea Physic Garden near London), wealthy citizens supported these establishments as well. Botanical gardens also spread to European overseas colonies. In 1653, for example, a botanical garden was set up in the Dutch colony at the Cape of Good Hope. During the nineteenth century, the Royal Garden near London (today known as Kew Gardens) and the Jardin du Roi in Paris (which was renamed Jardin des Plantes after the revolution and became part of the Muséum national d'Histoire naturelle) evolved into centralized state institutions that became deeply involved in the worldwide exploration and colonial exploitation of natural resources.[52]

As the activities of these natural history institutions became more and more global, a problem came to the fore that had not presented itself as long as they had been dedicated to the continuation of local pharmaceutical traditions. To be able to communicate about plants and animals across regional and cultural divides, naturalists had to settle on criteria that would allow for the reliable identification of plants and animals wherever they were encountered. In 1686, the English botanist and natural theologian John Ray therefore proposed to rely exclusively on characters that "originate from the seed of the same plant."[53] Characters that could be manipulated by external factors (such as the color of flowers, which could be changed by watering them with dyes) and naturally arising differences (such as those related to the sex of an organism) were not considered sufficient to distinguish species. Such variable characters identified varieties at most. As discussed in the previous chapter, Carl Linnaeus generalized this criterion roughly fifty years later with his species definition. And a few years later, Georges Buffon—in a section of his *Histoire naturelle* dedicated to the mule—introduced a related criterion for zoology by proposing that animals that were able to produce viable and fertile offspring should count as one species, whatever differences they showed in other respects. What constituted a species, Buffon proclaimed, was "neither . . . the number nor the collection of similar individuals, but the constant succession and renovation of these individuals."[54]

The impetus behind these species criteria was that they could be put into practice in botanical gardens. Botanists who imported plants from all over the world could eliminate all environmental variation on their plots by establishing a homogeneous cultivation regime. This had the result of accentuating the constancy of species differences. The reproduction of species could thus, at least in principle, be tied to the behavior of individual traits across generations. Yet organisms did not always conform to the ideas of naturalists. In 1744, a student drew Linnaeus's attention to a wild population of common toadflax (*Linaria vulgaris*) in which a few specimens exhibited a radically different morphology: the flower parts of these individuals were not arranged in the usual bilaterally symmetrical pattern, but rather in a radial fashion (fig. 3.3). As both varieties were otherwise similar and grew at the same closely confined spot in the wild, one had to assume that they were of common descent. And yet they showed a pronounced difference in flower morphology, normally a constant character. Linnaeus called this new variety *Peloria* (derived from the Greek word for "monster") and compared its discovery to that of the regenerative potential of the freshwater polyp that had caught and fueled the imagination of European naturalists at the same time. He later succeeded in reproducing *Peloria* in the botanical garden at Uppsala, further proof that it was a distinct and constant variety.[55]

Many observations of this kind were reported in the course of the biogeographic and taxonomic exploration of floras and faunas around the world during the eighteenth century. These observations led to the development of two strategies for explaining the observed variation patterns. The first of these explanations relied on the two common assumptions that organisms that spread into new environments were altered by the conditions they encountered and that such changes would also be transmitted to their offspring in the long run. Eighteenth-century naturalists referred to such processes as "degeneration," but often without implying that they were pathological (this only became part of the nineteenth-century meaning of the term). A degenerating influence was also, and especially, ascribed to the human practice of domesticating plants and animals. Buffon was one of the first naturalists to speculate that degeneration in this sense could even lead to the formation of new

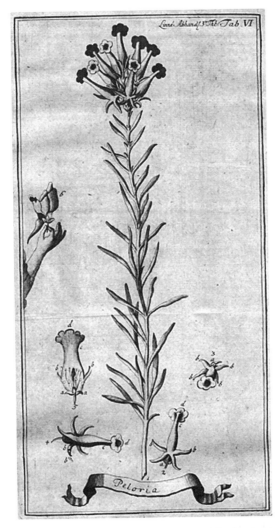

FIGURE 3.3 *Peloria*, a mutation of the common toadflax (*Linaria vulgaris*). Eighteenth-century copperplate. From *Carl Linnaeus Auserlesene Abhandlungen aus der Naturgeschichte, Physik und Arzneywissenschaft*, edited and translated by E. J. T. Hoepfner, 3 vols., Leipzig: Adam Friedrich Böhme, 1776–1778, vol. 3, 175.

species. Moreover, with the theory of evolution that Jean-Baptiste de Lamarck outlined in his *Philosophie zoologique* (1809), the inheritance of acquired characters became a standard element in the set of mechanisms that biologists employed to explain the formation of new varieties and species.[56] It also formed the basis for the many attempts to acclimatize exotic plants and animals to Europe in the late eighteenth and nineteenth centuries.[57] Societies like the French Société Nationale d'Acclimatation were established for the purpose of exploring this possibility, and such societies often entertained their own experimental gardens and farms.[58]

The second strategy for explaining heritable variation focused on the process of hybridization, the production of offspring by cross-fertilization between members of two different species. Linnaeus, in particular, believed that hybridization could produce new species. In 1759, he carried out the first carefully controlled hybridization experiment by crossing two color varieties of the showy goat's-beard (*Tragopogon pratensis*). The results of these experiments were submitted to the St. Petersburg Academy of Sciences in the context of a prize competition for studies on the sexuality of plants. This in turn inspired the botanist Joseph Gottlieb Kölreuter, who was active in St. Petersburg at the time, to carry out a large series of hybridization experiments with many different plant species. Contrary to Linnaeus's contentions, Kölreuter wanted to demonstrate that parental characters did not enter into new combinations in hybrids, but rather that the characters fused, which meant that one species could be gradually transformed into another by repeated backcrossing.[59] Subsequently, other naturalists in Europe occupied themselves with the question of hybridization, among them Carl Friedrich Gärtner, an apothecary in the Swabian town of Calw, as well as Charles Naudin in France and William Herbert in Britain. Gärtner's book, which was based on decades of experiments with about seven hundred different plant species and published in 1849, contains the most exhaustive pre-Darwinian synthesis on questions of plant hybridization. The general conclusion that he drew from his researches was that "types of hybrids" were not "vague and variable, but constant, obeying certain invariable laws of formation [*Bildungsgesetze*]"; they were "brought about constantly and regularly [*gesetzmäßig*] from the same factors again and again."[60] Only a decade later, this conclusion provided the starting point for Gregor Mendel's famous experiments with peas.[61] In zoology, a comparable experimental tradition was lacking, although Buffon carried out hybridization experiments with dogs as well as goats and sheep on his rural estate at Montbard. Keeping large numbers of animals usually entailed too many difficulties to make this a viable experimental program for naturalists.[62]

The experiments just outlined, together with the countless observations of the geographic distribution and variation of plants and animals that were reported in the taxonomic literature, resulted in a number of conceptual innovations. Again, the process resembles that which we described for medicine in the previous section. When Gärtner published his voluminous tome in the mid-nineteenth century, it was well known that inheritance affected some characters in a mosaic-like fashion—that is, that parental characters were passed on independently of one another and recombined in offspring—while other characters blended in offspring, thus assuming a middle ground between the parental varieties. It was also well known that some characters showed a tendency to dominate in offspring, while others receded and remained latent only to reappear in later generations; that the offspring of hybrids showed a tendency to "revert" or "regress" to either of the original forms; and that it was often of no consequence for the product of a hybridization whether the hybridized varieties were represented by the maternal or the paternal organism. Even though metaphors of inheritance continued to be used sparingly—in Gärtner's book we find the term once, while Mendel, in his short essay of 1866, used it twice—the literature on hybridization already defined an epistemic space that was focused on the transmission of individual properties.

One might expect that naturalists would have readily adopted the knowledge about the propagation of plants and animals that gardeners, farmers, and breeders had accumulated over centuries. We have already mentioned the extensive body of literature on the rearing of falcons and dogs that had its origins in the medieval period and that contained deliberations about how the "nobility" of certain "races" was passed on to offspring through the seed.[63] Yet until far into the nineteenth century, this literature treated the whole complex of factors that could be used to improve domesticated plants and animals. As much attention was paid to the nourishment and general care of creatures as to the importance of keeping local breeds "pure" or "upgrading" their "blood" occasionally by crossing them with imported breeds. "Blood" and "soil" remained inextricably entwined in this discourse.[64] "There are as many genera, species, and varieties," the baroque horticulturalist Petrus Laurembergius therefore claimed with reference to wine, "as there are regions, cities and towns."[65] Cultivated varieties largely remained tied to specific localities in the early modern period.

In this respect, an important development occurred in the late seventeenth and eighteenth centuries. Some breeders began to specialize in the production of plant seed and animal stock whose quality was determined not only by certain desired traits, but also by their ability to pass on these traits reliably to offspring reared at distant locations.[66] This ability, sometimes termed "prepotency," was especially in demand among gardeners who cultivated vegetables and fruit in the surroundings of rapidly growing urban centers, especially in England and the Low

Countries, and among agriculturalists who produced raw materials such as wool and cereals for manufacture and industry. Thus, as the cattle breeder John Saunders Sebright wrote in 1809:

> Regard should not only be paid to the qualities apparent in animals, selected for breeding, but to those which have prevailed in the race from which they are descended, as they will always show themselves, sooner or later, in the progeny: it is for this reason that we should not breed from an animal, however excellent, unless we can ascertain it to be what is called *well bred*; that is, descended from a race of ancestors, who have, through several generations, possessed, in a high degree, the properties which it is our object to obtain.[67]

In the course of the eighteenth century, breeders began to organize themselves in professional associations, and they began to exchange their breeds across long geographic distances in order to test their "inheritance capacity" in offspring. A sensational precedent in this respect was established by the Swedish industrialist Jonas Ahlströmer, a contemporary and close associate of Linnaeus. For centuries, fine merino wool had been a monopoly of Spain, where huge flocks of sheep were driven from their summer grazing grounds in the Pyrenees to the lowlands each autumn. Shepherds, with the help of bureaucrats, sought to ensure that the individual flocks did not interbreed during their annual journeys. This unique way of keeping sheep—unique both ecologically and in its avoidance of outbreeding—made it seem unlikely that merino breeds could be established anywhere else but in Spain. In 1723, however, Ahlströmer managed to import a few merino rams, to cross them into local Swedish sheep races, and to create a breed that both gave wool of high quality and endured the Swedish climate. Many sheep breeders followed his example, and by the end of the eighteenth century, merinos were kept in Saxony, Prussia, England, and France.[68]

Animal as well as plant breeders thus carried out transplantation and hybridization experiments that were quite similar to those that naturalists conducted in their gardens and menageries. And, like naturalists, they began to develop a technical vocabulary to describe their procedures and the phenomena they observed. Expressions such as "improvement," "breeding in-and-in," "disposition," "strength of inheritance" or "procreative power," "fixedness of character," "hereditary defect," and "regression" became common parlance among breeders. This common ground notwithstanding, the relationship between practical breeders and naturalists remained complicated and strained for some time. Institutional and social barriers prevented the development of a unified perspective on inheritance until well into the nineteenth century. On the one hand, breeding practices and their products were not seen as providing adequate materials for serious botanical and zoological study. One reason for this was that breeding produced an infinitude of varieties, and naturalists in the Linnaean tradition felt they had to abstract from these to describe "natural" species.[69] On the other hand, breeders like Robert Bakewell in England—whose "Dishley" sheep, bred for meat production, achieved record prices on the market—focused their efforts on individual, marketable traits and tried to "fix" these alone. This focus was alien to naturalists, whose interests remained in the origin, stability, and possible transformation of what they considered to be "true" or "good species," species that differed from one another in many respects rather than in individual traits.[70] It was only in the last third of the nineteenth century that researchers like Gregor Mendel began to adopt the kinds of questions and methods that some of the more advanced breeders had occupied themselves with for almost a century already.[71]

The eighteenth century was also the era in which naturalists like Georges Buffon and physicians like Johann Friedrich Blumenbach began to elaborate a "natural history of mankind." In chapter 5 we will examine the important role that racial anthropology and eugenics played in the emergence of biological theories of inheritance. For now, we only want to take a brief look at the sources that anthropologists relied on when they classified humans on the basis of external physical characteristics such as skin color or the shape of the skull. As we will see, these anthropological sources throw into relief the historical conditions that shaped the epistemic space onto which metaphors of inheritance could be projected.

The concept of human races was not simply an invention of Enlightenment anthropologists, albeit claims to this effect are frequently made. The founders of racial anthropology rather drew on travel reports that described a peculiar system of social castes that had taken shape in the Spanish and Portuguese colonies of Latin America during the seventeenth century. Known as *las castas*, this system regulated how legal and social status were distributed among different sections of colonial society where "miscegenation" among the corresponding groups was common. The system operated on the basis of a classification of the population according to skin color and other physical characteristics as well as geographic origin. Children resulting from mixed marriages were positioned in this scheme according to their filiation. During the eighteenth

**FIGURE 3.4** Mexican *castas* paintings from the eighteenth century. (A) *De Español y Negra Mulata.* (B) *De Español y Alvina Negro Torna atras.* With kind permission from the Museo de América.

century, the system of *castas* also found expression in a rich pictorial genre in which sets of pictures were arranged in serial or tabular form (fig. 3.4). Each individual picture showed a mixed couple and its child, and each bore an inscription that stated the components entering the mixture—that is, each parent's *casta*—as well as the result—the child's *casta* (fig. 3.4A).

"From a Spaniard and a Black woman [results] a Mulatto"; "From a Spaniard and a Mulatta [results] a Morisco"; "From a Spaniard and an Indian woman [i.e., a Native American woman], [results] a Mestizo"—

in this way, all sorts of combinations were labeled with some seventy terms. Buffon offered the following translation of this system in a section of his *Natural History* that was dedicated to human races: "*D'un blanc et d'une négresse sort le mulâtre à longs cheveux*" (From a white [man] and a black woman issues the mulatto with straight hair).[72] And, as mentioned earlier, this work was Immanuel Kant's starting point for speaking about the inheritance of racial traits in his anthropological essays.

The *castas* system is of interest above all because it highlights that race and inheritance did not simply complement each other. To put it differently, the "races" distinguished in the system were not simply different species whose stability was beyond doubt. Skin color, it is true, provided a manifest trait for the differentiation of *castas*. But skin color was also a highly variable characteristic par excellence whose significance was far from evident.[73] Indeed, black skin color was not associated with Africa until the early modern period, when it was also attributed to direct descent from Ham, one of the sons of Noah.[74] Skin color did not become an object of anatomical and physiological studies and a subject of speculation about its historical origin until the end of the seventeenth century.[75] Much like hereditary diseases, differences in skin color were regarded as individual, contingent deviations rather than essential characteristics. Even in the *castas* system, the decisive criterion for the allocation of an individual to a particular caste was not provided by skin color alone, or indeed by any other physical trait. For determining the *casta* of a person, it was sufficient to know the *castas* of his or her parents. Owing to this analytic structure, the *castas* system could absorb a wide array of phenomena while remaining stable in its basic outlines. The system even accounted for some of the more capricious patterns of heredity, as exemplified by a special caste in the system, the *torna atras* (literally "throw-backs"). The *torna atras* issued from a Spaniard and an *alvina*; that is, a white, blonde, and blue-eyed woman who had one black great-great-grandmother. And despite the skin color of its parents, the child was usually depicted as "dead" black (fig. 3.4B).

Direct experimentation was out of reach in anthropology. Nevertheless, the discipline became a hotbed of debate about heredity. On the one hand, anthropologists had recourse to what appeared to be "natural experiments," such as cases of "wolf children" that had grown up in isolation or in the wild.[76] On the other hand, the exponential growth of anthropological reports from regions outside Europe created what Philip Sloan has called a "veritable 'laboratory of human nature.'"[77] The system of *castas* described above, for example, constituted "a vast field of 'pre-Mendelian' investigation."[78] It was the empirical basis of what became a universal scheme of racial distinctions after Linnaeus had promulgated it to a wider audience of scholars in his *Systema naturae* (1735), which led to the development of various further theories about the origin of racial differences within the human species.

Like the concept of caste, its correlate in cultural history, the concept of human races subsequently took on broader biopolitical significance. The unequal distribution of wealth, power, and social opportunity was increasingly explained and justified via references to differences in the physiology of human reproduction. The same is true for the opposition of *nature* and *nurture*. Emerging in discussions of so-called wolf children, these notions inspired speculation about the origin and (self-) reproduction of genius around 1800 and later fueled the Victorian ideology of the "self-made man"—long before Galton published his book *Hereditary Genius* (1869).[79] In all these settings, whether literary or scientific, the crucial point was to find ways to relate the specific and the individual in conceptions of elementary dispositions. While dispositions or *Anlagen* manifested themselves in selected individuals only, their sum was seen to constitute the species as a whole. In these contexts, the puzzle consisted in finding an explanation for something that appeared like an oxymoron: the reproduction of difference.[80]

The erosion of institutional, social, and cultural barriers, such as those that separated naturalists from breeders or medical practitioners from naturalists and anthropologists, later played a crucial role in the ascent of a discourse of heredity. In the course of the nineteenth century, it became possible to view the specific and the individual, the normal and the pathological—in short, the regular and the deviant—as determined by the same set of natural laws of organic reproduction. To quote Carl Friedrich Kielmeyer once again, the "life of the species" became the center of attention. The scope of nineteenth-century biology widened from a focus on the individual being to bring the life of a manifold of individual beings into view. As we will see, heredity soon took a privileged place in answers to the question of what it was that connected this life of the many. Borrowing from Foucault's archeology of knowledge, we might say that the discourse of heredity, once it had achieved the status of positive knowledge in separate domains, now stood at the "threshold of [its] epistemologization."[81]

# 4 First Syntheses

As we have seen, phenomena of heredity had not gone unnoticed before the end of the eighteenth century. At the same time, the fact that knowledge of heredity was scattered over several more or less separate domains indicates the absence of a general concept of biological heredity that could have established coherence between the observations made in these different domains. Such a general concept of biological heredity developed only in the first half of the nineteenth century. As we noted at the beginning of the previous chapter, Carlos López Beltrán has identified this change in a telling linguistic shift. Whereas usage of the adjective "hereditarian"—at least in the context of medicine—can be traced into the late Middle Ages, the noun "heredity" appeared on the scene only in the late eighteenth century, in the writings of Kant among others, and its systematic use commenced only around 1830. "Heredity" (*hérédité*) initially gained currency among French physicians and physiologists but soon spread through wider scientific circles in Europe. By the 1880s, the term finally acquired the status of a dominant and ubiquitous catchword in the life sciences.

According to López Beltrán, the nominalization of the concept of heredity signifies the coming into being of a "structured set of meanings that outlined and unified an emerging biological concept."[1] If heredity is to be understood as a dispersed epistemic space in the previous century, this space began to be theoretically articulated in subsequent developments during the nineteenth century. By the end of the century, heredity had been condensed into a number of objects of experimental research. In parallel, another shift took place: the erosion of a cluster of traditional distinctions concerning the similarities between parents and children or ancestors and descendants. Among them were distinctions being made between species-specific and individual, paternal and maternal, and normal and pathological similarities, as well as similarities concerning the left versus the right side of the body, which were all seen as due to distinct causes.[2] In the course of the nineteenth century, these distinctions gave way to a generalized concept of heredity. That concept thus increasingly captured elementary characters or dispositions that came to be seen as the general substratum of the particular forms and appearances of life, be they left- or right-sided, pathological or normal, maternal or paternal, individual or specific.

The emergence of such a generalized concept of biological heredity depended on the concatenation of phenomena that were initially located and studied in disparate domains of knowledge. As described in the last chapter, considerable institutional barriers stood between these domains. In the course of the eighteenth century, some of these barriers were overcome, not least through the philosophy of the Enlightenment era. Contemporary philosophical thought, which began to emancipate itself from the three traditional university faculties of theology, jurisprudence, and medicine, often indulged in sweeping claims that were made without much regard for experimental corroboration or disciplinary rigor, both of which later came to characterize the nascent natural sciences. However, it was precisely the pretension to direct, unmediated explanation that allowed eighteenth-century philosophers to leap across the boundaries of disparate domains of knowledge. This can be seen in exemplary fashion in the work of the French philosopher and scientist Pierre-Louis Moreau de Maupertuis.

In his pseudonymously edited work *Vénus physique*, published in 1745 and subsequently reprinted in numerous further editions, Maupertuis brought together his reflections with the observations of breeders and physicians and the results of breeding studies with dogs and other animals from his own private *ménagerie*. Moreover, he included genealogical information that he had collected about a family in Berlin whose members were now and then born with supernumerary fingers or toes. His work had been sparked, Maupertuis reports in the first edition, by the exhibition in a Parisian salon in 1744 of a "four or five year-old child" that "possessed all the characters of a negro, and whose very white and pale color only augmented its ugliness." With respect to this child, which he called a "little monster," Maupertuis made a remark that appears to anticipate the generalized concept of biological heredity of the nineteenth century: "Whether one regards this whiteness as a disease or some kind of accident, it will be nothing but a hereditary variety (*variété héréditaire*), that will confirm itself or vanish in the course of generations."[3] Further, Maupertuis also considered the variegated garden plants as members of this category of "hereditary varieties."

It would nevertheless be inappropriate to think of Maupertuis or of contemporaries like Buffon—who had obtained some of the dog varieties that figured in his *Histoire naturelle* in 1753 directly from Maupertuis's breeds—as historical "forerunners of Charles Darwin."[4] Maupertuis saw "no danger" in contending that hereditarian phenomena pointed to the fact that organic matter, in addition to its "physical" properties of extension and movement, displayed "a certain degree of intelligence, desire, aversion, and memory."[5] His remarks were thus largely related to debates about the *generation* of living beings, and they were embedded in a broad ideological dispute about the existence of vital forces and the authority of the natural sciences.[6] Similarly, Kant's engagement with the origin of the human races can be said to occupy a strategic position in his philosophical *oeuvre*, a position that in this case reflects his aversion to the assumption that vital forces exist in nature, an assumption that Maupertuis, in contrast, entertained.[7]

The different cultural domains in which heredity became apparent in the form of separate phenomena were, moreover, not united by the overwhelming influence of one domain, even philosophy, on the others. Rather, one might compare the process of unification to a domino effect: conjunctures took place between domains such that a mobilization in one area became effective under the specific conditions of another. For example, the development of a social class whose wealth predominantly rested on mobile property went along with the spread of activities such as collecting or breeding plants for purposes of private enjoyment or public representation.[8] This widespread passion in turn kicked off sustained efforts to import and acclimatize more and more exotic plants to Europe for economic purposes.[9] Breeders who had successfully established marketable varieties of plants and animals then became role models for the self-made man. The bourgeois individual envisioned himself as capable of permanent advancement.[10] It is therefore not the case that a homogeneous "culture of heredity" suddenly emerged around 1800. Rather, local, social, and institutional networks were reconfigured step by step, and it was only as a result of these reconfigurations around the middle of the nineteenth century that a field of connected phenomena presented itself to, and could be analyzed as such by, theories of heredity.

In the life sciences, the correlate of this general social dynamic was the genesis and development of theories of evolution, with their encompassing historical dynamization of the living. Indeed, Charles Darwin's book *On the Origin of Species* (1859) did not only present a new theory of evolution. Although falling short of presenting a general theory of heredity, it was also one of the first widely noticed works that exposed heredity as the central problem of biology. The way in which Darwin introduced heredity is highly revealing in terms of contemporaneous presuppositions. At the time of writing, Darwin still fought against "theoretical writers" who doubted that heredity really existed. Drawing on "Dr. Prosper Lucas's treatise, in two large volumes," and the empirical knowledge of breeders, Darwin argued:

> When a deviation appears not unfrequently, and we see it in the father and child, we cannot tell whether it may not be due to the same original cause acting on both; but when amongst individuals, apparently exposed to the same conditions, any very rare deviation, due to some extraordinary combination of circumstances, appears in the parent—say, once amongst several million individuals—and it reappears in the child, the mere doctrine of chances almost compels us to attribute its reappearance to inheritance.[11]

This quotation shows that Darwin did not want to focus on the frequent similarities between ancestors and progeny when he talked about heredity. These similarities were easily explainable by the fact that similar causes had contributed to successive generations at conception and during development. Far more specifically, when he spoke about heredity, Darwin was interested in those cases in which rare but distinctive aberrations were reproduced under conditions that did not produce the same aberrations in other individuals. These cases could obviously not be explained by the prevailing conditions of life. For Darwin, heredity and its correlate, variation, were thus capricious processes that did not necessarily lead to adaptation. Through variation, differences were produced again and again, even under identical external conditions, whereas heredity conserved peculiarities even if the surrounding conditions changed. Both processes could eventually result in better adaptation to the environment, but this outcome was by no means necessary. Even the "divergence of characters," the elementary principle on which Darwin's theory of the origin of species rested, was—as he himself stressed—not at all a "necessary contingency; it depends solely on the descendants from a species being thus enabled to seize on many and different places in the economy of nature."[12]

In contrast to generation theorists of the eighteenth century, Darwin did not depend on rare reports about "curiosities," as the sources mentioned in the quotation above illustrate. In particular, Prosper Lucas's encyclopedic two-volume treatise on heredity, his *Traité philosophique et physiologique de l'hérédité naturelle*, represents the preliminary culmination of a by then rich medical tradition of studying hereditary diseases, psychic ones in particular.[13] In addition, the accumulated experience of breeders precipitated a wealth of publications in the first half of the nineteenth century, which Darwin studied meticulously.[14] Moreover, in both bodies of literature, the temporal dimension of the epistemic space of heredity came to the fore, even if the emphasis was different. Both physicians and breeders regarded hereditary varieties as due to events that resulted in more or less permanent deviations from a given norm. Whereas physicians saw them as nonadaptive deviations (diseases), breeders regarded them as adaptive deviations with respect to their breeding goals.

In the next chapter, we will consider the role of medicine as a driver of nineteenth-century discourses on heredity. To breeding, on the other hand, we will come back in chapter 6. For the rest of this chapter, we want to focus on the theoretical debates that articulated the epistemic space of heredity in the second half of the nineteenth century. Our basic claim is that heredity became a general biological problem when organisms acquired a genuine "history"; that is, when the forms of life ceased to be fixed by assumed species boundaries. For if species were neither prefigured from the time of creation (Linnaeus), nor seen as manifestations of the permissible combinations of organic matter (Buffon, Maupertuis), nor regarded as steps in a progressive development governed by general laws (Lamarck, Erasmus Darwin), then two related questions assumed urgency: What was the mechanism by which species changed? And what laws, if any, maintained them in a transiently stable state that lasted for several generations at the very least? Hereditary variations indicated that life possessed an autonomous quality that was neither compatible with the view that organisms had been designed for an everlasting environment nor with the view that they possessed a capacity for unlimited plasticity or perfectibility. As Jean Gayon remarked pointedly, "To be firmly established, natural selection required the development of an experimental science of variation and of heredity which quite simply did not yet exist."[15] Evolution and heredity thus form two sides of the same coin. It is therefore not by chance that it was Darwin's ideas about heredity—culminating in his provisional "hypothesis of pangenesis"—that became the reference point for multiple speculative attempts to come to terms with the phenomena of heredity in the later nineteenth century.

## Darwin, Galton, and the Phenomenon of Regression

In the opinion of Prosper Lucas—whose *Traité* Darwin cited more than once—it was obvious that "the general mistake of all old as well as modern theories consists in equating generation and inheritance."[16] However, Lucas made another distinction that would play no further role in Darwin's theory of evolution. He distinguished a "law of heredity" (*loi d'hérédité*), responsible for the constancy of the specific type, and a "law of innateness" (*loi d'innéité*), responsible for the hereditary transmission of individually acquired or spontaneously arising defects and interfering with regular hereditary processes.[17] An altered character that appeared in an individual could thus be transmitted to its progeny without affecting the hereditary type, even though the altered character was "innate" from the perspective of the progeny. "It is clear that with respect to the specific type it is always the innate that vanishes and the inherited that stays," Lucas therefore claimed.[18] In Darwin's theory of evolution, which no longer recognized the concept of type that underlay Lucas's deliberations, this distinction had become obsolete. Darwin's theory recognized only a sum of potentially variable characters as well as the laws that regulated their "correlation" and thus limited their independent transmission.[19]

Darwin was especially fascinated by the phenomenon of intermittent hereditary characters; that is, characters that disappeared in one generation and reappeared in one of the following generations. As early as the first chapter of the *Origin*, he avowed:

> The laws governing inheritance are quite unknown; no one can say why the same peculiarity in different individuals of the same species, and in individuals of different species, is sometimes inherited and sometimes not so, why the child often reverts in certain characters to its grandfather or grandmother or other much more remote ancestor; why a peculiarity is often transmitted

from one sex to both sexes, or to one sex alone, more commonly
but not exclusively to the like sex.[20]

This quotation shows very clearly what caught Darwin's attention. It
was the "independent life" of qualities whose distribution among progeny could obviously not be explained by external circumstances but had
to be attributed to a hidden mechanism. The epistemic space in which
such peculiar characters circulated could no longer be regarded as restricted to the individual relation between parental progenitors and their
children.

In 1868, Darwin published a two-volume sequel to the *Origin* entitled *The Variation of Animals and Plants under Domestication*. He
had intended it to be part of the *Origin*, but it ended up occupying him
for another ten years. Here, Darwin first assembled all the evidence he
could come by about variations and their transmission in the variegated
literature of breeders, physicians, and natural historians. Then, in chapter 27, he tried to "connect by some intelligible bond" the observations
that he deemed to be most important in the following areas: sexual procreation, graft-hybrids, xenia (alterations produced on the mother plant
by foreign pollen), development, the functional independence of the elements or units of the body (cells, as we would call them today), and finally, variability and inheritance. He identified this bond in the title of
the chapter, "Provisional Hypothesis of Pangenesis," and introduced it
with the following words:

> It is universally admitted that the cells or units of the body in-
> crease by self-division or proliferation, retaining the same na-
> ture, and that they ultimately become converted into the various
> tissues and substances of the body. But besides this means of in-
> crease I assume that the units throw off minute granules which
> are dispersed throughout the whole system; that these, when sup-
> plied with proper nutriment, multiply by self-division, and are
> ultimately developed into units like those from which they were
> originally derived. These granules may be called gemmules. They
> are collected from all parts of the system to constitute the sexual
> elements, and their development in the next generation forms
> a new being; but they are likewise capable of transmission in a
> dormant state to future generations and may then be developed.
> Their development depends on their union with other partially
> developed or nascent cells which precede them in the regular

course of growth. . . . Gemmules are supposed to be thrown off
by every unit, not only during the adult state, but during each
state of development. . . . Hence, it is not the reproductive organs
or buds which generate new organisms, but the units of which
each individual is composed. These assumptions constitute the
provisional hypothesis which I have called pangenesis.[21]

Thus, according to Darwin's hypothesis of pangenesis, as already explained at the end of chapter 2, not only were all relevant characters
of an organism gathered in the germ cells as gemmules, but so were
countless such gemmules stemming from more remote ancestors. Gemmules could be transmitted in a dormant state for generations until they
were activated again by circumstances that could not yet be specified.
For Darwin, this hypothesis resolved one of the most pressing problems
of inheritance, a problem that haunted breeders in the form of reversion or atavism and to which Darwin returned again and again. Like
the spontaneous and unpredictable appearance of sports, phenomena
of regression provided glimpses of the autonomy, if not capriciousness
and eccentricity, that characterized life and its evolution.[22] Likewise,
the sometimes stronger and sometimes weaker expression of particular characters could be attributed to the variable quantities or penetrating powers of gemmules. Finally, the problem of development could be
equally well grounded in the material substrate of inheritance, since the
accumulated gemmules represented all the stages of this process. This
hypothesis reflects a general feature of theories of heredity in the second
half of the nineteenth century: while they increasingly tended to separate
the phenomena of inheritance and development, they also continued to
try to explain both processes in a unitary fashion.

Darwin's half-cousin Francis Galton, whose explicit comparisons
to bureaucratic and political processes we exposed in chapter 1, also
played a significant role in elaborating a theory of heredity. Until the
early 1860s, Galton had mainly worked on geographic and meteorological problems, but impressed by the *Origin*, he turned to the problem of inheritance for the decades to come. An extensive series of blood
transfusion experiments with rabbits, carried out from 1869 to 1871,
led Galton to the conclusion that his cousin's idea of a pangenetic circulation of gemmules throughout the body did not withstand scrutiny. His
reflections in this area led him from a physiological to a statistical theory
of heredity and finally to the "law of ancestral heredity." Here we are
interested only in the formulation of Galton's theory of heredity, which
began to appear in print in the mid-1860s.

In the essay "Hereditary Talent and Character," published in 1865, Galton hypothesized that we are "no more than passive transmitters of a nature we have received, and which we have no power to modify."[23] This emphasis on the autonomy of the hereditary process is an enduring feature of Galton's concept of heredity, and it is one respect in which he diverged from Darwin's theory of inheritance. Jean Gayon has drawn attention to the fact that Darwin at times called his "theory of descent with modification" a theory of "inheritance with modification."[24] Galton adopted this emphasis on inheritance, but in addition, he claimed that it was a process largely independent of its individual manifestations:

> We shall therefore take an approximately correct view of the origin of our life, if we consider our own embryos to have sprung immediately from those embryos whence our parents were developed, and these from the embryos of *their* parents, and so on for ever. We should in this way look on the nature of mankind, and perhaps on that of the whole animated creation, as one continuous system, ever pushing out new branches in all directions, that variously interlace, and that bud into separate lives at every point of interlacement.[25]

Like Darwin, Galton conceives of the hereditary process as a system of budding germs, but these germs are not a direct product of the bodies of the parents. Rather, ancestors and descendants are the product of a common substrate of germinal material, the "stirp", on which the hereditary process is based and through which it manifests itself in the development of individuals. In his paper "A Theory of Heredity" of 1876, which is central to our argument, Galton starts from the premise—which he shared with Darwin—that organic bodies are composed of "organic units" of some kind that possess particular properties and are to some extent independent from one another. He then divides these units into "inborn" and "acquired" ones and argues that only the former are relevant to his theory of heredity. In this context, Galton emphatically rejects Darwin's hypothesis of pangenesis, a theory that, after all, also provided a convenient explanation for the inheritance of somatically acquired characters.

Galton's own theory is grounded in four postulates. First, a germ corresponds to each independent unit of a body. Second, the sum of the germs that constitute the "stirp" of the fertilized egg is much larger than the sum of the germs that actually develop in an individual. Third, the latent germs that do not develop form a "residue," the most vital part of which enters the germ cells of the organism. Fourth, Galton assumes, with Darwin, that the germs possess specific forces of attraction and repulsion that guarantee a regular formation of structures and an orderly development, thus making the assumption of a central vital force superfluous.[26] "I look on the sexual elements," Galton wrote, "as directly descended from the 'residue,' and"—in contrast to Darwin—"do not suppose the gemmules to travel freely."[27] Galton thus concludes that "direct descent"—in the sense of direct parenthood, or generation in the premodern sense—"is wholly untenable."[28] It is exactly this everyday meaning of descent, as Galton had already contended in his essay "On Blood Relationship," published in 1872, that so far hampered a true understanding of the more complex, often bewildering phenomena of heredity. The "true hereditary link" would not connect "the parents with the offspring, but the primary elements of the two, such as they existed in the newly impregnated ova, whence they were respectively developed. . . . [W]e gratuitously add confusion to our ignorance, by dealing with hereditary facts on the plan of ordinary pedigrees—namely, from the *persons* of the parents to those of their offspring."[29]

In "On Blood Relationship," Galton also distinguished between "patent" and "latent" elements. Although this distinction should not be equated with the relation between dominant and recessive genetic elements, it nevertheless defined an important element of the epistemic space that became constitutive for the genesis of classical genetics. As in Darwin's work, the curiosity of "reversion," despite its comparative marginality and unpredictability, led Galton to a major distinction that became decisive for the conceptualization of the hereditary process in the long run: the clear division between the transmission of hereditary elements on the one hand and their role in the development of the organism on the other.

*Heredity and Cell Theory*

Almost all hereditary theories before 1900—here bundled together as "first syntheses"—were coupled to reflections on evolution or development or both. Transmission of hereditary dispositions, evolutionary descent, and organismic development formed a theoretical unity.[30] However, as foreshadowed in Darwin's and even more so in Galton's theories of heredity, this unity began to dissolve as cell theory moved into the center of considerations of heredity and as phylogeny and ontogeny became disassociated and began to be represented as discrete processes. In his paper "From Heredity Theory to 'Vererbung,'" Frederick Churchill stresses that the theory of heredity reached a "watershed" in the 1880s owing to progress in cytology, albeit cell theory itself had become mainstream some forty years earlier.[31] François Duchesneau similarly speaks of a "historical delay" in connecting cell theory and heredity.[32]

As early as 1858, Rudolf Virchow speculated in his *Cellular Pathology* that it was probable "in a high degree" that the nucleus plays an extraordinarily important part within the cell, "a part . . . less connected with the function and specific office of the cell, than with its maintenance and multiplication as a living part."[33] Subsequent to these early remarks about a division of labor between nucleus and cytoplasm, a number of cytological observations in the 1870s added to the growing acceptance of the materiality of hereditary factors and, in particular, lent plausibility to the idea that they were located in the cell nucleus. Let us only mention Eduard Strasburger's and Oscar Hertwig's research on the fusion of nuclei upon fertilization, or Walther Flemming's observation of a longitudinal division of the "nuclear particles" (*Kernkörperchen*), for which the expression "chromosomes" caught on following a suggestion by Wilhelm Waldeyer in 1888.[34] Moreover, in 1887, Oscar and Richard Hertwig reported that they had succeeded in removing the nuclei from sea urchin eggs by vigorous shaking. The enucleated eggs could be fertilized and showed signs of cell division.

Theodor Boveri took up this experimental thread. From his experiments with the horse roundworm *Ascaris megalocephala*, he concluded in 1888 that the chromosomes were to be seen as "autonomous individuals retaining their autonomy in the resting nucleus as well."[35] During work carried out at the Zoological Station in Naples a year later, he succeeded in fertilizing the enucleated eggs of the sea urchin *Sphaerechinus granularis* with nucleus-carrying sperm from another genus, *Echinus microtuberculatus*. Some of the products of this experiment developed into an early larval stage that resembled larvae of *Echinus*. This experiment established that the nuclei determined the form of the growing larvae, not the cytoplasm of the egg. The larvae exhibited the characteristics of the species from which the nucleus was derived. With this result, cell research abandoned its predominantly descriptive orientation and entered an experimental phase.[36] Boveri's experiments did not go unnoticed: Thomas Hunt Morgan translated his report into English only four years later.[37] Finally, August Weismann postulated that a reduction division of the nuclei of the gametes had to take place before fertilization, a process he was subsequently able to observe microscopically. These are

only a few of the findings that followed one another in quick succession
in the two and a half decades between 1870 and 1895, not least as a result of new microscopic preparation and staining techniques.

For the early 1880s, Churchill has documented an unusual surge in
cytological reports that either carried the concept of heredity explicitly in
their title or reported studies of the phenomena of heredity at a cellular
level. In addition to those mentioned earlier—the botanist Strasburger
and the zoologists Hertwig and Weismann—Moritz Nussbaum and Albert Kölliker were among the authors of these reports. They all shared a
certain vision of hereditary "continuity" that "preserved the organized
material of transmission through the disruptive and complex events of
differentiation."[38] Cell theory, as promulgated by Virchow—characterized by the principle of *omnis cellula e cellula*—had finally put an end
to the radical ideas of epigenesis dating from the late eighteenth and
early nineteenth centuries. Life now appeared to be inevitably bound
to a structure—the cell—that could sustain itself only through its own
proliferation: a structure that persisted by dividing itself. This theory
provided a basis on which the discourse of heredity could build, insofar
as the continuous development of cell structures, and of the nucleus in
particular, suggested that the cell should be seen as bearer and transmit-
ter of a material that was preserved in transmission. The historical "delay" in connecting cell theory with the discourse of heredity diagnosed
by Duchesneau can therefore also be interpreted as a premature conjuncture of fields that would be elaborated and exploited later only by
classical genetics. The Mendelian theory of heredity—in the form it took
after 1900—is indeed unthinkable without this conjuncture. For Mendel, too, as we will see in chapter 5, cell theory provided a decisive point
of departure.

### The Synthetic Drafts of Nägeli, Bernard, de Vries, and Weismann

In Germany, Carl Wilhelm von Nägeli worked out a comprehensive theory of heredity in the late 1870s and the early 1880s, and he summed up
his thoughts on a "mechanical-physiological theory of descent" in a
monumental book (*Mechanisch-Physiologische Theorie der Abstammungslehre*) published in 1884. The title indicates that Nägeli, too,
framed his ideas on heredity in the context of a theory of descent, but he
considered his theory an alternative to Darwin's doctrine of evolution.
In the framework of his "molecular physiology"[39] Nägeli drew a consequential distinction. He differentiated two distinct parts in the "stereo-
plasm" of the organism: the "trophoplasm" formed the body substance,

whereas the "idioplasm" represented the hereditary substance. According to Nägeli, the idioplasm consisted of a "modification of the albuminates"[40] that organized themselves into "micelles" of a quasi-crystalline order—somewhat like the starch in the starch granules that one could find in plants, as the botanically versed Nägeli added. At the same time, he emphasized that "the configuration of the idioplasmatic system" was "not a geometrical, but a phylogenetic task."[41] In other words, its structure could not be derived from first—material—principles, but was a historical product. The specificity of each idioplasmatic system lay in the configuration of the cross section of its bundle of strands, in which "all of the ontogeny with all of its peculiarities must be contained as a disposition."[42] The idioplasm, Nägeli explained, "is therefore in a sense the microcosmic image of the macrocosmic (fully grown) individual." But this, as he immediately added, "does not of course imply that the micelles of the idioplasm correspond to, say, the cells of the mature organism or possess an analog order. On the contrary, these two orders are fundamentally different."[43]

According to Nägeli, the idioplasmatic substance of the organism thus constituted a separate system. Within this system, "matter had organized itself into units of the same order," and these units "could be compared among each other and measured by one another."[44] At least in principle, if not with the technical means actually at hand at the time, this order could thus be studied empirically and did not have to be derived from "hypothetical and unknown smallest things."[45] Implicitly, this was a polemic response to Darwin's talk about "gemmules." Yet, like the gemmules, the elements of Nägeli's idioplasm had the capacity to grow by the apposition of further micelles and to multiply without losing their phylogenetic configuration.

To illustrate the relation between idioplasm and trophoplasm, Nägeli resorted to the analogy of a piano: "We must therefore imagine that the idioplasm unfolds the dispositions for different organs similarly to the way a piano player brings the consecutive harmonies and disharmonies [of a piece of music] on his instrument to an expression. For every a, as for every other tone, he strikes again and again the same chords. The adjacent groups of rows of micelles in the idioplasm are therefore comparable to chords, each of which represents another elementary phenomenon."[46] The idioplasm thus consisted of material, individualized inductors of elementary organic phenomena; they, and not the individuals composed from them, were the ultimate bearers of the evolutionary potential, a position that Nägeli affirmed by means of another remarkable image: "If all circumstances are favorable," he wrote in a paper of 1856, "a disposition must be able to build itself up further and further through a series of generations, as capital grows to which the annual interest is added."[47]

Nägeli assumed that the idioplasm "extend[s] throughout the whole organism as a seamless web," the knots of which he suspected to be located in the cell nuclei. This "web" took its most compact form in the fertilized egg.[48] But just as the idioplasm spanned the individual body as a whole, so it spanned the succession of generations. For him, a genealogy in its entirety was basically "a single, continuous individual consisting of idioplasm."[49] Much like Galton, Nägeli presumed that this image turned the common view of heredity on its head: "For it is not that the parents bequeath part of their characters to their children, it is on the contrary the same idioplasm that first forms the parental body corresponding to its essence and a generation later the very similar body of the child also corresponding to its essence."[50] The hereditarian system possessed its own powers, and it governed the individual organism. In comparison to this system, the individual tended to be ephemeral.

Nägeli thus maintained a sharp distinction between two material regimes. The idioplasm was the continuous genealogical embodiment of a progressive phylogenetic acquisition of ever more complex structures. The trophoplasm was a transitory shell that carried the evolving idioplasmatic system through the generations. In contrast to Darwin's heap of gemmules, Nägeli's idioplasm, although composed of units of varying chemical composition, had the character of a system. It was "strung together in bands of lower and higher divisions."[51] Above all, its substance was distinct from bodily matter. Among all those who developed theories of heredity before 1900, only Nägeli assumed that the two plasms were constituted by two chemically distinct systems.

At this point, it is worthwhile to briefly consider the contributions to theories of heredity by the French experimental physiologist Claude Bernard.[52] Unlike his earlier physiological studies, Bernard's late work in this area was not based on experiments, and he himself considered it to be speculative in nature. In an 1878 series of lectures on the general phenomena of life (*Leçons sur les phénomènes de la vie communs aux animaux et aux végétaux*), Bernard distinguished between a "chemical synthesis" and a "morphological" or "organizing synthesis" in the organism. The former consisted of physiological functions that could be reduced to physicochemical processes and were amenable to experimentation. The latter consisted of the concatenation, subordination, specific positioning, and temporal succession of different physiological functions and consequently provided the basis of any generation and reproduction of organic forms. Of one thing Bernard was certain: the starting point for morphogenesis was the fertilized egg, which had to contain "some kind of formed and substantial element [common to] the consecutive generations,"[53] an element that "is transmitted, is enduring, but not like an organ belonging to the individual as its bearer, rather as an element that belongs to the ancestor and that, in the economy of the present being, represents something like an atavistic parasite."[54] In his *De la physiologie générale*, he referred to this element as a "trace" (*empreinte*) or as a "recollection" (*souvenir*),[55] and he regarded it as acting "in the manner of a preconceived idea that transmits itself via atavism or organic tradition from one being to the next."[56] In the *Leçons*, he later described it as "virtually inscribed" (*virtuellement inscrit*) or acting as an "assignment" (*consigne*).[57] "The work of morphology," Bernard continued, "is a pure *repetition*; it has its reasons not in a force acting momentarily at any time; it has its reasons in a preceding force. There is no morphology without precursor."[58] Inheritance was thus the essential factor in the "morphology of life."[59] Twenty years later, the American biologist Edmund Beecher Wilson elaborated on Bernard's distinction in his influential book *The Cell in Development and Inheritance* of 1896. In contrast to Bernand, however, control over "morphological synthesis" reside firmly in the nucleus of the cell for Wilson.[60]

Let us finally discuss two prominent synthetic conceptions formulated less than a decade after Churchill's cytological "watershed" of the early 1880s, both of which made clear references to earlier work by Darwin and Galton. In his *Intracellular Pangenesis* (1889), the Dutch botanist Hugo de Vries undertook to reconcile Darwin's hypothesis of pangenesis with the latest developments in cell theory toward the end of the nineteenth century. For de Vries, the independence and free miscibility of hereditary dispositions was of decisive importance, as it had been for Darwin previously. On this point he left no room for doubt: "Independence and miscibility are therefore the most essential attributes of the hereditary factors of all organisms."[61] He deemed any further speculation about a stable connection between, or even an overarching architecture of, the hereditary dispositions superfluous. His "pangenes" were autonomous units of life. Although they were composed of physical atoms or chemical molecules, the specificity of their composition was "historical" in nature.[62] According to de Vries, however, this historical nature manifested itself in independent units, and—unlike Nägeli—he did not refer to an overarching system into which they were bound. In particular, pangenes were endowed with the two basic characteristics of life: they could grow and multiply on their own.

At the center of *Intracellular Pangenesis* stood the cell. For de Vries—as for Virchow—the morphological distinction between nucleus and cytoplasm that cytologists had established reflected a functional distinction, a division of labor: "The function of the nucleus is transmission, that of the cytoplasm, development."[63] All the different pangenes of an organism were represented in the nucleus. As long as they were stored in the nucleus, they were inactive. Depending on the developmental conditions in which the organism found itself, the pangenes became selectively active in the cytoplasm where they multiplied: "Therefore, a transmission of the hereditary characters from the nucleus to the cytoplasm must in some way take place."[64] The pangenes were thus not only the hereditary dispositions, but also the living molecules that built up the differentiated body via selective multiplication. For this purpose, they had to be retrieved from the nucleus—"in some way," as de Vries put it. During cell division, in contrast, they were passed on from nucleus to nucleus in their totality.

In opposition to Nägeli, who referred to plant and animal breeders as "practitioners" of whom no theoretical progress could be expected, de Vries—like Darwin before him—frequently referred to the work of breeders, such as Louis de Vilmorin in France, Wilhelm Rimpau in Germany, or the breeders of the Svalöf experimental station in Sweden, in order to bolster his most important claim; to wit, that the hereditary dispositions were discrete and represented discrete characters.[65] In particular, he hoped that an analysis of the emergence and fixation of variations would allow him to go beyond a purely theoretical synthesis and open up an experimental approach to heredity. We will consider in chapter 6 how he later achieved this goal.

Finally, August Weismann's synthesis vis-à-vis those of his contemporaries Nägeli and de Vries is to be considered. Weismann developed his theory of heredity in his essay on the continuity of the germ plasm ("Die Continuität des Keimplasmas als Grundlage einer Theorie der Vererbung," 1885) and in his extended work *The Germ-Plasm: A Theory of Heredity* (1892). Like Nägeli, Weismann distinguished between "germ-plasm," "hereditary plasm," or "idioplasm," on the one hand and the body plasm, or "soma," on the other. The former ensured the continuity of inheritance through the generations; the latter was the origin of the cells of the individual developing organism. But whereas Nägeli, in addition to this functional distinction, asserted that the idioplasm consisted of specific "albuminates," Weismann—like de Vries in this respect—subscribed to the assumption that the germ plasm was essentially composed of the same organic molecules that were active in

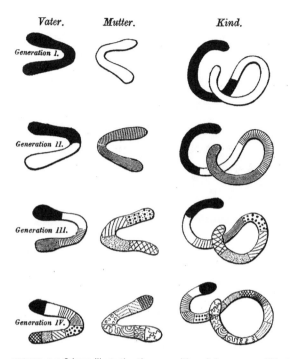

FIGURE 4.1 Scheme illustrating the composition of chromosomes ("idants") from individual elements ("ids"). From August Weismann, *Das Keimplasma* (Jena: Fischer, 1892)
fig. 19.

the rest of the protoplasm, or soma. For Weismann, however—now in
contrast to de Vries, but in accordance with Nägeli—the makeup of the
germ plasm was a complicated edifice, a "fixed architecture, which has
been transmitted historically,"[66] and that, according to his theory, integrated the "biophors"—the elementary "bearers of vitality"[67]—in "determinants," the determinants in "ids," and the ids in "idants." One can
roughly equate the last of these entities with the morphologically visible
chromosomes (fig. 4.1).

Weismann thought that the biophors, in a manner comparable to
de Vries's pangenes, formed a depository in the nucleus, and that parts
of this depository could, in the course of ontogenesis, selectively diffuse
into the cytoplasm, where they would be used as "material" for the construction of differentiated cells.[68] "These biophors," he claimed, "constitute *all* protoplasm—the morphoplasm which is differentiated into the
cell-substance, as well as the idioplasm contained in the nucleus."[69] In
Weismann's theory, the germ plasm was deployed in a regular fashion in
the course of development from the egg to the adult organism, following

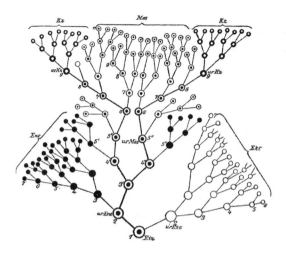

FIGURE 4.2 Schematic representation of the germ line of the worm *Ascaris nigrovenosa*. From August Weismann, *Das Keimplasma* (Jena: Fischer, 1892), fig. 16.

a "mechanism existing in the idioplasm."[70] At the same time, the complete historical architecture of the germ plasm remained spatially separated in the germ line from the differentiating body and was therefore unaffected by its modifications.

Whereas the germ plasm was preserved in its totality in the germ line, it irreversibly and differentially unfolded in the body during ontogeny (fig. 4.2). The nucleoplasm in development, Weismann wrote in 1885, thereby adding to the arsenal of metaphors we encountered so far,

> may be to some extent compared with an army . . . . The whole army may be taken to represent the nucleoplasm of the germ-cell: the earliest cell-division . . . may be represented by the separation of the two corps, similarly formed but with different duties: and the following cell-divisions by the successive detachment of divisions, brigades, regiments, battalions, companies, etc.; and as the groups become simpler so does their sphere of action become limited.[71]

In contrast to Nägeli, Weismann thus believed that the plasm of the nucleus, which was passed on in the germ line, did not consist of a separate substance. However, in comparison with the soma, it was subjected to a different regime, insofar as it was passed on continually and in its totality through the generations, thus retaining its potential to become unfolded into a complete new individual. Weismann's repudiation of the inheritance of characters that the organism might have acquired in the course of its life is based on this separation of two regimes. And it rests on the parallel assumption that no way existed for differentiated body cells to feed back into the germ cells. In Weismann's image, this would have been tantamount to a battalion performing the function of an army—an impossibility in his eyes.

The countless references to "Weismann's legacy" often overlook the fact that Weismann by no means implied that the nuclear plasm that was passed on in the germ line was unable to acquire new qualities. He assumed spontaneous variability of the germ plasm, and he allowed for the possibility that it could be directly influenced by external environmental factors.[72] His experiments with butterflies, for example, took for granted such a direct—and directed—influence by the environment. This assumption implied a manipulability of the variation of the germ plasm, or a direct action of "external influences" (*Mediumeinflüße*) on the germ plasm, as he put it.[73] Accordingly, Weismann assumed that heritable differences in the wing color of the butterfly *Polyommatus phlaeas*, whose home range extended from the north to the south of Europe, could be ascribed to the differences in climatic conditions and supply of nutrition, particularly to different regional temperatures. In his butterfly cultures, he thus worked with temperature stimuli and investigated whether he could experimentally induce the effects that in nature had been provoked by an environmental gradient and fixed by selection. In contrast to those of de Vries, Weismann's experiments were consequently not inspired by the hybridization practices of breeders, but by physiological considerations relating to the expression of characters and the induction by external influences of heritable variations in the germ line. However, these experiments remained inconclusive and contributed little to Weismann's theory of heredity, which stands as an impressive edifice that towers over his experimental achievements by far.

## "Hard" and "Soft" Inheritance

It appears thus that the efforts to create something like a unifying theory of heredity in the second half of the nineteenth century extended throughout the biological disciplines, spreading from natural history (Darwin), to statistics (Galton), to physiology (Bernard), to botany (de Vries, Nägeli), to zoology (Weismann), to cytology, and to medicine at large. At the end of this chapter, the task that remains is to explain what prompted all these sweeping and at times overzealous speculations about heredity. We would suggest that this excess of theory reflects not only the perceived importance of a new subject, but also a more general, and at the same time deeper, sense of uncertainty that frequently found expression in the sociopolitical comparisons that these authors adopted. So far, historians of science have mainly narrated the history of nineteenth-century theories of heredity along the lines of oppositions such as "soft" versus "hard"[74] or "blending" versus "non-blending" inheritance.[75] They have thus adopted a narrative that culminates in—or degenerates into, depending on one's standpoint—the development of classical genetics.

These oppositions do not always capture and some times even occlude, the diversity of theoretical, and at times purely hypothetical, visions found in this period. A closer look shows that general speculations on the mechanism of heredity were not organized according to these polar positions until the end of the nineteenth century. If they were articulated at all, then they remained rather loosely defined. Among physicians, for instance, the belief that diseases of the parents could press their "stamp" on the children remained widespread. And even after Weismann's germ line principle began to gain acceptance, nothing stopped medical doctors from invoking the hypothesis that substances such as alcohol could "poison" the germ line. This implied that heritable changes—at least in the form of damage—could actually be irrevocably acquired.[76] Physicians were not alone in entertaining such ideas. As we have seen, not even Weismann ruled out the possibility that the environment could have an influence on the germ line, and he even investigated this possibility experimentally. For the sake of argument, one might say that this reflects a conception of heredity that is both "soft" and "hard." It is "soft" insofar as it assumed that environmental factors can effect a change in the hereditary substance, and it is "hard" insofar as it assumed that this change, once it happened, was irrevocably transmitted in the germ line—an obvious analogy to original sin, as it were. Similarly, the social Darwinism of the late nineteenth century was full of Lamarckian ideas, the influential social philosophy of Herbert Spencer in particular, but these sentiments obviously did not attenuate the racist convictions that were often promulgated by this intellectual current.[77]

Other theoretical proposals, such as Darwin's pangenes and Galton's stirp, likewise allowed for both the idea that parental characters could amalgamate or "blend" in the outward appearance of progeny and the assumption that the hereditary substance was composed of "non-blending," enduring elements. It thus appears that the separation of "hard" and "soft" as well as "blending" and "non-blending" inheritance could become serious positions of debate and contention only after Mendelism began to catch on in 1900. With respect to the notion of "blending inheritance," Theodore Porter has reminded us that Darwin knew no expression with this particular meaning. For Darwin as for Galton, blending always related to the visible characters; that is, to the result of inheritance, not to the process of inheritance. Galton consequently used the expression "blended inheritance" specifically with respect to the exterior appearance of certain hybrids, not with respect to a putative mechanism within the hereditary material.[78]

At the end of the nineteenth century, questions about the material constitution of the hereditary substance and about its causal powers had an importance that was far more fundamental than the exact nature of hereditary mechanisms. These questions were accordingly debated with greater intensity than they subsequently would be in classical genetics. The question of highest priority concerned the mutual relations of the organic elements involved in the hereditary process and their relation to a possible overarching organic system. Were the elements, as postulated by Darwin and de Vries in particular, largely autonomous, such that they could freely recombine and individually unfold under appropriate circumstances? Or were they, as assumed by Nägeli and Weismann, dominated by an overarching system that had grown historically and that determined where and when they were "called" into action?

Like cell theory more generally, theories of heredity thus dealt with the hierarchical relationship between the parts and the whole. This focus also facilitated an interpenetration of biological and political visions. Renato Mazzolini has analyzed what was at stake in these visions in detail for Virchow's concept of the organism.[79] For Virchow, the organism represented "a kind of social arrangement of parts, an arrangement of a social kind."[80] Nevertheless, Virchow firmly opposed attempts by conservative contemporaries to turn his metaphors into a naturalizing justification of an authoritarian "state organism." In his 1849 essay on physicians' relationship to the state ("Der Staat und die Ärzte"), he argued that the state is "never and will never be an organism. It is nothing but a complex of organisms. The so-called state organism prospers thus best if the development of the individual is best granted."[81] According to Virchow, the "monarchic principle" of a "vital force" dominating the organism as a whole had to be replaced by the principle of a "generative force" that had to be "attributed to the cells as hereditary and peculiar."[82]

If there is a dichotomy that is characteristic for this period, then it is the question of whether inheritance was to be conceived of as a force or as a material structure. As Jean Gayon has shown, this debate remained charged throughout the nineteenth century.[83] Those who favored the view that inheritance was a force could easily affiliate themselves with epigenetic conceptions, as paradigmatically articulated in Johann Friedrich Blumenbach's concept of a "formative drive" (*Bildungstrieb* or *nisus formativus*). From an epigenetic perspective, such a force was generative. In the context of heredity, its function shifted to maintaining the specific type. Prosper Lucas, for example, characterized his "law of inheritance" in exactly this way. Darwin, in contrast, spoke of a "force of inheritance" or a "power of inheritance" in a rather unspecific sense. At the same time, he pointed out "how feeble, capricious, or deficient" this force could be.[84] Throughout the nineteenth century, breeders typically thought of heredity as a force that accumulated and could be reinforced over generations—or weakened by neglect.[85] In Galton's "law of ancestral heredity," finally, it was identified as a measurable force or "tendency," as we will see in the next chapter.

Authors who indulged in energetic or panpsychic views also favored the conception of heredity as a force. One of the most prominent among them was the monist and propagator of Darwinism in Germany, Ernst Haeckel. In his *Perigenesis der Plastidule* of 1876, he envisioned the germ plasm as consisting of molecules characterized by certain states of vibration. The cells dividing during ontogenesis were exposed to different external conditions, and the states of vibration changed accordingly. On this basis, Haeckel postulated a strict parallelism between ontogenetic and phylogenetic development. His "biogenetic basic law" claimed that the same mechanism of the "inheritance of acquired characters" was in operation in both developmental processes. Haeckel furthermore understood heredity as an essentially conservative force, insofar at least as each organism, in its individual development, recapitulated its phylogenesis. In his predilection for the creation of neologisms, he called this process "palingenesis." Variations interrupted this process, constituting something like a "history of frauds" (*Fälschungsgeschichte*, or "cenogenesis").[86]

This short and incomplete inventory reveals that different concepts can be covered under the umbrella term "force." But despite the presence of all kinds of conceptions of hereditarian forces, the alternative vision that phenomena of inheritance were connected with material structures gained increasing prevalence. At the same time, many decisive questions remained open or gave rise to controversy: Was the source of the hereditary material the ancestral organism itself, or something that somehow transmitted "itself" independently from the individual organism?

To what extent and under which conditions could acquired characters become hereditary? Did the germ plasm consist of a peculiar substance, and where would it be located? Was it bound to particular cellular structures such as the nucleus? What was the elementary makeup of the hereditary substance, and how did its units relate to one another? Could they undergo a process of fusion or blending, or would they never intermingle? Were the hereditary units particulate and essentially independent from one another, or did they constitute an indissolubly complex system? How would they mold the future organism—via direct and lifelong determination, or by guiding the first steps in development only? Finally, which particular roles did the two sexes play in the process of inheritance?

As far as the last question is concerned, a massive break occurred in the second half of the nineteenth century. In past theories of generation, it was almost universally assumed that the contribution of the two sexes to the product was asymmetrical, reflecting, not least, the contemporary legal and political relations between men and women.[87] If such an asymmetry was not located in the very act of conception itself, as Aristotle or the "ovists" and "animalculists" of the seventeenth and eighteenth centuries held, it was at least regarded as a given with respect to the formative influence of the pregnant woman, including her "imagination," on the embryo.[88] In particular, the sex of the child was ascribed to the nutrition and behavior of the pregnant woman well into the nineteenth century.[89] In the second half of the nineteenth century, this asymmetry was problematized, in spite of cytological research that yielded evidence to the effect that extreme differences in size and form existed between sperm and eggs. Against this background, the shifting of the hereditary processes into the nucleus of the cell and, with the beginning of the twentieth century, onto processes of exchange among chromosomes can be read as a desperate effort to single out a level at which the female and the male enjoyed equality despite the overwhelming evidence to the contrary.[90] However, the resulting "nuclear monopoly" implied that the female body (as had been the case in the views of Harvey) was removed from the sphere of influences shaping the offspring and came to be conceptualized as part of what is commonly called "the environment." In addition, the cytoplasm of the fertilized egg was similarly excluded from the realm of heredity, although it was "transmitted," via the mother, to an extent that greatly exceeded the paternal contribution.[91]

Despite intense debate, late-nineteenth-century wisdom offered no generally accepted answers to the questions about heredity sketched in the preceding paragraphs. Some concurrence can nevertheless be discerned among the biologists discussed in this chapter. After Darwin, reflections on heredity gradually revolved less around the relation between ancestors and progeny—that is, a vertical, diachronic relation between particular individuals—and more and more around the synchronic relation of populations and generations to a common hereditary substrate. This loss of importance of the temporal ancestors in favor of a common stock of hereditary dispositions defining relationships among contemporaries appears to be connected to, not least, the fact that in the context of late-nineteenth-century euphoria over social progress, heredity was associated increasingly with the future instead of the past and with the sustainable expansion of tradition instead of with its justification and explanation.

Insofar as the past played a role in the discourses on heredity in the cultural and life sciences, it appeared either as a threat to the present in the form of a creeping "degeneration" or as a "heritage" that had to be appropriated anew by each generation. Whereas the trope of heredity migrated from the legal realm into the life sciences during the eighteenth century, the biological concept of heredity started to fuel cultural and political visions at the end of the nineteenth century, as we will see in the next chapter.[92] In the realm of biology proper, animal and plant breeding began to provide a crucial model, oriented as these activities were toward goals to be achieved in the future. It is therefore not surprising that early Mendelian genetics entertained such close connections with breeding and agricultural concerns, as this reflects a general trend in the biology of the epoch. Genetics was seen as a decisive step to "induce" phenomena of heredity "forward, to future generations."[93] From a Mendelian perspective, to be discussed in detail in chapter 6, the genetic constitution of an organism could neither be deduced from the distant history of its ancestors nor established for the organism itself in the present; it could only be derived from its progeny.

# 5

## Heredity, Race, and Eugenics

The speculative hereditarian theories that we unfurled and examined in the preceding chapter—often esoteric in character and addressing a scientifically literate audience—were the tip of an iceberg. Human inheritance, the threats and promises that it held out for the health of nations, and the policies proposed and implemented to promote or fend off the effects of heredity were at the heart of many political debates in the period between the publication of Darwin's *On the Origin of Species* and the end of World War II. At first sight, the relationship between these debates and the science of heredity as it emerged toward the end of the nineteenth century seems straightforward. Eugenics and "racial hygiene"—the latter expression became prevalent in the German-speaking countries—advertised themselves as disciplines that would apply the new knowledge of heredity to solve pressing medical and social problems, and they did their best to raise public awareness about the relevance of the new science in providing such solutions. But a closer look at the history of eugenics and racial hygiene reveals that the connection between the science of heredity around 1900 and its application to human heredity is looser than implied by this rhetoric, albeit no less intriguing.

For one thing, the prominence eugenics enjoyed around 1900 has to be seen against the background of two longterm historical developments that were largely independent of con-current developments in biology. On the one hand, the European nation-states and the United States, as well as the so-called Western off-shoots, Canada, Australia, South Africa and Argentina, were undergoing dra-matic demographic changes. These changes were first summarized by the American population scientist Warren Thompson under a general model that has since become known as "the demographic transition."[1] The essential parameters of this model can be summarized as follows: an increase in life expectancy is followed by a decrease in birth rate, with the time lag between these developments producing rapid net popula-tion growth initially; migrants from rural areas flood into urban centers; and agricultural and industrial productivity increase substantially, fol-lowed by a general increase in living standards. A small European nation such as Denmark provides a good example of the magnitude of these changes. Between 1801 and 1901, 360,000 Danes moved from the coun-tryside to Copenhagen, 143,000 in the 1880s alone. Between 1901 and 1950, the population of rural Denmark dropped by another 850,000. At the same time, the overall population of Denmark grew from 929,000 to 4,281,000, and all this despite massive emigration to the Americas.[2] The United States provides the prime example of "explosive" popula-tion growth, of course. In 1870, its population hardly exceeded that of the German Reich, but only twenty years later more people lived in the United States than in all European nation-states, including Russia, taken together.[3] A comparison of birth rates in 1890 and 1911 in the German Reich finally indicates how radically reproductive behavior changed in the course of the demographic transition: while in 1890 the birth rate was 40.9 per thousand inhabitants, it had dropped to 28.2 by 1911.[4]

The second long-term historical development against which the in-tense interest in heredity around 1900 has to be seen is the increasing so-cialization and politicization of medicine. There are striking differences among countries as far as the establishment of public health systems is concerned. In France, the process began in the early nineteenth century; in the German Reich, public health insurance was introduced in 1883 as part of the social security reforms under Bismarck; the National Health Service in Great Britain was created after World War II; and the United States has only recently adopted a form of nationwide health care. De-spite these differences, however, some general trends hold true: the pro-vision of medical care was progressively professionalized and special-ized, at the expense of traditional forms of self-medication and healing that were increasingly denounced as "quackery"; administrative and po-litical positions in the health sector were more and more occupied by university-trained physicians rather than lawyers; state administrations were prepared to interfere with individual lifestyles as well as social and family life for purposes of preventing disease;[5] and disciplines that focused on whole populations as their subject, such as demography, epidemiology, hygiene, and social medicine, emerged and would soon receive growing resources.

The demographic transition and the emergence of public health systems were closely intertwined, although the relationship between these developments was far from deterministic. The decline of the birth rate—initially in high-income segments of society, later in segments with lower income as well, and finally even in rural areas—and the catastrophic hygienic conditions that caused grave health problems among the urban proletariat were perceived as critical symptoms of a "degeneration" of national populations by scientific and political elites alike, especially during World War I.[6] Health thus became a target of state intervention, and public health care one of the primary means of achieving, in the words of the historian of medicine Paul Weindling, "a cohesive and integrated society during the upheavals of industrialization."[7] Responses to these perceived threats differed between countries. In German-speaking countries, for example, a liberal professional elite dominated public medicine until World War I. Its representatives stood up for the sovereign rights of the medical profession and opposed all interference by the state in the private relationship between approbated physicians and their patients.[8] The path toward the crimes under the National Socialist regime, committed in the name of eugenics, racial hygiene, and population politics, was far from straightforward, as Weindling has shown in detail. It was rather riddled with intense ideological conflicts about the relationship between individuals, professions, other social groups, and state authorities. The authoritarian and technocratic "solutions" that some physicians, biologists, and intellectuals promoted vehemently in the name of eugenics and racial hygiene continually had to overcome strong and vociferous resistance.

Eugenics and racial hygiene should therefore not be seen as parts of a monolithic political movement either. The past three decades have seen the publication of a wealth of research into the history of eugenics in different national contexts, which has revealed a surprising variety of programmatic positions and coalitions. Eugenically motivated initiatives existed not only in Germany, but also in North America and Great Britain, France, the Scandinavian countries, Russia and later the Soviet Union, South America, and some African and Asian countries that were under colonial rule.[9] In addition to these national initiatives, some eugenicists organized themselves internationally in the first decades of the twentieth century.[10] Pleas for eugenic population policies were not issued only by right-wing racists. Karl Kautsky, for example, a leading figure in the German Social Democratic movement, put out a book in 1910 that bore the title *Vermehrung und Entwicklung in Natur und Gesellschaft* (Propagation and Development in Nature and Society). He proposed that measures of "artificial selection" should replace the processes of "natural selection that the struggle for existence produces" in every society.[11] Still, socialists with eugenic leanings usually rejected forced sterilization and euthanasia, putting their hope in education and voluntary birth control instead. And in contrast to right-wing racial hygienists, they rarely argued against an extension of better hygienic and therapeutic conditions to all members of society on the grounds that such measures would only promote the propagation of the idle and "unfit."[12]

Despite the variety and popularity of eugenic arguments around 1900—"eugenics" was one of the most frequently used entries in the *Reader's Guide to Periodical Literature* of 1910[13]—eugenicists never succeeded in forming a true mass movement. Nominal membership in the British Eugenics Education Society, founded in 1907, never exceeded 1,700, and the American Eugenics Society, which emerged from a coalition of local and regional associations in 1923, had only slightly more than 1,000 members.[14] Much the same was true in Germany, where the Society for Race Hygiene registered 425 members in 1913 (111 of whom were female) and 1,300 in 1930. These organizations, moreover, were clearly dominated by members of the professional middle class: physicians, university professors, engineers, teachers.[15] As the sociologist Donald MacKenzie has argued, engagement with eugenic problems was one way to strengthen the social status of this group. Members of eugenic organizations presented themselves to politicians and state authorities as experts who could contribute to the rational solution of burning social problems.[16] The rapid rise to dominance of Mendelian genetics after 1900 was in part due to the promise it held for scientifically legitimated social technologies, a point to which we will return in the next chapter.

Absolute numbers and the social composition of eugenic associations indicate that the pervasiveness of eugenic arguments cannot simply be attributed to political propaganda that was organized centrally or "from above." Some associations dedicated to the "improvement of heredity" were little more than agencies for arranging marriages among like-minded people. They were a small and often eccentric part of a much broader public discourse.[17] Eugenic arguments were employed in all sorts of contexts, and these arguments remained much the same regardless of whether those who employed them aimed at social reform, or even revolution, or whether they anticipated the decline of Western civilization, seeking measures to prevent it. We have already mentioned the working-class movement and attempts to reform health care systems as one relevant background. The women's rights movement was another important context for the mobilization of eugenic arguments. Demands for sexual and reproductive autonomy, and thus basic political rights, were frequently supported by eugenic considerations.[18] "More children from the fit, less from the unfit," the American feminist Margaret Sanger, for instance, proclaimed in 1919, "that is the chief issue of birth control."[19] Debates about prostitution were permeated with eugenic ideas owing to the prevalent view that venereal diseases attacked the hereditary material as well. Likewise, promoters of abstinence denounced tobacco and alcohol as "racial poisons," while social reformers attacked modern lifestyles, modern urban life in particular, as a source of racial degeneration. That eugenic keywords and slogans were taken up by racist and nationalist movements seems self-evident, but the very same slogans were often adopted by those who fought for the political emancipation of minorities who came under attack by those movements, like African Americans or Zionists.[20]

The enactment of sterilization laws in Germany, North America, and Scandinavia in the late 1920s and early 1930s could therefore rely on a broad public consensus. Eugenic views were socially acceptable, and eugenic demands seemed to reflect widespread worries about the health of future generations. Moreover, coercive measures and purely biological reasons for sterilization generally played a minor role, except in Nazi Germany as well as in some states in the United States and Canada.[21] Usually, sterilization required the formal consent of the individual concerned, or that individual's closest relatives, and the decision was based on whether the "patient" in question was deemed able to raise children, a question that even in those days could hardly ever be answered based on biological considerations alone. This may explain why in countries such as Sweden sterilization laws could remain in place without raising public concern until the 1970s.[22]

Most historical studies of eugenics do not scrutinize the concepts of heredity that underwrote these movements. Instead, they usually focus on the ideological and political impact of eugenic doctrines. If any effort is made to characterize notions of heredity, the well-worn dichotomy of "hard" versus "soft" is usually employed. After all, it seems that if one grants the inheritance of acquired characteristics, then there is at least some room for measures that seek to promote the "hereditary health" of the population through improved living standards. Conversely, rejecting the inheritance of acquired characters seems to imply that only strict selection and targeted breeding will contribute to eugenic aims. Indeed, neo-Lamarckian positions were particularly common among eugenicists with socialist leanings who sought to reform society in order to raise living standards for the less privileged and in Catholic contexts in which birth control and euthanasia were rejected for religious reasons. This has been demonstrated in great detail for the Soviet Union, France, and Latin America.[23] Yet more recent studies have shown that scientific, political, and religious positions often intersected in other, more surprising ways. Some right-wing Lamarckists believed that noxious living conditions caused irreparable damage to the germ line and that this damage warranted radical eugenic countermeasures. And some of those who defended Weismann's neo-Darwinism or Mendelism argued that mechanisms of "hard" inheritance, which left no room for environmental influences, did not support the idea that social phenomena like criminality or poverty could be reduced to biological causes. "Not every Lamarckian was democratic and egalitarian, nor every convert to Weismannism a reactionary," finds historian Diane Paul. "Scientific theories are socially plastic; they can be and frequently are turned to contradictory purposes."[24]

Most of the popular eugenics literature did not spell out a particular conception of heredity at all. Nor did it show its colors as far as the assumed mechanism of heredity was concerned. By and large this literature simply argued for medical or bureaucratic interventions into human reproductive behavior with the aim of reducing or increasing the quantitative proportions of social groups of putatively different economic or social value ("negative" and "positive" eugenics). Given this "demographic" perspective, as one would call it today, eugenicists required little more grounding for their proposals than the age-old adage that like begets like. Thus they convinced themselves and others that elimination of the "feeble-minded," "paupers," "criminals," "vagrants," "Jews," "colored people," or more generally of "unfit" and "alien" population elements was an adequate instrument to avert the "degeneration" of the body politic and to strengthen one's own "nation" or "people." It is clear today, as it was then, that such biologistic reifications denied social status, and with it fundamental human rights, to certain groups. Their acceptance across society depended on several ideological conditions, which occasionally came together to form a more or less compact, scientocratic *Weltanschauung* including social Darwinist elements. One component in this mixture was that national identity was seen to rely on the natural sciences, which were regarded as possessing an exclusive monopoly on culturally and socially relevant technical expertise; another was that "progress" was taken to mean the survival of one's own "nation" or "race" at the expense of others. At the same time, eugenic calculations were economic in nature. The aim was not to let the "struggle for existence" reign supreme, but, quite the contrary, to organize and channel it in such a way that the "right" kind of people—as a rule, those who enjoyed the same social status as oneself, whether in terms of class, nationality, or "race"—would ultimately be on the side of the winners. The deep-rooted anxieties that nourished these calculations, and which in turn were fueled by the results of these calculations, not only dominated the late nineteenth and early twentieth centuries, but make themselves felt even today, albeit more subtly, in debates about immigration, criminality, and health. The "ordinary" nature of the eugenics enterprise is also reflected in the efficiency and "rationality" with which eugenically and racially motivated population policies were implemented. The executives who planned, administered, and carried out National Socialist extermination campaigns were primarily trained economists, geographers, and demographers, rather than anthropologists driven by a fanatical concern for racial purity.[25]

Nevertheless, it is indisputable that biology and medicine furnished some of the key concepts and guiding principles of eugenics and racial hygiene, and conversely, that the movement of eugenics and racial hygiene created specific preconditions (a variety of institutions, practices, and working tools) for the establishment of a science of heredity— genetics—at the beginning of the twentieth century. In this chapter we focus on two important sets of techniques developed in the second half of the nineteenth century that genetics later relied on: the measuring devices and statistical techniques that racial anthropologists employed to quantify human variability, and the rigorous methods for the analysis of kinship that became increasingly significant in medical studies of human heredity. These techniques are of particular relevance for understanding the history of heredity because they promoted the idea that populations could be analyzed in terms of differential elements—traits, dispositions, genes—that were redistributed and recombined in each generation, thus giving rise to a new set of individuals. What was at stake here was thus not the truism that like begets like, but a very specific question put forward as early as 1837 by Cyrill Franz Napp, the abbot of the monastery in Brno at which Mendel later carried out his famous experiments: "What is it that is inherited and how is it inherited?"[26]

## Race and Anthropometry

The representation of human history as a merciless struggle for survival among different races was certainly one of the most poisonous innovations of the European Enlightenment.[27] The crimes against humanity planned and executed in the areas of Europe that were under the control of National Socialist Germany during World War II have so fundamentally discredited the application of the concept of race to humans that it is difficult to approach the complex history of this concept and its relationship to theories of inheritance with critical distance.[28] The gut reaction of most historians is to treat race simply as a social myth, invented for the ideological justification of oppressive practices, and this is also appropriate in most contexts. Yet any serious attempt to unravel this history has to recognize that "race" was not only a speculative idea but was empirically grounded as well, however technically or socially mediated this grounding was. We have already encountered two of its empirical roots in chapter 3 of this book: animal breeding and the system of *castas*, which was used in colonial Latin America to classify people of mixed indigenous, European, and African descent for purposes of administration and jurisprudence.

A third empirical root is less well known, but no less significant. In the first half of the nineteenth century, a number of European historians began to refer to the political factions that struggled for dominance in the English Civil War (1642–1649) and the French Revolution (1789) as different "races."[29] In doing so, they relied on historical sources that documented a corresponding self-identification of the warring factions. In the case of the European nobility, the self-identification as belonging to a "race" of its own can even be traced back to the revival of ancient notions of "noble blood" in the late medieval period. "It seems to me," the French historian Augustin Thierry mused in 1835, "that something of the conquest of barbarians still bears weight today in our country, despite the temporal distance; and that we can descend, stage by stage, from what we are living through today all the way down to the invasion of a foreign race into the core of Gaul and its domination over the native race."[30] He was referring to the invasion of the Roman provinces of Gaul by the Germanic tribe of the Franks in the fifth century AD, which Thierry saw as the origin of the conflict between bourgeoisie and nobility that had reached its pinnacle in the French Revolution. In a letter to Friedrich Engels, Karl Marx could therefore refer to Thierry as "*le père* [the father] of class struggle."[31]

Both the *castas* system and the reinterpretation of social and religious conflicts in early modern Europe as resulting from an underlying "race struggle" demonstrate that the notion of human races was initially by no means tied to physical or biological characteristics alone. Throughout the nineteenth century, the term "race" was used synonymously with other terms like "nation," "people," or "tribe." It thus covered groups, both under their own description and under the descriptions of others, that differed not only in their physical traits, but in their language, mentality, material culture, religion, and social organization. There was no "clear line between cultural and physical elements or between social and biological heredity," writes the historian of anthropology George Stocking about the nineteenth century. "'Blood' was for many a solvent in which all problems were dissolved and all processes commingled."[32]

The ideological efficiency of this "solvent" is revealed particularly well in frequent debates about "vanishing tribes" in the nineteenth century. An address by James Cowles Prichard to the British Society for the Advancement of Science in 1839, published in the journal of the British and Foreign Aborigines' Protection Society, provides a good example of its amphibolic nature. "Wherever Europeans have settled, their arrival has been the harbinger of extermination to the native tribes," Prichard stated plainly.[33] Like many others, however, Prichard did not assume that the inferiority of "native tribes" could be reduced to biological differences. More important was the technological and cultural superiority of European colonizers over native populations, which manifested itself, according to Prichard, in agriculture, literacy, military organization, and Christianity on the one side as well as in self-defeating mentalities and practices—a lack of self-control, tribalism, and cannibalism, which, according to common prejudice of the time, reigned among the "savages"—on the other.[34]

This optimistic view of colonial expansion—optimistic, that is, from the perspective of the Europeans—was the discursive reverse of the degeneration fears that at the same time took hold of European elites with respect to their own societies. Prichard's address is exemplary in setting aside all moral or civilized scruples by portraying the destruction of other cultures and people as the victory of culture writ large over autochthonous nature. The Scottish geologist Charles Lyell similarly consoled himself in his *Principles of Geology* (1832–1833) about the more sordid sides of colonial conquests: "Yet, if we wield the sword of extermination as we advance, we have no reason to repine at the havoc committed . . . . We have only to reflect, that in thus obtaining possession of the earth by conquest, and defending our acquisitions by force, we exercise no exclusive prerogative. Every species which has spread itself from a small point over a wide area, must, in like manner, have marked its progress by the diminution, or the entire extirpation, of some other."[35] Lyell's *Principles* were an important inspiration for Darwin, who made quite similar remarks in his *Descent of Man* (1870–1871).[36] To put it somewhat cynically, the extermination of "native tribes" provided the most concrete example of the principle of natural selection that the nineteenth century provided.

Discourse about "vanishing tribes" was supported by the same unspecific notions of inheritance as the discourse on degeneration that unfolded simultaneously. Often, no more than a general idea of reproduction—the "golden chain of improvement," as Johann Gottfried Herder expressed it in his *Outlines of a Philosophy of the History of Man* (1784–1791)[37]—was adduced to explain the persistence of different languages, cultures, and racial types. Yet in this context, a focus on the transmission of isolated traits that could be described and measured with precision became more and more prominent. Prichard, for example, identified a revealing way of saving the "tribes" doomed for destruction, at least for the virtual world of science:

> In the meantime, if Christian nations think it not their duty to interpose and save the numerous tribes of their own species from utter extermination, it is of the greatest importance, in a philosophical point of view, to obtain much more extensive information than we now possess of their physical and moral characters. A great number of curious problems in physiology, illustrative of the history of the species, and the laws of their propagation, remain as yet imperfectly solved.[38]

The interest in the diversity of human physiological and psychological traits that is evident in this quote can be traced back to the appropriation of the Latin American *castas* system by European naturalists in the eighteenth century. As outlined in chapter 3, skin, hair, and eye color played a role in visual representations of the *castas*, but these features remained embedded in a multiplicity of other references—such as dress, occupation, and emotional character—to the social and cultural status of the depicted persons. In his *Systema Naturae* of 1735, Linnaeus transformed the *castas* system into an abstract classification scheme that dif-

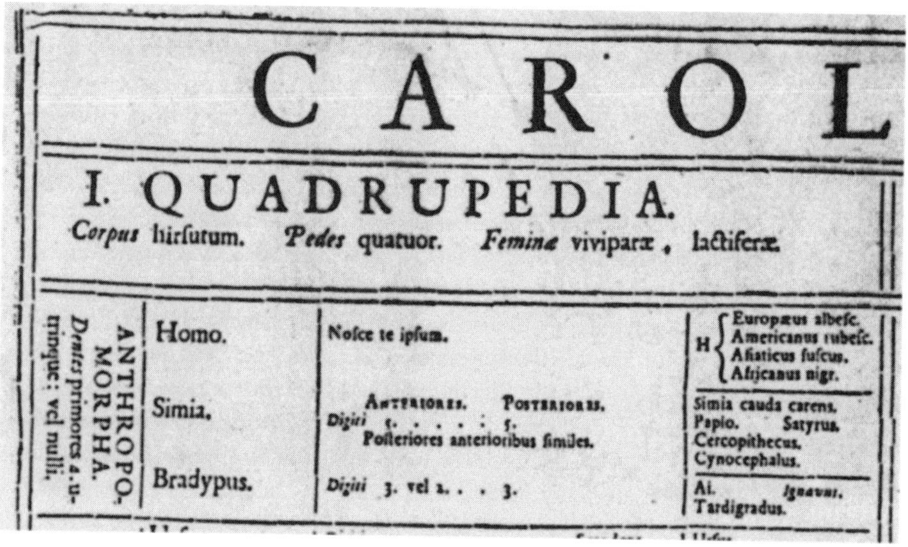
CAROL

I. QUADRUPEDIA.

*Corpus* hirsutum. *Pedes* quatuor. *Feminæ* viviparæ, lactiferæ.

| | | | |
|---|---|---|---|
| ANTHROPOMORPHA. *Dentes* primores 4. utrinque: vel nulli. | Homo. | Nosce te ipsum. | H { Europæus albesc. Americanus rubesc. Asiaticus fuscus. Africanus nigr. |
| | Simia. | *Digiti* Anteriores. 5. Posteriores. 5. Posteriores anterioribus similes. | Simia cauda carens. Papio. Satyrus. Cercopithecus. Cynocephalus. |
| | Bradypus. | *Digiti* 3. vel 2. . . 3. | Ai. *Ignavus.* Tardigradus. |

**FIGURE 5.1** Detail from Carl Linnaeus, *Systema naturae* (Leiden: De Groot, 1735), listing four human races.

fered from its sources by extreme reduction and a claim to universality. For Linnaeus, the human species was partitioned into four "varieties": the "whitish European," the "reddish American," the "tawny Asian" ("lurid" in later editions), and the "blackish African" (fig. 5.1). Subsequent editions of the *Systema Naturae* connected this quaternary scheme with the medical doctrine of four temperaments: the red skin color of Americans was associated with a choleric disposition, the white skin color of Europeans with a sanguine disposition, the (now) "lurid" skin color of Asians with a melancholic character, and the black color of Africans with a phlegmatic temperament.[39] The particular sources from which Linnaeus developed this classification are not known. What is known, however, is that Latin American *savants* had established similar associations between physical characteristics and temperament. In fact, complexion had already been used as a diagnostic sign for underlying temperament in the medieval period, albeit without raciological intent.[40]

Renato Mazzolini has proposed that eighteenth-century physical anthropology was founded on the question of how differences in skin color come about.[41] There is much evidence that supports this thesis, but it also presents a historical puzzle. A white, a red, or a dark complexion had long been regarded as a reliable indicator of health, which in turn was thought to be determined by factors such as climate, temperamental disposition, and individual lifestyle. Complexion was thus not regarded as a character that was set at birth and remained constant throughout life. Despite his schematism, Linnaeus had no qualms about treating variation in skin color like variations in other highly variable characteristics such as body weight and stature.[42] Color, moreover, was generally regarded with suspicion by naturalists because it varied substantially within species and was, in the last analysis, contingent on subjective judgments. Even Johann Friedrich Blumenbach, who is commonly regarded as one of the "fathers" of physical anthropology, maintained in his *On the Natural Varieties of Mankind* (1775) that human skin color was of accidental origin and varied with climate and lifestyle.[43] Of course, intermarriage also contributed to color variation, as acknowledged by Buffon in a description of Mexico City in his *Histoire naturelle*. "In the town of Mexico," he wrote, "there are Europeans, Indians from north and south America, African Negroes, Mulattoes, and mongrels of every kind; so that we see men there of every shade between black and white."[44] Most early anthropologists hence agreed that skin color was far from being a sure sign of racial belonging.

Influenced by Blumenbach, physical anthropologists therefore began to take an interest in "harder," apparently more "constant" characters than color. Skeletal features—especially of the skull, which was assumed to have a close connection with physiognomy and brain anatomy—attracted the attention of anthropologists in particular.[45] The Dutch anatomist Petrus Camper, who was also a practicing sculptor, saw the facial angle as an infallible sign of the position of a "nation" on a developmental scale defined by cultural achievement, and in 1842 the Swedish anatomist Anders Adolf Retzius introduced the cephalic index (the ratio of the maximum head width and maximum head length), which would occupy physical anthropologists for almost a century to come. On the basis of his index, Retzius made a distinction between a brachycephalic (shortheaded) and a dolichocephalic (long-headed) type. He further proposed that the former was representative of a European Stone Age population that in its pure form had survived only in small pockets (Basques, Finns, and Lapps), while the latter descended from an Indo-Germanic "race" that had invaded Europe during the Bronze Age. Such broad-brush theories drew immediate criticism from, for example, Paul Broca, the French founder of the influential Société anthropologique, who devoted much energy to developing instruments for more refined skull measurements.

During the nineteenth century, the number of measured variables increased continually, as did the human varieties that were distinguished on this basis. Measurement techniques and instruments were standardized, and they became so widespread that anthropometry around 1900 bears comparison to other "big sciences" like astronomy or meteorology. Anthropometric data of all kinds were collected for entire cohorts of military recruits in countries such as Italy and Sweden, and hundreds of morphological parameters could be elicited from a single human skull. Toward the end of the century, the scope of such studies was extended to psychological and cognitive properties for example, in the "psychology of nations" (*Völkerpsychologie*) of the German psychologist Wilhelm Wundt and in the "intelligence tests" developed by Alfred Binet and Lewis M. Terman at the beginning of the twentieth century.[46]

The trend toward operationalization and mechanization of data collection in the field within anthropology—the overpowering drive toward "mechanical objectivity," which it shared with most scientific disciplines throughout the nineteenth century[47]—is interesting for our history of knowledge of heredity in two respects. On the one hand, it promoted an analytic perspective, even in contexts in which anthropometric measures were used as signs of underlying holistic or typological entities, such as temperament, race, or degree of civilization. Under this analytic gaze, the human body, cultures, and languages increasingly emerged as objects that were composed of simpler elements—traits or dispositions, elements of style, phonemes, and grammatical structures. On the other hand, this perspective inadvertently brought to the fore what looked like autonomous laws that governed the distribution of such elements in populations. The development of mathematical statistics in the eighteenth and nineteenth centuries was indeed closely linked to demographic and anthropometric projects.[48] We cannot offer a full history of statistics here, but some episodes are worth looking at because they illustrate how the analytic measurement of populations led to the discovery of "laws of inheritance."

The main protagonist, once more, is Francis Galton. In 1861, Galton met the French astronomer Adolphe Quetelet, who in 1835 had begun to transpose the statistical theory of errors from his home discipline to the study of "average man" (*l'homme moyen*), a creature that supposedly represented the "type" of its nation, in all its physical and moral properties, and as such formed the object of Quetelet's "social physics" (*physique sociale*). Galton was an admirer of Quetelet's work but disliked his methodological focus on averages. "Some thorough-going democrats may look with complacency on a mob of mediocrities," Galton once wrote, "but to most other persons they are the reverse of attractive."[49] Galton was more interested in the tails of statistical curves. In *Hereditary Genius* (1869), for example, he attempted to demonstrate that more talented individuals were to be found among the relatives of extraordinarily gifted musicians, lawyers, and politicians than would be expected under the assumption of a purely stochastic distribution of these traits. He swept aside the obvious objection that such clustering was attributable to education, family tradition, and nepotism, rather than inheritance, by claiming that such factors played only a small role in liberal England, at least when it came to truly extraordinary talent.[50]

Galton's interest in the inheritance of extreme traits also surfaced in his experiments with sweet peas, which he carried out from the mid-1870s onward. In the spring of 1876, he sent packets of pea seeds to a number of acquaintances. Having weighed the seeds beforehand, he asked his helpers to cultivate seeds that fell into different weight classes separately. After receiving the crop, Galton analyzed the results according to two parameters, the weight of the harvested seeds and the weight of the seed they had been grown from. When Galton compared the two resulting distributions, a phenomenon emerged that he called "regression," thus adopting an expression from contemporary theories of inheritance. The offspring of peas that deviated from the average, or "type," of a population also deviated from that type, but to a lesser degree. Offspring thus seemed to "regress" to the mean of the overall population, and the degree to which they did so was proportional to the degree to which the parental seeds had been deviant. In other words, the heavier or lighter the parental seeds had been, the more strongly their offspring tended to revert back toward the population mean.[51] Galton thought that regression had the function to preserve a stable type. Accordingly, he later developed ideas of discontinuous evolution that were taken up by the early geneticists Hugo de Vries and Wilhelm Johannsen.

The mathematician Karl Pearson proposed another interpretation of these findings. Inspired by Galton's studies, he developed the mathematical apparatus needed to describe statistical relationships in populations in a series of papers that appeared from 1894 to 1904 in the *Philosophical Transactions of the Royal Society* under the title "Mathematical Contributions to the Theory of Evolution." This body of work established Pearson as the founder of what came to be called the "biometrical school" of heredity. Pearson assumed that regression resulted from a "correlation" of the trait under study with a number of other traits that, on average, could be expected to lie closer to the population mean.

Both Galton and Pearson assumed what they called a "law of ancestral inheritance" in their conjectures. According to this law, the germ plasm of an organism consisted of heritable elements that were derived not only from the parents but from more distant ancestors as well. The law further assumed that the relative fraction of ancestral elements in the

FIGURE 5.2 Scheme illustrating Francis Galton's "law of ancestral inheritance." From Francis Galton, "A Diagram of Heredity," *Nature* 57 (1898, p. 293).

germ plasm diminished in proportion to the number of generations that separated ancestor and descendant. For example, when Galton first proposed this law, he surmised that one-half of the *stirp* derived from the two parents, one-fourth from the four grandparents, one-eighth from the eight great grandparents, and so on (fig. 5.2).[52] Pearson, on the other hand, believed that these proportions could vary for different traits. In his "Biometrical Laboratory" at University College in London, which from 1911 bore the name Galton Laboratory for National Eugenics, Pearson worked out a statistics of correlations with the aim of measuring the "strength of inheritance" for particular traits.[53] This work laid the foundations of modern multivariate statistics.

As these examples illustrate, the distribution patterns that emerged as the statistical relationships between isolated traits were investigated became objects of research in their own right. Conversely, the emerging regularities seemed to warrant the assumption that the traits chosen for study were indeed clearly specifiable, heritable traits. In the research of Galton and Pearson, there was no well-defined boundary between the development of mathematical algorithms on the one hand and biological research on the other. The terms "regression" and "correlation" initially derived their meaning from research into the presumed laws of inheritance, and they referred to fundamental biological forces presumed to determine the transmission and development of traits in organisms. Today, both terms merely designate mathematical relationships between two frequency distributions, from which causal propositions can be derived only with great caution and by drawing on additional information. Perhaps the easiest way to make this point is by looking at the early history of the concept of regression. Galton himself soon realized that regression could equally be observed if one compared parents with their children; that is, the parents of children with extreme traits also tended to "regress" to the population average. However, a biological interpretation of this "tendency" was never contemplated, for this would have implied that children somehow influence the makeup of their parents.

Moreover, as much as the statistical procedures described above could legitimate and reify social and cultural stereotypes, they could also be used as instruments for the critical examination of stereotypes. The collection of large quantities of information about traits in frequency tables and curves, for example, soon revealed that the variability of traits went far beyond the discrete types that were recognized in everyday contexts. The distribution of the value of a trait petered out at its tail ends, often resulting in substantial overlap between types (fig. 5.3). For many traits, it became evident that the difference between two averages that defined supposedly discrete types carried hardly any weight compared with the broad variation within each type. Moreover, upon closer inspection, some trait distributions showed two or more peaks, an indication that the population in question was not homogeneous, but a composite of heterogeneous but overlapping types. And, more often than not, studies of correlation, which—like Galton's studies of sweet peas—always involved comparisons of frequency distributions resulting from classifying the same raw data according to different criteria, revealed that a presumed relationship between two variables did not exist at all. The positivist tendency toward ever more fine-grained specification, classification, and reclassification of data thus brought about a dialectical development that not only solidified but also subverted customary categories of race, class, and gender. The distribution and transmission of many traits, if considered in isolation, seemed to be subject to laws that were largely independent of the "types" whose recognition had originally inspired their study.[54]

One of the earliest anthropologists who realized the critical potential of statistical procedures was Franz Boas. He had studied anthropology with Rudolf Virchow and the geographer Adolf Bastian in Berlin and emigrated to the United States after a legendary expedition to the Canadian Arctic in 1883–1884.[55] Boas came to dominate American

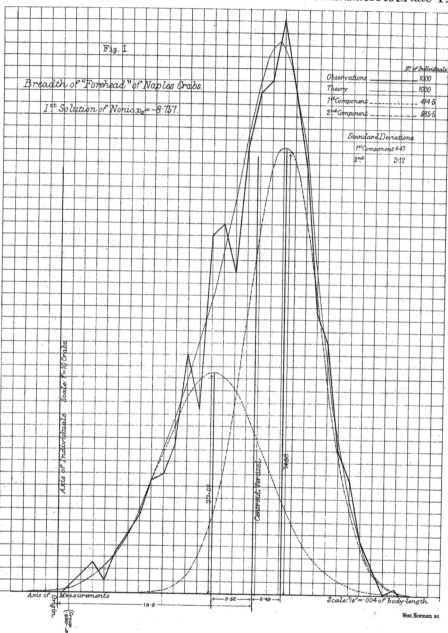

**FIGURE 5.3** Curves from Karl Pearson und Walter F. R. Weldon representing trait distribution in a crab population. From Karl Pearson, "Contributions to the Mathematical Theory of Evolution," *Philosophical Transactions of the Royal Society of London* 185 (1894), pl. 7.

anthropology for more than fifty years and today is habitually referred to as the "founding father" of modern cultural anthropology. Boas himself, however, insisted throughout his career that physical anthropology was an integral part of the broader discipline of anthropology. Until the beginning of World War I, he undertook large-scale studies of physical anthropology, and some of his students continued this line of research after the Great War. One of Boas's first contributions to this field was a short article that appeared in the journal *Science* under the title "Mixed Races." In this article, he demonstrated that the cephalic indices of two ethnic groups of Native Americans—the "Oregonian Athapascans" and "Southern Californians," as he called them—did not, as was widely expected for anthropometric traits in general, blend in the offspring of intermarriages. Instead, he found that offspring were of a "mixed type," with their cephalic indices clustering around the average of one or the other of the parents.[56] In the mid-1890s, Boas began to adopt the methods Pearson had developed for the study of statistical correlations, and one of his subsequent findings caused considerable consternation among physical anthropologists, including Pearson. He showed that the two variables that constitute the cephalic index—head length and head width—did not correlate with each other, and thus apparently did not point to some common cause such as shared ancestry or physical type.[57]

Boas's anthropometric investigations, which we will explore further in the next section, can be regarded as thought experiments involving repeated reclassification and reanalysis of raw data from a variety of angles. The overall results of these investigations led Boas to subject the concept of race to a critical revision in a monograph published in 1911 under the title *The Mind of Primitive Man*.[58] This book already assembled all the arguments against the view that race was a biologically—and politically—meaningful category that were laid out forty years later in the so-called UNESCO Statement on Race of 1951, which is often portrayed as the most manifest sign of the irrevocable collapse of racial anthropology as a science after World War II. In the 1930s and 1940s, Boas himself had contributed to the decline of racial anthropology through the ceaseless promulgation of his views, both in the form of popularization and among his peers. Yet he did not believe that his scientific investigations had had only this negative result.[59] The title of the German version of his 1911 book, published in 1914, was *Kultur und Rasse* ("Culture and Race"), and this title probably better reflects Boas's scientific project. Rather than just debunking the category of race, his aim was to understand the exact relationships between the many elements whose variation constitutes human diversity, including physical characteristics.

As it turned out, many of these elements, contrary to deep-rooted expectations, were only loosely correlated, and Boas regarded this as a positive result. It indicated that the present distribution of cultural artifacts, linguistic particularities, and physical characteristics was the result of historical processes of circulation among cultures rather than of their evolution within cultures.[60] Language, culture, and bodily traits obeyed different laws in their transmission from generation to generation and in their diffusion across geographic space. Still, they could be viewed as domains that in their sum made up humanity. Boas therefore continued to study the physical, linguistic, and cultural characteristics of Native Americans with unfaltering methodical rigor and fixation on details.[61] His student Alfred L. Kroeber, in contrast, concluded that language and culture constitute a "superorganic" domain of their own. This implied a strict division of labor between cultural and physical anthropologists. Interestingly, Kroeber supported his argument by resorting to Weismann's theory of the germ plasm. He argued that culture and language, in contrast to physical properties, were indeed "acquired" and that the modalities of their transmission were therefore fundamentally different.[62] In both Boas's and Kroeber's hands, however, Herder's "chain of improvement" had fallen apart into separate, albeit tangled, strands.

Finally, we want to make some comments about a notorious pair of concepts that came to the fore in contestations about the relationship between nature and culture around the turn of the twentieth century: "nature versus nurture." These terms were again first used by Francis Galton, and they also provided the title for a programmatic essay by Pearson, which derived from an earlier plenary address at the opening of the annual meeting of the Social and Political Education League in 1910.[63] For both Galton (in his earlier writings) and Pearson, however, the dividing line between nature and nurture differed from the one today's biologists tend to adhere to. For them, nature and nurture did not correspond to two organic registers, one of which was responsible for the transmission of traits and the other for their development. Instead, what they had in mind were two different sets of causes that acted statistically on descendants: on the one hand, the "indefinitely numerous small causes"[64] that reached down from the past into the present, resulting in variations that had a measurable average effect on descendants; and on the other hand, the myriad of causal factors that affected every organism during its lifetime, often instantaneously. As the nature-nurture dichotomy is mobilized in present-day debates about the evaluation of results in human genetics (especially with respect to sociobiology and evolutionary psychology), be it for affirmative or critical purposes, one has to bear in mind that the frame of reference has shifted to a genetic understanding of inheritance. While it made perfect sense for Galton and Pearson to compare "nature" and "nurture" quantitatively, as both categories included causal factors of a similar kind, such comparisons make little sense today because genes and environmental factors are held to be categorically distinct. To say that 90 percent of the variance of a certain feature is genetically determined does not therefore warrant the conclusion that environmental factors (or "nurture") may have no dramatic effects in this respect.

*Genealogy and the Analysis of Kinship*

In the first decade of the twentieth century, a fierce and bitter debate erupted between Karl Pearson and William Bateson, chief promoter of the institutionalization of Mendelian genetics in Britain at the time. This debate has often been interpreted as a clash of mutually exclusive paradigms in which the Mendelians won the upper hand over the biometricians in the end.[65] Apart from the personal and political differences that also fueled the debate—Bateson was more conservative, while Pearson had radical nationalist and socialist leanings[66]—such an interpretation overlooks two pertinent facts. On the one hand, the methods that the biometricians developed to measure statistical correlations between traits continue to be used for the study of heritability to this day, especially in twin research.[67] On the other hand, while it is true that biometricians and Mendelians endorsed different theoretical perspectives, these perspectives were far from mutually exclusive. This was pointed out as early as in 1902 by the Scottish biometrician Udny Yule. The law of ancestral inheritance, Yule warned, "is a law applying to aggregates and predicates nothing concerning the individual."[68] Biometricians like Pearson regarded heredity as the sum of the causal ancestral influences that affect the members of a given generation stochastically. This understanding of heredity implied the law of ancestral inheritance, since the number of ancestors doubles with each generation until the ancestors come to represent the whole "race." The law of ancestral inheritance could thus be used to calculate the probability that a descendant would resemble its ancestors with respect to a certain trait. However, this line of reasoning does not license conclusions about the actual physiological contributions that individual ancestors made to the constitution of their descendants.[69]

The statistical concept of heritability provides an example by means of which Yule's argument can be clarified. The ability to predict with a certain probability that children will resemble their parents with respect to a trait warrants no conclusions with regard to the question of whether the similarity is attributable to common genetic heritage or a common environment. In order to answer the latter question, one has to resort to genetic experiments such as those instituted by Mendel. These experiments do not consider the sum total of ancestral influences, but rather infer the genetic constitution of individual ancestors by observing the distribution and transmission of particular, well-defined genetic factors in their offspring.[70] Once the behavior of such factors had been ascertained experimentally, it was indeed possible, as Yule and Pearson himself later demonstrated mathematically, to account for their effect on populations by means of the law of ancestral inheritance.[71]

The debate between biometricians and Mendelians shows that the discourse of heredity was not consolidated only by the application of statistical methods. The analysis of concrete genealogical relationships between ancestors and descendants—in short, the analysis of kinship—was equally important. Dramatic developments in the analysis of kinship occurred during the nineteenth century, although these developments have not received the same attention from historians of biology as the concurrent developments in statistics. Below, we want to identify some of the decisive conceptual steps that were taken in the analysis of kinship and that proved to be relevant to the history of the concept of heredity.[72]

One important step on the way to thinking in terms of heredity was a major extension of the concept of generation that had already begun to take hold toward the end of the eighteenth century. As we have seen in chapter 2, "generation" had traditionally been used as a synonym of "procreation," the momentous act of creating a new being. This is also true for the metaphorical meaning that the term had taken on in the context of genealogy and chronology. Historical epochs and degrees of kinship were often defined by the number of successive procreative acts in lines of linear descent, thus defining dynasties, for example. But around 1800, as Ohad Parnes has pointed out, a new concept of generation began to assert itself not only in the life sciences but in the social and cultural sciences as well. Generations were now understood as collectives, encompassing humans or other organisms that were born at roughly the same time. "New was the assumption," Parnes emphasizes, "that generations are identifiable units within society, and that the members of generations share significantly more than merely the simultaneity of existence." Most notably, succeeding generations were identified as the carriers of sovereignty in political theories such as those put forward by Thomas Jefferson, who even likened generations and the relationships between them to those of "distinct nation[s]."[73]

The significance of this new concept of generation for the emerging discourse of heredity may appear paradoxical. "Generation," in the sense of a "cohort"—the more technical term in use among sociologists since Karl Mannheim's groundbreaking study "The Problem of Generations" of 1928[74]—or a group of individuals born at about the same time, ignores concrete kinship relations. Generations in this sense do not succeed one another like individuals in a line of descent, as any society at a given time will consist of individuals belonging to diverse age groups. Nor do members of a generation share a uniform degree of relatedness; they are rather related to one another in widely different degrees. Yet, according to Parnes, precisely these features of generations presented new opportunities of accounting for the economy of transmission. To what degree properties—whether in the biological or socioeconomic sense—were conserved and to what degree they varied from one generation to the next could now be described in exact quantitative terms. Moreover, as a result of its inherent ambiguity, the concept of generation allowed for a variety of appropriations. As Parnes sets out in detail, social philosophers like Auguste Comte, John Stuart Mill, and Karl Marx portrayed the reproduction of society as a succession of generations in which one generation is productive at any given time and is replaced by the next. Biologists such as Carl Nägeli, Alexander Braun, and notably Gregor Mendel, on the other hand, relied on such economic and political images to articulate a new understanding of biological reproduction that adopted a similar bird's-eye view of evolving populations.

As Parnes emphasizes, the decisive point about this new understanding of generation was that one could now think about the transformation of whole societies as well as species within a time frame accessible to experience. Generations were the units onto which the heritage of previous generations fell and which, thus finding themselves in a *new* starting position, could make sovereign use of this heritage and finally pass on an enriched inheritance to the next generation. The wider concept of generation thus placed the idea of inheritance within a system of diachronic and synchronic coordinates that allowed one to think of its transformation as a regular process of change. The "complex web of hereditary transmissions"[75] that constituted the relationships between generations

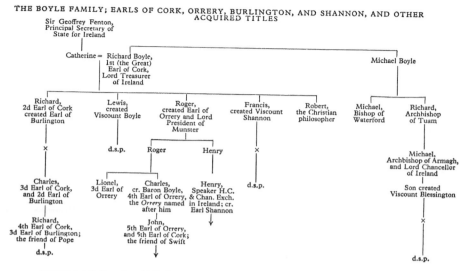

FIGURE 5.4 Pedigree from 1892. From Francis Galton, *Hereditary Genius: An Inquiry into Its Laws and Consequences* (London: Macmillan, 1869), p. 204.

and the laws that were assumed to govern these relationships became objects of scientific scrutiny.

A concomitant change in the meaning of kinship also contributed to the new context of understanding hereditary transmission. We have seen in chapter 2 that premodern notions of kinship were deployed flexibly, often either to exclude close relatives from hereditary succession on formal grounds or to confer entitlement to an inheritance on more distant relatives. Relations of parentage were therefore often reduced to unbranching lines of descent that omitted female relatives and collateral lines. This tendency can still be gleaned from Galton's *Hereditary Genius*: many (though certainly not all, as is sometimes claimed) of the numerous "records of families" in the book exclusively registered male family members and their relationships. However, the genealogical diagrams that accompany the text (fig. 5.4) also reflect the conceptual change that interests us. The diagrams were structured by generations and included collateral relatives, such as uncles, nephews, and cousins. This structure shows that Galton sought to document kin relationships within each generation in a comprehensive and precise manner.

Vernacular kinship terminology is often insufficient as a basis for such a precise analysis of kinship. The English word "uncle," for example, has two genealogical meanings: it can designate one's father's brother but also one's mother's brother. Galton dealt with this problem by introducing a notation system that symbolized seven fundamental

| Description of persons. | Relationships in Seneca. | Translation. |
|---|---|---|
| **LINEAL LINE.** | | |
| 1. My great grandfather's father........ | Hoc'-sote........ | My grandfather. |
| 2. " great grandfather's mother........ | Oc'-sote........ | " grandmother. |
| 3. " great grandfather........ | Hoc'-sote........ | " grandfather. |
| 4. " great grandmother........ | Oc'-sote........ | " grandmother. |
| 5. " grandfather........ | Hoc'-sote........ | " grandfather. |
| 6. " grandmother........ | Oc'-sote........ | " grandmother. |
| 7. " father........ | Hä'-nih........ | " father. |
| 8. " mother........ | No-yeh'........ | " mother. |
| 9. " son........ | Ha-ah'-wuk........ | " son. |
| 10. " daughter........ | Ka-ah'-wuk........ | " daughter. |
| 11. " grandson........ | Ha-yä'-da........ | " grandson. |
| 12. " granddaughter........ | Ka-yä'-da........ | " granddaughter. |
| 13. " great grandson........ | Ha-yä'-da........ | " grandson. |
| 14. " great granddaughter........ | Ka-yä'-da........ | " granddaughter. |
| 15. " great grandson's son........ | Ha-yä'-da........ | " grandson. |
| 16. " great grandson's daughter........ | Ka-yä'-da........ | " granddaughter. |

**FIGURE 5.5** Detail from one of Lewis H. Morgan's tables of kinship terminology. From Lewis H. Morgan, *Systems of Consanguinity and Affinity of the Human Family*, Washington 1871.

kinds of kin: father, grandfather, uncle, nephew, brother, son, and grandson. In addition, he used typographical means to highlight the gender of a person and to indicate the side of a family: italics indicated female gender, whereas capital letters referred to paternal and lowercase letters to maternal kin. Thus, although the abbreviation "*u*" was derived from "uncle," it denoted a maternal aunt. "The notation I employ," Galton claimed, "gets rid of all this confused and cumbrous language. It disentangles relationships in a marvellously complete and satisfactory manner, and enables us to methodise, compare, and analyse them in any way we like."[76]

Anthropologists and physicians took the analysis of kinship one step further. In 1871, only two years after the publication of *Hereditary Genius*, the American lawyer Lewis H. Morgan published his *Systems of Consanguinity and Affinity of the Human Family*. In this colossal work, Morgan documented and analyzed kin designations in languages from all over the world, collected with the help of the Smithsonian Institution, various missionary organizations, and the U.S. State Department (fig. 5.5). To facilitate the translation of kinship designations from different languages, Morgan had developed a "new instrument of ethnology" with which he equipped the numerous diplomats, missionaries, and naturalists who had agreed to help him. This "schedule," he claimed, was "sufficiently full to describe every known relationship, and yet arranged upon such a method as to be simple and intelligible."[77] In a fashion not dissimilar to Galton's notation, Morgan's schedule represented all kinship relations by combining a limited set of elementary relations:

mother of, father of, son of, daughter of, brother of, sister of, husband of, and wife of. Simple as well as complex relationships could thus be translated by concatenating these elementary terms, from the simple "my son" to "my mother's mother's mother's brother's son's son's son." Thus the schedule could account for no less than 273 kin relations, covering relatives up to the fifth degree.

By comparing the kinship terminologies extracted from all sorts of languages—from Swedish to Tamil, from Celtic to Iroquois—with their translations into vernacular English and with his own analytic terminology in countless tables appended to the book, Morgan was able to demonstrate that the systems of kinship found in different natural languages obeyed very different principles. But Morgan did not pursue a solely philological enterprise. His schedule, he maintained, possessed a "numerical . . . character" and rested "upon an ordinance of nature." It could therefore manifest "family relationships" that "exist in virtue of the law of derivation, which is expressed by the perpetuation of the species through the marriage relation. A system of consanguinity, which is founded upon a community of blood, is but the formal expression and recognition of these relationships."[78] The analysis of kinship systems, Morgan believed, provided direct access to the evolutionary history of humankind, bringing to light the different stages it had passed through as well as the different branches into which humankind was unfolding in the present.[79] Ultimately, kinship systems could be used as an index for racial and cultural diversity because they were an immediate expression of the hidden transmission mechanisms that supported the cultural and biological reproduction of every single society.

Morgan is now seen as one of the fathers of social anthropology. His work stood at the end of a line of research that saw the publication of several other classic works of the formative period of this discipline in the 1860s. In *Mother Right*, the Swiss scholar Johann Jacob Bachofen postulated in 1861 that humanity had passed through three evolutionary stages: a primitive, promiscuous stage, followed by a matriarchal and finally a patriarchal stage. That same year, Henry James Sumner Maine, who had been a consultant for the British colonial government of India for some time, drew attention to the close but complex interdependencies between marriage rules and territorial claims and proposed to distinguish between "stationary" and "progressive" societies on this basis. Four years later, John F. McLennan introduced the analytic distinction between exogamy and endogamy in *Primitive Marriage*. Also in 1865, Edward Tylor—a lawyer by training just like Morgan, Bachofen, Maine, and McLennan—clarified the relationship between cross-cousin marriage, exogamy, and matrilocal residence in his *Researches into the Early History of Mankind and the Development of Civilization*. In short, the complicated rules that govern marriage, parentage, and inheritance in every society became an object of almost obsessive anthropological scrutiny, just like the physical variability of mankind. Overall, these studies showed that kinship systems did not simply represent relations of blood. They rather incorporated and used these relations to construct autonomous systems of cultural exchange and transmission. This was further evidence for what Kroeber later called the "superorganic" nature of culture.[80]

The studies discussed in the preceding paragraph were cited widely in the nineteenth century. Darwin discussed them in his *Descent of Man* of 1871, as did Friedrich Engels in his *The Origin of the Family, Private Property, and the State* of 1884. The results of researches into kinship in early social anthropology thus fed indirectly into the most important intellectual and ideological currents of the late nineteenth century.[81] Owing to the exceedingly complex and variable nature of the topic, however, the literature remained esoteric and was mainly written for a specialist audience. In contrast, contemporary medical and psychiatric literature had a more immediate effect on the broader understanding of kinship. In this context, kinship was analyzed to uncover causes for "familial degeneration," and eugenic interests combined with an almost omnipresent interest in genealogy within the growing middle classes. This combination of interests led to the collection of genealogical data on a hitherto unprecedented scale.

Information about family relationships entered the realm of psychiatry in two closely related ways. On the one hand, such personal details were collected for administrative purposes in asylums, and digests of this information filled institutional statistics from the early nineteenth century.[82] On the other hand, psychiatrists early on promoted the idea that psychic and mental diseases could be attributed to a hereditary disposition (called diathesis) and that the unimpeded spread of these conditions would cause sweeping degeneration.[83] Their case histories therefore usually included information about the incidence of mental diseases, mental retardation, suicides, and alcoholism among the relatives of a patient. Contemporary institutional records correspondingly distinguished between patients that suffered from mental diseases that were "directly," "indirectly," "atavistically," or "collaterally" inherited, depending on whether their parents, their grandparents, or collateral relatives were affected by a comparable mental disorder. And while many psychiatrists attached high etiological significance to hereditary factors, heredity was

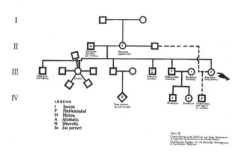

FIGURE 5.6 "The near blood-kin of a feebleminded woman sterilized by the State of California." Family pedigree from the Eugenics Record Office, ca. 1935. Courtesy Cold Spring Harbor Laboratory.

only one among many aspects that physicians recorded in what they called "developmental" histories of mental diseases.[84]

From the mid-nineteenth century, it became more and more common for medical and psychiatric publications to include genealogical diagrams to document the distribution of diseases among members of a family. The best-known examples of this are the diagrams produced at the Eugenics Record Office (fig. 5.6), which was founded in 1910 by the biologist Charles B. Davenport at the Station for Experimental Evolution in Cold Spring Harbor, with support from the Carnegie Institution.[85] These diagrams mark the end of a long history of standardizing genealogical representations for the purposes of medical studies of heredity, and they are therefore especially well suited to highlight the differences that separate modern from more traditional forms of representing genealogy.[86]

Two traditional ways of drawing up ancestor relations existed: ancestor tables (*Ahnentafeln*), which listed the parents, grandparents, great-grandparents, and so forth of a given individual; and pedigrees (*Stammbäume*), which charted the children, grandchildren, great-grandchildren, and so forth of an individual. Although such diagrams quickly grew to encompass an ever larger number of individuals, they consistently omitted collateral relatives and thus represented the ancestry or descendancy of one particular individual only. The pedigrees that came into use among late-nineteenth-century physicians and eugenicists, in contrast, remained curiously open towards all sorts of relationships. They betray an attempt not to single out only particular relations, but rather to capture the whole network of kin relationships that surrounds any given person. Although structured by the linear succession of generations, typical genealogical diagrams from around 1900 therefore exhibit a remarkable degree of two-dimensionality that is reminiscent of schematic drawings of electrical circuits rather than the linear structures that are otherwise common in genealogy.[87]

When Davenport and his assistant Harry McLaughlin, who had originally been trained as an agricultural scientist, collected genealogical diagrams at the Eugenics Records Office, their aim was to look for evidence of Mendelian inheritance patterns in particular families. The scope of their data collection activities was widened not only through the participation of hospital directors, but also through the contributions of many amateurs, who either sent in pedigrees of their own families or conducted household surveys as voluntary "field workers" for the Eugenics Record Office. More than scientific curiosity supported this collecting enterprise. Family research had become something of a popular pastime among the middle classes in the second half of the nineteenth century, as evidenced by the foundation of many local and national genealogical associations. Like the production of more inclusive pedigrees in medical contexts, these genealogical pursuits had the aim of creating complete charts of family relations, including collaterals on both paternal and maternal sides. The allure of such studies frequently resided in the hope of identifying a prominent historical figure as a relative. Sometimes the hope also lay in making contact with remotely related persons who were still alive.[88]

The data that accumulated as a result of such genealogical activities raised the possibility of creating an "inventory of the blood" of entire nations. In Germany, the historian Ottokar Lorenz first noticed this possibility, which he pursued in cooperation with psychiatrists. At the time, traditional genealogy was regarded at the German universities as an ancient but somewhat outdated subject, a handmaiden of history proper. Seeking to rehabilitate genealogy, Lorenz expounded his vision of a "scientific genealogy" in his *Lehrbuch der gesammten wissenschaftlichen Genealogie* ("Textbook of All of Scientific Genealogy"), published in 1898, in which he also showed himself rather skeptical of eugenicists' talk about degeneration and expressed reservations about Galton's statistical methods. In a loose reference to chemistry, he claimed that genealogy was "the element of historical events." The existence of nations and classes with peculiar characteristics—but also of familial idiosyncrasies like the Habsburg lip, whose irregular, "atavistic" reoccurrence in a family line had always caught the eye of genealogists—could be explained by a simple genealogical principle, for example. If one traced the ancestry of an individual in an ancestor table, the number of forebears would normally not grow geometrically, because marriage among close relatives like cousins would lead to a "loss of ancestors" (*Ahnenverlust*) and hence to a narrowing down of the biological heritage. Lorenz's general conservative credo was that "love best comes to fruition with loss of ancestors and with social equality."[89]

Apart from promoting his theory of ancestor loss, Lorenz also called on psychiatrists and historians to join forces in an enterprise dedicated to the systematic collection of genealogical data, and psychiatrists like Emil Kraepelin and Robert Sommer eagerly heeded this call. In 1904, the Central Office for German History of Persons and Families (*Zentralstelle für deutsche Personen- und Familiengeschichte*) was founded, and many hospitals and psychiatric institutions began to collect genealogical data about their patients, albeit problems with data standardization soon became apparent. A research institution that had the capacity to collate and process empirical data on a national scale then came into being in 1917 with the Genealogical-Demographic Department of the German Research Institute of Psychiatry in Munich, which received financial support from the Rockefeller Foundation and the Kaiser Wilhelm Society.[90]

In the debates about adequate standards and methods for the creation and analysis of genealogical records that accompanied these institutional developments, Germany reached a crucial turning point that has recently been reconstructed by the historian Bernd Gausemeier. The accumulation of genealogical data en masse raised the possibility of turning genealogy into a tool not simply for outlining the parentage of individuals, but rather for the statistical analysis of whole populations. Initially, the idea was "to go more into the depth than the breadth," as the psychiatrist Wilhelm Strohmeyer put it, and the resulting genealogical activities resembled those of the Eugenics Record Office at Cold Spring Harbor.[91] In the early 1920s, however, Ernst Rüdin, the director of the Genealogical-Demographical Department of the German Research Institute of Psychiatry, began to pursue a radically different approach, drawing on results derived by the Stuttgart physician Wilhelm Weinberg and the English mathematician Godfrey Harold Hardy. In 1908, Weinberg and Hardy had independently shown that the frequency of genetic alleles in a population remains constant in the absence of evolutionary selection or mutations. Building on the so-called Hardy-Weinberg law, Rüdin shifted the focus of his work on the heritability of schizophrenia from understanding pedigrees to a reconstruction of whole generations. To simplify somewhat, his method consisted of recording the incidence of schizophrenia among only the parents and siblings of a patient, but doing so for cohorts of several thousand patients. By observing whether the frequency of the illness among siblings corresponded to the frequencies derived from the Hardy-Weinberg law (25 percent where both parents were healthy, 50 percent where one of the parents was schizophrenic), he could then conclude whether or not schizophrenia behaved like a recessive Mendelian trait.[92] In a single analytic stroke, two generations were thus represented as if they had been the product of a large number of independently executed Mendelian breeding experiments.

Franz Boas similarly employed an intriguing concept that epitomizes the analytic-constructive perspective that underwrote these early studies in human population genetics: the concept of the "family line." On the basis of a large-scale anthropometric study carried out for the U.S. Immigration Commission, Boas was able to demonstrate in 1910 that physical characteristics like the cephalic index or height changed slightly but significantly in the American-born children of immigrants of various ethnic origins or "races." Racial characteristics thus appeared to be generally unstable.[93] Boas and his assistants had analyzed the raw data, gathered from about eighteen thousand individuals in New York schools and on Ellis Island, in a manner that was similar to the methods Rüdin later employed. They grouped parents together with their children into units that Boas called "family lines." As Boas later emphasized repeatedly, there was a tension between the concepts of race and family line. The "races" of everyday language—in his immigrant study, Boas had distinguished between Eastern European Jews, Sicilians, Neapolitans, Czechs, Poles, and Scots—usually consist of a conglomerate of many different family lines, and conversely, a single family line may be composed of individuals belonging to distinct races, as Boas liked to point out to his German colleague Eugen Fischer. In a study of the "Rehobother Bastards," a small "mixed" African-European population in South West Africa, Fischer had shown in 1913 that "racial" traits like eye color also followed Mendelian patterns of inheritance. For Fischer, the Rehobother Bastards constituted a paradigmatic case of a "mixed race."[94] Boas, in contrast, pointed out that this locally constrained as well as endogamous population approached "purity of race," for it consisted of only a small number of family lines, which intermarried again and again.[95]

Such genealogical and statistical analyses enabled later population geneticists to unravel the internal mechanisms of human heredity. However, direct and targeted interventions in human reproduction in order to lay bare the essence of heredity remained impossible. Practitioners of the new science instead had to rely on past events and processes and to analyze them retrospectively as if they had been experiments—a method of Darwin's choice as well.[96] There was no lack of material for such research. For example, as Veronika Lipphardt has shown, many studies on the "biology of the Jews" around 1900 followed this approach. Not all of these studies were inspired by the prevalent anti-Semitism. Jews were considered an ideal object of study above all because a dominant "biohistorical narrative" maintained that Jews had remained a "pure race" over centuries as a consequence of their religiously motivated endogamy and their involuntary isolation in European ghettos.[97]

In general, such biohistorical narratives were ideologically ambiguous and could be instrumentalized for a variety of purposes. For the most part, they were used to shape political identities and to assert one's belonging to a certain social group, nation, or race. The nineteenth-century debate between "polygenism" and "monogenism" in anthropology is especially revealing with regard to the ideological uses of biohistorical narratives. From today's perspective, this debate appears strangely academic, if not entirely wrong-headed, because insufficient stress was laid on the apparently obvious common ground. Polygenists presumed that human races were distinct species, whereas monogenists cited the fertility of people of mixed race as evidence to the contrary—namely, that all humans belonged to one species. Yet most participants in the controversy—those who favored the view that human races were distinct species as much as those who held that human races were varieties of the same species—believed in the common descent of all species and thus implicitly admitted that all humans, in the last analysis, had to share a common ancestor.[98] Hence the ideological import of the debate did not reside in its implications for the existence, or not, of distinct races. Rather, the conflict between polygenists and monogenists was explosive because it was closely tied to political discussions about the legitimacy of slavery and colonial rule.[99]

Ideological motivations, however, fail to explain how the analytic perspectives described in this chapter came into being. It is notable that such perspectives, which ultimately dissolved races and nations into collections of loosely correlated but countable and predictable elements, were endorsed not only by scientists who were skeptical about the epistemic legitimacy of the race concept from the outset, such as Boas, but also by outspokenly racist scientists like Fischer and Rüdin. In order to understand the ascendancy of these analytic approaches, one has to recognize that anthropologists in general, whether racist or not, wanted to understand processes of societal transformation in order to provide instruments for the political control of such processes.[100] According to a famous aphorism by Tylor, anthropology was to be understood as a "reformer's science."[101] Given such aims, anthropologists could not rest content with a mere description of the status quo. Only if populations were conceptualized as dynamic aggregates of independent elements of varying degrees of stability could one hope to impinge on their developmental trajectory. The epistemic space that was opened by statistical and genealogical analyses of heredity was significant for many researchers because it allowed for the identification, investigation, and manipulation of "heritable" elements. In contrast to conventional accounts, according to which eugenic discourses resulted in a "reification" of social relationships, we suggest that the analytic procedures described in this chapter are more similar to the chemical process of fractionation, which is used to separate the constituents of complex chemical mixtures—for the sake of "purity," but also in the hope of enabling new, powerful combinations.[102]

Advances in population genetics in the 1930s and 1940s pushed the fractionation of race so far that after World War II, and after all the atrocities committed in the name of "race," human geneticists could self-confidently maintain that races did not really exist at all.[103] This claim focused on a typological concept of race that was now often attributed to "the man on the street"—prejudiced as he was, in the eyes of academics. Scientists pretended to entertain a more dynamic conception of human populations, from which "races" emerged only temporarily through underlying processes of mutation, selection, selective mating, geographic isolation, and hybridization. The Russian-American population geneticist and evolutionary biologist Theodosius Dobzhansky summarized this view in a powerful formula: "Race," he claimed, is "not a state of being, but of becoming."[104] More generally, categories of seemingly obvious biological and social significance such as race were dissolved in favor of a "rationalization of sexual life."[105] The gene—the concept that became the center of gravity of the life sciences for more than a century—was the result of this process of dissolution of everyday categories and of the concomitant condensation of the epistemic space of heredity into a new epistemic object.

# 6

## Disciplining Heredity

During the second half of the nineteenth century, the life sciences split into more or less separate areas of activity. Physiology was increasingly separated from comparative anatomy and morphology, new areas such as cytology and microbiology came into being, and natural history was transformed into evolutionary biology. Each new discipline identified its own subject matter and developed and deployed its own canon of methods. At the same time, this differentiation in the landscape of biological research created a need for overarching perspectives. Claude Bernard's 1878 lectures on "the phenomena of life common to animals and plants" were one early response to this challenge, and a growing number of contributions subsequently embraced the notion of "general biology" in their titles.[1] Whereas a science of life—that is, biology—had established itself around 1800 alongside with, and in contrast to, physics and chemistry with the aim of analyzing the diversity of organized and reproducing bodies, biology around 1900 no longer defined itself in contradistinction to the world of the nonliving. Instead, biology now centered on the definition of a level that allowed it to bring all forms of life under the same theoretical perspective. One of the consequences of this development was the introduction into biological research of what came to be called "model organisms." To put it pointedly, biology now pursued its agenda by focusing on particular organisms whose highly idiosyncratic qualities facilitated the experimental characterization of the most general properties of living beings.

This profound transformation around 1900 was especially noticeable in the discussion of hereditary phenomena. For half a century, heated debates and far-reaching speculations had animated the field, in the main inspired by evolutionary and developmental considerations, cytological observations, and anthropological as well as sociopolitical and cultural concerns about human populations. Then, in the time span of little more than the first decade of the twentieth century, an experimental discipline sui generis crystallized around a small number of model organisms, first plants such as *Pisum* (the pea) and then animals such as *Drosophila* (the fruit fly). The key events in this process have been described more than once, and its overall development has long taken a place in the history of scientific revolutions.[2] On the occasion of the Third International Conference on Plant Hybridization in London in 1906, the British biologist William Bateson proposed the name "genetics" for the new discipline, and he established the *Journal of Genetics*, together with Reginald Punnett, in 1910. In 1909, the Danish botanist Wilhelm Johannsen, who taught plant physiology at the University of Copenhagen, coined the term "gene" for the unit around which the new discipline revolved. As Johannsen saw it, the development of genetics was based on and brought about by a combination of two experimental practices: the production of "pure" plant or animal lines on the one hand and the hybridization of the resulting pure lines on the other. Both practices had roots in applied animal and plant breeding, but neither had been used systematically as an experimental tool for examining the succession of characters from one generation to the next before 1900. The only exception was the work of Gregor Mendel, who had also developed a system of symbols that allowed him to evaluate the results of his experiments statistically. But the rules that he observed in his pea generations—insofar as his findings were acknowledged at all during the last third of the nineteenth century—were not endorsed as a general model of the process of inheritance. As we will see, Mendel himself was hesitant to generalize his results.

In chapter 5, our aim has been to capture the pervasiveness of nineteenth-century thinking about biological heredity in medical and biopolitical contexts. In this chapter we examine how the epistemic space of heredity condensed into a number of circumscribed epistemic objects with the establishment of experimental genetics. Research was increasingly carried out on plant and animal models that either had been already used in agriculture or showed favorable experimental characteristics, such as small size and fast reproduction. In the eighteenth century, as we have seen in chapter 3, academic institutions like botanical gardens played a decisive role in the production and identification of phenomena in the realm of subspecific variation that then became the focal points of a biological discourse of heredity. Toward the end of the nineteenth century, conjunctures between university biology and the biotechnologies of the second industrial revolution reconfigured the epistemic space of heredity.[3] The resulting new research practices were pursued at ambitious scales. Biochemistry, microbiology, and the techniques of pure breeding and hybridization contributed to the mapping out of a domain in the phenomena of life that further subverted the traditional dichotomy of species and individual.[4] With the emergence of genetics, biology arguably assumed the character of an experimental discipline that was not only a basic science, but also held tremendous prospects for application—a process that repeated itself in the second half of the twentieth century, albeit under different circumstances and at another scale, with the rise of molecular gene technology and genomics.[5]

Our analysis of these developments casts doubt on the common picture of scientific revolutions as paradigm shifts that take place abruptly within relatively isolated scientific communities. To begin with, we will concentrate on Gregor Mendel and investigate what the novelty of his work actually consisted in and what he nevertheless had in common with his contemporaries. We will then consider the history, in the second half of the nineteenth century, of animal and plant breeding on the one hand and bacteriology on the other. As we aim to show, a number of elements that were later associated with Mendelism derive from these contexts. The subsequent sections treat the history of what is generally called "classical" genetics in the wake of Mendel's work. It will become clear that this discipline was far from monolithic; it was marked by conceptual ambivalences and by research practices that changed continuously. It is thus better understood as a loose ensemble of research strategies and technologies that both helped to forge a disciplinary identity and were soon adopted by neighboring areas such as embryology and evolutionary biology.

## Gregor Mendel and the Hybridists

Many commentators have deplored the fact that Mendel's experiments, which were published in volume 4 of the *Verhandlungen des Naturforschenden Vereines in Brünn* ("Transactions of the Natural History

Association in Brno") in 1866, were not noticed by contemporary biologists, even though they could have helped Darwin, for instance, to a sound theory of heredity—the missing link in his theory of evolution. However, Mendel himself would have seen all this differently: he presented his work with peas as "detail experiments" in the long-established but narrow research tradition in hybridization that we discussed briefly in chapter 3.[6] Following the discovery of "constant varieties" by Linnaeus, among others, botanists had studied the role of hybridization in the possible generation, or "transmutation," of species. Subsequently, experimental studies of hybridization became more and more extensive. Carl Friedrich Gärtner, for example—whose *Versuche und Beobachtungen über die Bastarderzeugung im Pflanzenreich* ("Experiments and Observations on Hybrid Generation in the Plant Realm") of 1849 had become the point of reference for work in plant hybridization in Mendel's time—had conducted more than ten thousand separate experiments on more than seven hundred plant species. Mendel presumably came into contact with this tradition during his studies in Vienna, where his lecturers included Franz Unger, a plant geographer and early proponent of cell theory. Unger wanted to work out a "physics" of the plant organism by formulating "laws of development" on the basis of transplantation experiments and quantitative biogeographic and paleobotanical studies. Unger himself was influenced by Carl Nägeli, with whom Mendel was to exchange a series of letters after the publication of his hybridization paper.[7]

Although Mendel's work was part of this tradition, his experiments also had idiosyncratic features. How Mendel's experiments related to the hybridization practices of contemporary botanists is best seen by comparing his work with that of his French colleague Charles Naudin, who conducted experiments in Paris at the same time as Mendel in Brno. One of the most skilled plant breeders in the middle of the nineteenth century, Naudin worked at the Muséum national d'Histoire naturelle, where he had the overall responsibility for horticulture after 1854. At the beginning of the 1870s, Naudin moved to the French Pyrenees, where he conducted experiments on the acclimatization of exotic plants in a garden that had been created for that purpose. In 1876, he became director of the botanical garden founded by Gustave Thuret in Antibes.

The work that we consider here was Naudin's response to a prize competition set by the Parisian Académie des Sciences, which called for the "study [of] plant hybrids under the perspective of their fertility and the perpetuity or the non-perpetuity of their characters."[8] Perpetual hybrids, if obtainable, were of paramount importance for agriculture. As we have seen in chapter 3, similar prize competitions concerning hybridization also motivated the studies of Kölreuter and later Gärtner.[9] Naudin assessed the perpetuity of the results of his plant crosses by following them over several generations and meticulously recording their characters. His experiments were broad in scope and included a wide range of vegetables and ornamental plants of interest to French farmers and gardeners. His report lists more than forty different crossings, but he rarely derived more than a few dozens of offspring from each cross. Moreover, like most plant hybridizers before him, Naudin concentrated less on individual, precisely identifiable characters than on the habitus of the descendants by comparing the overall character of parents and progeny. As an empirical regularity, Naudin observed that the first generation of descendants from a cross appeared rather uniform, one parental form being, as a rule, "preponderant" over the other, whereas in the second generation the hybrid form began to "dissolve," as some individuals resembled one of the two original parental forms again. Naudin explained this behavior by assuming that a "specific essence" inherent in each species determined its character and that this essence reasserted itself against forced mixtures in the long run.[10]

In contrast to contemporaries like Naudin, Mendel reported the results of an eight-year-long series of crossing experiments with a few constant varieties of a single species, *Pisum sativum*. Moreover, he focused on a small number of defined character pairs, and he strictly controlled the artificial, experimental fertilizations of his plants. As he declared in the introductory remarks of his report, he had undertaken to formulate a "generally applicable law of the formation and development of hybrids."[11] According to his paper, Mendel first carefully chose parent plants that exhibited "constantly differing traits," then focused on those whose progeny retained full fertility. Second, as he had set himself the task of determining the "numerical proportions" of forms in subsequent generations, he cultivated a large number of hybrids, as well as their subsequent progeny, with a view to deriving statistical ratios.[12] Third, he used algebraic symbols to represent his assumption that any two alternative characters were based on distinct factors that were transmitted independently through the germ cells. When he was working with several character pairs, the symbol system helped him to represent the resulting permutations exhaustively. More importantly, in the case of dominant-recessive character pairs—the exclusive focus of his report—this notation allowed him to distinguish between the hereditary constitution of a plant and its appearance. Finally, Mendel assumed that the hereditary factors that determined the characters were freely exchangeable and randomly distributed during the formation of the germ cells. In contrast to his contemporaries, whose views we described in chapter 4, Mendel assumed that the egg and pollen cells each carried only one of a pair of factors that determined the character of the offspring. The separation of factors was thus sharp and unequivocal for each individual descendant, but because it occurred randomly, a rule for character distribution could only be determined statistically for the population of descendants as a whole.

As far as the scope of his laws was concerned, however, Mendel's argument was very similar to Naudin's. The latter had held that the "specific essences" of his crossed species were not permanently miscible and that hybrids were therefore not permanent. Even if only at the level of individual factors, Mendel similarly concluded, on the basis of his experiments, that "the arrangement between the conflicting elements" in the fertilized cell was a transient one that did not last "beyond the life of the hybrid plant"; whereas identical factors of individuals of the same species or variety became "entirely and permanently mediated" upon fertilization and entered a "viable union."[13] For Mendel, it therefore was, and remained, a peculiarity of hybrids that they appeared uniform in the first generation, exhibiting the dominant character throughout, and that they segregated in a ratio of 3:1 in the second generation with one-fourth of plants showing the recessive character again. Despite all this, it is clear from Mendel's correspondence with Nägeli in Munich, and from his second botanical publication on hybridization experiments with *Hieracium* (hawkweed), that he remained fundamentally interested in the possibility of "constant hybrids." His *Hieracium* hybrids appeared to point to the possibility of a stable compromise taking place between the opposing parental factors that entered a hybrid union, and at the same time to cast doubt on the general applicability of his law for hybrids.[14]

*Pedigrees and Pure Cultures: Breeding in Agro-industrial Contexts*

In his inaugural lecture at the Collège de France, Michel Foucault referred to Mendel as a "true monster," claiming that he "spoke of objects, employed methods and placed himself within a theoretical perspective totally alien to the biology of his time." According to Foucault, Mendel thus did not move "within the true" (*dans le vrai*) of the biological discourse of his time.[15] With this interpretation, Foucault set himself apart from those historians of science who time and again have shown themselves to be astonished by the fact that Mendel's contemporaries were apparently unable to recognize the importance of his results. However, in doing so, Foucault reinforced another cliché, namely, that of Mendel as a cloistered monk—of the garden variety, one might say— who was not part of any scientific network. Yet we have already seen that Mendel's experiments were linked to a well-established tradition of research and that they attracted the attention and interest of no one less than the renowned botanist Nägeli in Munich. Moreover, during his scientific education at the Universities of Olomouc and Vienna, Mendel had acquainted himself with combinatorial mathematics, with the principles of physical experimentation, and as mentioned earlier, with contemporary cell theory and evolutionary theory. He was furthermore an active member of several associations that promoted the sciences as well as sheep and fruit tree breeding. These associations comprised scientists, scientifically educated public servants, industrialists, gentleman farmers, and breeders. Questions of heredity were discussed extensively in the journals of these associations. As Vítězslav Orel shows in his biography of Mendel, in the middle of the nineteenth century the Augustinian monastery in Brno, whose economic activities concentrated on textile industry and brewing, was closely associated with the agricultural development movement in Moravia, one of the most advanced gardening, culturing, and industrial landscapes of continental Europe at the time.[16]

If, in view of this context, one wants to maintain that Mendel was a "monster," then Foucault's term would have to be applied to his experiments. As described above, what was peculiar and unfamiliar about them was that Mendel combined elements that had not been brought together in this way before: combinatorial and statistical procedures, the methodology of physical experimentation, cell theory, and advanced breeding practices. By around 1900 the ground had become more receptive for such peculiar conjunctions and the results they gave rise to. Agricultural stations had been established all over Europe and North America toward the end of the nineteenth century. In the context of the broader movement that led to the establishment of institutes of technology, these stations conceived of themselves as being practice-oriented as well as scientific institutions. They thus provided a fertile site for interactions between university biology in the more traditional sense and the biotechnologies of the second industrial revolution. Animal and plant breeding, in turn, became accessible to influences from advanced academic disciplines such as microbiology.[17]

Let us first have a look at animal and plant breeding. For one thing, Mendel's experiments were influenced by ongoing debates in a local association that promoted sheep breeding as a way of elevating related economic activity (the Verein der Freunde, Kenner und Beförderer der Schafzucht ("Association of Friends, Experts and Promoter of Sheep-breeding"), among its members was Mendel's superior and mentor Cyrill Napp). The influence of this association manifests itself all the way down to some of the terminological choices in Mendel's publications.[18] There were, of course, good reasons why the Augustinian monastery was prominently represented by its abbot. The monastery was the biggest local economic enterprise. Through systematic inbreeding and out-breeding, Moravian sheep breeders had created so-called Merino breeds that produced wool of outstanding quality. One of the main points under discussion was the question of how best to balance inbreeding and hybridization. Both degeneration and uncontrollable variation had to be avoided. This concern is directly mirrored in Mendel's experimental protocol in that his experiments alternated systematically between self-fertilization (the most extreme form of inbreeding) and artificial crossing.

Throughout the nineteenth century, plant and animal breeders saw heredity predominantly as a "force" they could manipulate and control for their own purposes. As the markets for their products expanded—the first Saxonian sheep reached Australia in 1825 and played a key role in the establishment of the Australian national economy[19]—the stability of characters—their "heritability," in other words—gained in importance and became the ultimate goal of breeding.[20] These efforts also led to an important innovation in the breeding of agriculturally useful plants. Traditionally, breeders—mostly peasants and landowners—had simply selected seeds from those plants that they deemed to be of optimal quality and set them aside for sowing during the next season. The populations that resulted from this selection procedure—the so-called landraces—were heterogeneous and therefore exhibited particular resistance against extreme weather and parasites, but they were not suitable for quasi-industrial farming, which required standardized raw materials. As far as self-fertilizing plants are concerned, this difficulty was overcome through a procedure developed in the 1840s by the French family business Vilmorin, which is now called pedigree breeding or pure breeding.[21] The starting point was the seeds of an individual plant. The progeny were tested, and any variants that cropped up were repeatedly sorted out. Provided that the plants in question were self-fertilizers, variation due to hybridization was negligible. For the resulting pedigree cultures, there was, in Wilhelm Johannsen's words, "no doubt about the 'father.'"[22] Mendel similarly used a self-fertilizing plant and tested the "stability" of its varieties for several years before he fertilized them artificially in the course of his experiments. Mendel's work thus incorporated some of the breakthrough techniques recently developed by professional plant breeders.

Apart from the development of new breeding techniques, contemporary animal and plant breeding also involved the application of new forms of registration and calculation that were similar to those described in the previous chapter for anthropology and eugenics. Breeders' associations created central breeding registers and published detailed genealogies and statistical tables in their journals. There is a deeper reason for this parallelism between the realm of breeding and the human sciences: From the second half of the nineteenth century, animal breeders in particular often took note of and sometimes used contemporary strategies of family planning like cousin marriage. Conversely, as is well known, eugenicists indulged in breeding fantasies for which the practices of animal and plant breeding yielded concrete models.[23] At the agricultural stations that were established toward the end of the nineteenth century, the systematic and quantitative tabulation and inventorization of breeding experiments intensified to such an extent that the term "scientific breeding" began to take root. At the beginning of the 1890s, the Swedish Seed Breeding Association, for example, cultivated about two thousand different cereal strains produced by pedigree breeding on its experimental station in Svalöf. The development of each of these strains was monitored via quantitative assessments of its characters, sometimes by means of special measuring instruments. Year by year, the results were entered into a complicated "register" consisting of a "pedigree record," "field books," and a "journal of analysis."[24] It is no coincidence that Mendelism began to thrive in such institutions at the beginning of the twentieth century.[25]

The fruitful interaction between Mendelian genetics and applied breeding research probably hinged on the fact that the animal and plant breeders were not only the recipients of a new science, but actively involved in the production of new genealogical constructs. During the first decade of the twentieth century, two basic concepts were introduced into genetics that were related to such constructs: Johannsen's "pure line," which more or less corresponded to Vilmorin's "pedigree"; and the "clone," a concept introduced for asexually reproducing populations by the botanist Herbert J. Webber from the Plant Breeding Laboratory of the U.S. Department of Agriculture.[26] These concepts were organized not so much around the characteristics of progeny, such as constancy, as around the principle of genealogical construction employed. A "pure line," for example, was defined as a pedigree exclusively produced by self-fertilization.

The notion of a "pure line" epitomizes the development of such concepts and, at the same time, reveals a facet of a broader context that may be surprising. Johannsen, whose "genotype" theory we discuss below, began his scientific career—after an apprenticeship as an apothecary and ensuing studies in botany—in 1881 in the chemical department of the research laboratory of the Carlsberg brewery, where he investigated the role of nitrogen in the germination and the maturation of barley. In 1887, he became lecturer for plant physiology at the Royal Veterinary and Agricultural College in Copenhagen, but continued his collaboration with the Carlsberg Laboratory in the form of breeding experiments. It is in this context that he probably first applied the pedigree method, although in later writings he insisted that he had come across the concept of a pure line elsewhere.[27] Also at the Carlsberg Laboratory, Emil Christian Hansen adopted the methods of Louis Pasteur and Robert Koch for obtaining pure bacterial cultures in the early 1880s. Pasteur and Koch had developed these methods for the diagnosis and treatment of contagious diseases; Hansen applied the same methods to brewing yeast. Individual yeast bodies, isolated either by dilution or under the microscope, were introduced into sterile culture media. In this way, a "whole series of yeast species or races of varying practical value" could be produced in petri dishes, as Johannsen put it in a textbook on plant physiology he published with his teacher Eugen Warming in 1901.[28] Within a few years, two of Hansen's "races" had found their way into breweries worldwide: Carlsberg bottom yeast no. 1 and no. 2 facilitated the production of beer of relatively constant quality, whether in Denmark, Germany, North America, Australia, or India.[29]

This episode highlights some of the economic and social preconditions for the coming into being of genetics, such as the beginning of agro-industrial mass production of organic raw materials and foodstuffs, as well as of drugs and vaccines, toward the end of the nineteenth century. The Vilmorin business, for example, had some four hundred employees by around 1900. Division of labor and bureaucratic control played a growing role in these industries, which in turn furthered the development of an expert culture. Analysis, exactness, calculability, and predictability now had priority. The omnipresent catchword was "purity." Purity was an instrument of control, since the productivity of work processes could be measured in relation to the input of "pure" materials. And "pure" inputs facilitated the "fixation" of characteristics and the production of goods that could be identified, specified, and reproduced at any time or place. Instead of a seamless continuum of variations, one worked with discrete and stable life forms that could be recombined without fear of unpredictable interactions, which often occurred with "impure" products such as the traditional landraces.[30] "Purity" became a marketable quality of living beings. Concomitantly, at the end of the nineteenth century, sales-promoting purity promises were increasingly controlled and guaranteed by means of trademarks, quality seals, and patents.[31]

It therefore makes sense that the Mendelians compared their own young discipline favorably with synthetic chemistry, which had attracted attention during the previous decades through the creation of artificial dyes and drugs.[32] In this light, it is also unsurprising that two of the three so-called rediscoverers of Mendel's laws were active in the area of applied botany. Erich von Tschermak-Seysenegg had a background in plant breeding, and he became one of the first influential advocates of the new genetics in agriculture and gardening in the Austrian empire. After founding a plant breeding station in Groß-Enzersdorf, near Vienna, in 1903, he promoted the establishment of more than twenty experimental stations in Bohemia, Moravia, West Hungary, and Lower Austria, following the model of similar institutions he had encountered on his travels through France, Belgium, the Netherlands, Sweden, the United States, and Germany.[33] Hugo de Vries had studied under the plant physiologist Julius Sachs in Würzburg. During that time he had also written monographs on agriculturally relevant plants for the Prussian Ministry of Agriculture.[34] Moreover, de Vries was in regular contact with the microbiologist Martinus Willem Beijerinck, who had worked in the Nederlandsche Gist en Spiritus Fabriek (Dutch Yeast and Spirits Factory) in Delft before he moved to the Polytechnic School of Delft in 1895. Beijerinck inspired de Vries to study mutations and alerted him to Mendel's paper in 1899.[35] The biologists who embraced Mendelian genetics in the years after 1900 thus inhabited a world populated by a new form of living beings—living beings unlike those that Mendel's contemporaries had been accustomed to.

## The annus mirabilis *of 1900 and the Following Decade*

When Mendel published the results of his experiments with peas, he saw them as corroboration of a special law that applied to hybrids only. If one compares the "rediscovery" around 1900 with the situation around 1866, then the discontinuity lies precisely in the interpretation of Mendel's law. Even though de Vries's report on the "law of disjunction,"

as well as Carl Correns's publication on "Mendel's rule," still bore the
notion of "hybrid" and "race bastard" in their respective titles, both
clearly interpreted the rule as a *general law* of heredity.[36] That this was,
furthermore, immediately recognized by others is evident in William
Bateson's reaction to these publications, which begins with the following
sentence: "An exact determination of the laws of heredity will probably
work more change in man's outlook on the world, and in his power over
nature, than any other advance in natural knowledge that can be foreseen."[37] In the context of hybrid heredity, Mendel's triple achievement—
the combination of a statistical, quantifying treatment of data with a
combinatorial symbol system and the definition of a formal object, the
"factor," as the unit of the hereditary process—had not attracted much
attention. Now the same achievement was greeted enthusiastically for
throwing light on the hereditary process as such. In this light, the analytic distinction between bastardization as a special case of procreation
and the general process of species-bound sexual reproduction began to
vanish. The phenomenon of sexual reproduction of higher organisms
was transformed into a universal tool with the help of which one could
learn about the genetic constitution of the organisms involved. Again,
Bateson had recognized this as early as 1899, when he spoke of hybridization and cross-fertilization as a "method of scientific investigation."[38]
Eight years later, at the end of his inaugural lecture at the University of
Cambridge, he similarly stressed this point:

> For the first time *Variation* and *Reversion* have a concrete, pal-
> pable meaning. Hitherto they have stood by in all evolution-
> ary debates, convenient genii, ready to perform as little or as
> much as might be desired by the conjuror. That vaporous stage
> of their existence is over; and we see Variation shaping itself as
> a definite, physiological event, the addition or omission of one
> or more definite elements; and Reversion as that particular addi-
> tion or substraction which brings the total of the elements back
> to something it had been before in the history of the race.[39]

With this background in mind, the rapid spread of Mendelian experiments on heredity in the German-speaking countries, in Scandinavia, in
Great Britain, and in North America between 1900 and 1910 becomes
comprehensible; it reveals a more general shift in the knowledge regime
of heredity. In chapter 4 we examined the syntheses of theories of evolution and heredity in the nineteenth century. We spoke of an epistemic
*space* in which the biological conception of heredity took shape. What

we now observe around 1900 is the establishment of experimental systems in which the earlier, rather exceptional and capricious phenomena of "heredity" condensed into the generalized model of an epistemic object. Ultimately, the success of these systems, which would soon enjoy pride of place in the life sciences, can be explained via their limitation. In the words of the protozoologist and biotheoretician Max Hartmann, these systems made it possible for biologists to leave the space of "generalizing induction" and to attack the basic questions of general biology in an experimental form; that is, via "exact induction."[40] As a consequence of this shift, research on model organisms intensified and appeared in a new light. Yet, from the outset, the new experimental science of genetics also faced conceptual problems, which we now briefly introduce by way of Johannsen's and Correns's reflections on the nature of the hereditary units.[41]

Having contributed some key terminology to the field, Johannsen summarized in 1911 for the *American Naturalist* what knowledge had been established by genetics in the first decade of the twentieth century. As mentioned earlier, he thought that genetic research on heredity—in the narrower sense of "all true analytical experiments in questions concerning genetics"—was based on two experimental foundations: the breeding of pure lines, to which he himself (working with the princess bean *Phaseolus vulgaris*) had dedicated a great deal of his experimental work at the Agricultural College of Copenhagen from 1892 to 1905; and the crossing experiment, as introduced by Mendel.[42] As he pointed out in the preface to the first edition of his *Elemente der exakten Erblichkeitslehre* ("Elements of an Exact Theory of Heredity"), these experimental strategies made it possible to detach, through "more exact delimitation," a "narrower field" from the "large and indistinct domain of heredity."[43] According to Johannsen, purifying cultures and crossing them rendered the idea of hereditary "transmission" obsolete, specifically in the sense that the parents transmitted their personal characteristics to their offspring. For him, the development of these techniques meant that the opposite vision could take hold, according to which "the qualities of both ancestor and descendant are in quite the same manner determined by the nature of the 'sexual substances'—i.e., the gametes—from which they have developed."[44] Only Galton and Nägeli had previously put forward such a radical statement, but Johannsen did not cite them on this occasion. Instead, Johannsen compared the transition from the old transmission thinking to the new genetic thinking with the transition from the chemistry of phlogiston to the chemistry of oxygen, perhaps owing to his education as a pharmacist and his work as an assistant chemist under

Johan Kjeldahl at the Carlsberg Laboratory, where Emil Christian Hansen was purifying yeast cultures at the same time.

Johannsen defined the "genotype" as "the sum total of all the 'genes' in a gamete or in a zygote."[45] But although Mendelian crosses suggested that the gene concept "covers a reality," Johannsen remained convinced that, for the time being, "as to the nature of the 'genes' it is as yet of no value to propose any hypothesis"; in particular, he attacked the "morphologico-phantastical speculations of the Weismann school."[46] He was even inclined to bracket the concept of "gene" altogether, since he regarded it as an undue reification, and preferred to speak about "'genotypical' differences" or "genodifferences."[47] While Nägeli and Weismann had stressed tirelessly that the idioplasm or germ plasm possessed a historical structure, and that it was therefore to be understood as a historical entity, Johannsen conceived of the genotype just the other way round, as an "ahistoric" concept. For him, "genotype" referred to whatever remained identical in living beings through the generations and was therefore amenable to experimentation, just like the molecules in chemistry and the atoms in physics.[48] He thus likened genes to complex chemical substances, but at the same time he refused to speculate about their specific material constitution. In regard to the identification of genes, one might say, following Frederick Churchill, that Johannsen focused on the "vertical" analysis of heredity, but abstained from a "horizontal" analysis of the relation and interaction between the hereditary units.[49] It is in any case conspicuous that Johannsen restricted himself to the construction of pure lines and did not make any use in his experiments of the second cardinal technology of genetics; that is, hybridization.

Carl Correns stands out among the botanists who adopted hybridization as an experimental tool around the turn of the twentieth century. He had studied in Munich, where he was one of Nägeli's last students. Interestingly, Correns did not begin his hybridization experiments because he was interested in the hereditary behavior of hybrids. Instead, he wanted to solve the puzzle of xenia. In some plants, following fertilization with pollen from another variety, characters of the pollen-donating plant appear in the seeds and fruits of the maternal plant already, rather than only in individuals of the next generation. This phenomenon had earlier aroused Darwin's interest, and Correns hoped it would yield insights into the physiology of fertilization. In a short autobiography written when he was older, Correns remarked, "During the ten years which I spent as a *Privatdozent* in Tübingen, my works and publications mostly followed along paths already taken. . . . Besides that, as soon as I had the possibility to use a botanical garden through my Habilitation, I started, as an aside, to do various experiments which we would now qualify as 'genetic' and which had been going through my head already earlier." Later on in his autobiographical note, he spoke of these early trials as "skylarking" (*Allotria*).[50] But then, at first in addition to and later instead of the xenia, Correns became increasingly aware of and interested in statistical regularities that manifested themselves with respect to particular characters in the offspring of his hybrids. Between 1894 and 1899, he crossed a large number of corn varieties, and between 1896 and 1899, varieties of peas. He had chosen these two species because xenia had been described for them in the literature. However, at least in the case of *Pisum*, his search for xenia failed. Instead, at the same time as de Vries in Amsterdam and Tschermak in Vienna—who, like Correns, had set out to undertake and later departed from xenia studies—Correns found Mendel's ratios in the progeny of his bastards.[51]

Like Johannsen, Correns was part of a new generation of biologists who eschewed the sweeping theoretical syntheses of their forebears. They limited their reflections to more circumscribed phenomena and sought explanations for their particular experimental results. Nevertheless, his conclusions were consequential.[52] Looking to stake out a middle ground between the assumption that the idioplasm or germ plasm possessed an "architecture" (in the spirit of Nägeli and Weismann) and the idea that hereditary factors were freely miscible and exchangeable (following de Vries), Correns distinguished between "dispositions" and "characters"—anticipating, as it were, Johannsen's distinction between gen(otype) and phen(otype). The task of biologists interested in questions of heredity, according to Correns, was to "draw conclusions from the behavior of the characters"—which could be observed—"on the behavior of their dispositions"—which remained invisible.[53] The experiments in the wake of Mendel were directed at the invisible counterparts that underlay the visible phenomena. The goal was to make statements about "dispositions," not "characters."

Correns extracted two conclusions from his experiments: "(1) that for each character there is a *particular*, I would like to say an *individualized* disposition, and (2) that the individual dispositions in the germ plasm of a kin [*Sippe*] cannot be tightly bound to one another."[54] In addition, he assumed that the complementary dispositions residing in two germ cells did not mix in the resulting zygote, but rather remained individualized units that were separated in the course of the next reduction division of the chromosomes and hence could be distributed over different germ cells once again. This summary of Correns's views raises two important points. First, Correns's "dispositions" were not present in multiple copies in the cell, or in the cell nucleus, but in single, individual copies—in contrast to Darwin's and Galton's "gemmules," Nägeli's "micelles," de Vries's "pangenes," or Weismann's "biophors." Second, his units were not fixed and bound into a hereditary architecture of a higher order. But following de Vries in this respect meant to Correns that it had to "remain incomprehensible, as Weismann stressed against de Vries, and rightly so in my opinion, how the *developmental sequence* of the dispositions is determined. It is the recognition of this point that led Weismann as well as Nägeli to [their assumption of] a tight connection of the dispositions in the idioplasm."[55] Correns's dilemma was that if the independence of hereditary units from one another was assumed, then statements on the process of hereditary transmission no longer implied statements on the process of development because the latter depended on a sequence of events that had to obey a precise spatial and temporal order. Correns tried to resolve this difficulty by drawing on the observation that one single disposition could trigger different characters at different positions and in different developmental stages in an organism. This led him to the following proposal:

> In trying to resolve the obvious contradiction I have had recourse to a vision that I would like to disclose here, although I know that it will be received as a heresy. I propose to have the locus of the dispositions, without permanent binding, in the nucleus, especially in the chromosomes. In addition I assume, outside the nucleus, in the protoplasm, a mechanism that cares for their deployment. Then the dispositions can be mixed as they may, like the colored little stones in a kaleidoscope; and yet they unfold at the right place.[56]

Correns hence uncoupled transmission and development, probably even more radically than Johannsen, who never was willing to give up the idea of an underlying type at the level of the genes, as his terminology—genotype—indicates. Nägeli's and Weismann's solution had also been to burden the architecture of the germ plasm with the order of development. Correns instead defined a separate physiological space in which the effects of the hereditary dispositions could become manifest. However, the structure of this physiological space, as well as the relation between the two orders, remained, for the time being, obscure. Correns refrained from speculating about the proposed protoplasmatic "mechanism," although he became more and more interested in this question in his later research. For example, he hoped to learn more about the physiological and developmental effects of the hereditary dispositions through correlations of the following kind: As his experiments had shown, peas with green seeds gave rise to peas with yellow seeds if crossed with the yellow-seeded, dominant variety. But the seeds also became yellow if the peas were attacked by the larvae of the pea beetle—an observation, by the way, that had already been made by Mendel. Correns concluded that a "chemical influence" had to be responsible for this change in color, and he thought that such cases were particularly apt for experimental investigation. However, like Johannsen, Correns remained silent as far as the constitution of the hereditary dispositions themselves was concerned, though here he too favored chemical conceptions. In contrast to some speculations still current at the time, he refused to elevate the hereditary factors "almost to microorganisms." He may well have thought of Weismann's biophors when he declared that he himself would prefer to "collect building blocks for another conception."[57] Correns, like Johannsen, one could say, firmly expelled the living qualities that the hereditary units had once possessed in the grand theoretical designs of the nineteenth century.[58]

On the basis of his experiments, Correns became an early advocate of the view that the hereditary units were essentially freely recombinable (fig. 6.1). But he was also one of the first to describe the phenomenon of "coupling" of hereditary factors.[59] He tried to explain such cases of linkage with his own version of a chromosome theory of heredity. According to this theory, pairs of dispositions, which were ordered like pearls on a string, could sometimes be fixed in their position in groups and therefore were also bound to segregate together upon germ cell formation. If one examines Correns's research during the first decade after the "rediscovery" of Mendel's rules, one finds that he pursued the strategy of analyzing complicated cases of and apparent exceptions to Mendel's rules. They all corroborated the "nuclear" monopoly over the hereditary units and the fact that those units remained more or less linked during segregation. Correns showed that even the complicated sex determination patterns of *Bryonia* (bryony) could be explained by Mendel's segregation law. On the other hand, in experiments with variegated plants, Correns also hit upon phenomena such as chlorophyll variations that appeared to be transmitted via the cytoplasm; that is, through the plastids of the plasm. Even this finding did not shake his firm conviction that "for heredity proper the nucleus alone is responsible."[60] However, these observations enticed him to investigate the role of the cytoplasm in inheritance during

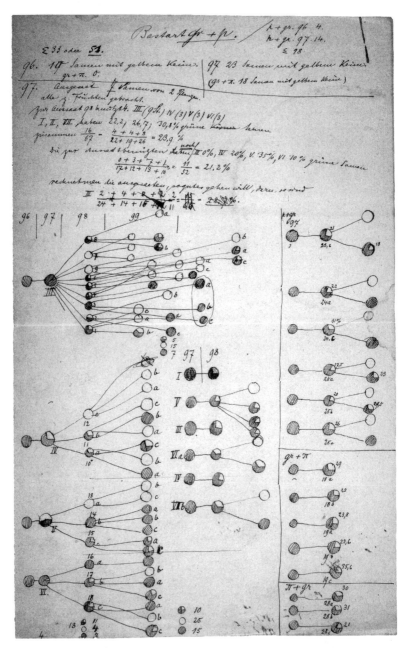

**FIGURE 6.1** Experimental protocol from Carl Correns's papers "Bastart gr + p" on the hybridization of the pea varieties "Grüne Folger" and "Purpurschote." Archive of the Max Planck Society, Carl Correns Papers, III. Abt., Rep. 17, Nr. 115, loose sheet, not dated. Courtesy Archive of the Max Planck Society.

the subsequent two decades, work that his successor Fritz von Wettstein at the Kaiser Wilhelm Institute for Biology, where Correns had worked since its foundation in 1913, would continue.

Let us come back to Johannsen for a moment. Although he introduced the gene concept, he was by no means convinced that all organismic characters, let alone the basic organizational features of a species, could be reduced to genes. "The entire organization may never be 'segregated' into genes," he mused.[61] Johannsen and Correns faced the same problem. In Mendelian experiments, it was possible to deal with hereditary dispositions—namely, genes—in an atomistic fashion. But neither Johannsen nor Correns were ready to leap to the conclusion that the organism was a mere mosaic of genes or of characters that were determined by genes. Having recourse to the terminology of the late eighteenth century, we could say that the "reproduction" of living beings lent itself to an analytic configuration, whereas their "organization" resisted the Mendelian grasp. Correns assigned organization to the cytoplasm that somehow triggered individual gene actions in a coordinated way. In principle, this organization could be subjected to physiological experimentation. Johannsen, in contrast, understood organization as an expression of the genotype. The cost of Johannsen's response was some ambiguity, however. Only certain genes and the associated phenotypic characters were subject to "segregation" and could thus be addressed by Mendelian experimentation. In contrast, the "type" itself—which Johannsen was not ready to give up because it warranted the organization—eluded the Mendelian experimental approach. "Personally I believe in a great central 'something,' as yet not divisible into separate factors," Johannsen wrote as late as 1923, in a critical review of the advances of genetics.[62]

Nevertheless, Johannsen did not refrain from defining heredity concisely as "the presence of identical genes in ancestors and descendants."[63] In fact, the inner nature of genes was of no consequence, as far as experiments with pure lines and the analysis of crosses between them was concerned. For that reason, the new discipline was often called "formal genetics." But in order to guarantee their formal functioning, a great deal of practical work had to be invested in the purification of the strains on the basis of which hybridization programs were built up. These research programs required genetic homogeneity, in the form of a supply of model organisms that were constructed in laboratories and experimental stations. "Pure cultures" of organisms became the most important material precondition for experimental work with organisms.

Only with their help could the "gene" assume the contours of a quasi-manipulable epistemic thing.

### Drosophila melanogaster *and the Reification of the Gene*

The expression "transmission genetics" caught on, despite Johannsen's opposition to the idea that heredity involved a transmission of parental characteristics to offspring. Hence research on heredity distanced itself from questions of evolution and embryonic development. Although transmission, evolution, and development remained related to one another, they were separated into three different dynamic regimes that apparently could be analyzed in their own right. The breakthrough—and the apotheosis, we might add—of transmission genetics occurred with the research program that was established and pursued with the model organism *Drosophila melanogaster*—the black-bellied fruit fly (fig. 6.2)—by Thomas Hunt Morgan between 1910 and 1930, first at Columbia University in New York and then at the California Institute of Technology in Pasadena.[64] The importance of the adoption of a model organism that could be reared relatively easily in mass culture within a confined space, and that underwent twenty to thirty reproduction cycles per year, can hardly be overestimated.[65]

After Morgan and his group had succeeded in assigning genes to groups linked to particular chromosomes, the four chromosome pairs of *Drosophila* became the focus of a comprehensive gene mapping endeavor. The project was based on two assumptions: first, that the genes were arranged linearly along the chromosomes; and second, that the frequency of recombination between homologous chromosomes—the breaking up of established linkage groups, in other words—made it possible to ascertain the relative positions of genes and to measure the relative distance between them. To be more precise, recombination emerged in these studies as a phenomenon that accompanied transmission. Over the years, more and more mutations appeared in mass cultures of *Drosophila*, thus adding to the arsenal of gene loci that could be mapped ever more densely. In this system, the gene was not only the unit of transmission, but also the unit of mutation and of recombination. The *Drosophila* geneticists experimented on a scale that was unheard of in biology. The logistics of their research were novel as well. Morgan assembled a whole research group around *Drosophila*—among them, in the founding generation, Alfred Sturtevant, Calvin Bridges, and Hermann Joseph Muller—and he encouraged other scientists all over the world to work with the fruit fly. A newsletter kept drosophilists au

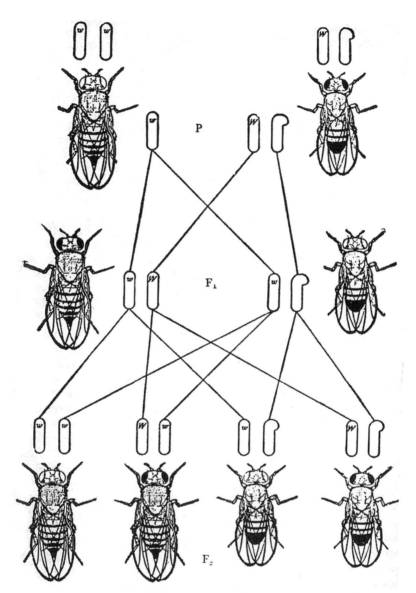

**FIGURE 6.2** Scheme illustrating sex-linked inheritance in crossing white- and red-eyed mutants of *Drosophila*. From Thomas Hunt Morgan, Alfred Sturtevant, Hermann J. Muller, and Calvin Bridges, *The Mechanism of Mendelian Heredity* (New York: Henry Holt & Co., 1915), fig. 10.

courant about the latest research results, and the community agreed to use a common terminology in order to foster efficient communication between laboratories.

The focus on a model organism that was biologically remote from humans and that had no immediate agricultural interest contributed to establishing genetics as an academic subject. But corn, or maize (*Zea mays*), was soon added to the genetic research agenda, as was a mammal, the mouse. Maize genetics created a bridge to agriculture, and experiments with mice later had a lasting impact on medical genetics.[66] In genetic crossing experiments, each of these model organisms was simultaneously an object of investigation and a precision instrument that was used to explore that object. As a prerequisite for this dual function, model organisms derived from pure cultures ideally differed in only a single, well-defined aspect of their genetic constitution, but were otherwise identical and could be assumed to react identically to the same conditions. A purely instrumental understanding of genes, whose variants (alleles) "made a difference" in the development of the organism, was therefore widespread among transmission geneticists.[67]

Genetically homogeneous populations—and hence model genetic organisms—were a decisive precondition for crosses fashioned on Mendel's experiments. Mendel himself had demanded from the varieties that entered his hybridization experiments that they exhibit "constantly differing" characters. This requirement had the consequence that practically no new characters could arise when these lines were crossed, except intermediate traits in the first hybrid generation. Only protracted crossing made it possible, or so experience suggested, to transfer a particular character exhibited by one variety to another variety, which was then bred pure to obtain a plant with a new combination of characters. However, the mass cultures required for Mendelian experimentation and for the construction of pure lines pursuant to Johannsen's methods also presented the possibility of observing the "spontaneous" appearance of completely new variants, which themselves could be selected and examined with respect to constancy—a practice not foreign to breeders, as we have already noted. These variants came to be called "mutations," especially following the publication of de Vries's *Mutation Theory*. The subtitle of this monumental two-volume work—*Experiments and Observations on the Origin of Species in the Vegetable Kingdom*—shows that de Vries claimed that mutations alone could underwrite the evolutionary process of species formation.[68]

The strains employed in these experimental cultures were also material resources in which genetic knowledge "packages" were maintained in a sort of hardware. Materialized genetic knowledge could thus be introduced into new experimental arrangements. This possibility led to the creation of experimental networks that helped researchers to coherently relate their research results to one another over long distances.[69] In this sense, the epistemic objects of genetics became increasingly reified and converted into tools. When Johannsen published his paper "The Genotype Conception of Heredity" in the *American Naturalist*, the protozoologist Herbert Spencer Jennings, then working with *Paramecium* cultures, immediately castigated Johannsen for his skeptical attitude. In the same issue, Jennings insisted, with respect to genotypes, "These things, whatever we call them, are concrete realities—realities as solid as the diverse existence of dogs, cats and horses."[70]

In parallel with the detection and collection of spontaneous variations in mass cultures, newly discovered sources of radioactivity, such as radium, and X-rays were being used already in the first decade of the twentieth century to elicit mutations. The botanist Albert Blakeslee at Cold Spring Harbor was one of the pioneers in applying radium to induce chromosome alterations.[71] But it took until the late 1920s for Hermann J. Muller and Lewis Stadler to succeed in producing viable X-ray–induced mutants of two common genetic model organisms, *Drosophila* and *Zea mays*. The so-called genetic target theory (*Treffertheorie*), developed by Nikolai Timofeeff-Ressovsky, Karl Zimmer, and Max Delbrück in Berlin around the same time, was also based on X-ray experiments with the fruit fly.[72]

The requirement for standardized model organisms soon found institutional expression. Plant gardens and animal stables that supplied such organisms were created. They operated either as commercial entities, such as the Jackson Memorial Laboratory for mouse breeding in Bar Harbor, Maine, set up by Clarence C. Little in 1929,[73] or with funding from research support organizations, such as the Notgemeinschaft für die Deutsche Wissenschaft (Emergency Association of German Science), that supported such "community work necessary for research."[74]

The search for variants that could be used in genetic experiments also prompted the introduction of new distinctions. Previously, Nägeli and later Morgan had already differentiated between hereditary variations and nonhereditary modifications. The work of Victor Jollos with the protozoon *Paramecium* alerted geneticists to what he called "permanent modifications" (*Dauermodifikationen*), relatively stable changes that were discussed in the context of cytoplasmic forms of heredity.[75] According to Jollos, these permanent modifications could be selected and handed down from one generation to the next, but they differed from nuclear modifications in that they eventually disappeared again, although often only after many generations. In extended cultivation experiments with the snapdragon *Antirrhinum majus* at the Agricultural College of Berlin, Erwin Baur—who later directed the first German research institute that carried the term "Vererbungswissenschaft" in its title—drew a distinction between the "factorial" mutations used in Mendelian experimentation and what he called "small mutations" (*Kleinmutationen*): tiny morphological or physiological changes that nevertheless persisted but could be recognized only by those who were experienced in handling a particular model organism.[76] His later success in breeding fruit varieties became legendary.

Overall, classical genetics developed a tendency to move toward higher and higher resolution. For Muller, the target theory of Timofeeff-Ressovsky, Zimmer, and Delbrück held out the hope of being able to drill down to the molecular dimensions of the gene. With this end in mind, he conducted ever more complex experiments with ever more sophisticated strains of *Drosophila* on a professional odyssey that took him from Austin, Texas, to Berlin, to Moscow, to Edinburgh, and finally back to the United States.[77] Though Muller himself fell short of attaining this level of resolution, Seymour Benzer finally mapped the fine structure of the genome of bacteriophage T4 in the late 1950s. It is worth noting that Benzer came to the conclusion that "gene" was a "dirty word," for his work demonstrated that the gene as a unit of function (*cistron*) by no means coincided with the unit of mutation (*muton*) and the unit of recombination (*recon*).[78] Barbara McClintock's investigations into the color variants of corn leaves, conducted at Cold Spring Harbor, likewise approximated molecular dimensions. Her recognition of "jumping genes" earned her, belatedly, the Nobel Prize in Physiology in 1983.[79]

In the final analysis, the gene mapping projects of classical genetics had the aim of identifying as many gene loci as possible, which in practice was possible only by taking advantage of allelic mutations. Physiological and developmental geneticists, on the other hand, were interested in resolving the cascades of gene actions involved in a particular metabolic or developmental process. They believed that complex characters that initially eluded Mendelian analysis would eventually be resolved with respect to their genetic components and thus become accessible for crossing analysis. This research program was only partially successful and thereby exposed the limits of experimental mutagenesis. Many organic changes, it turned out, that could be artificially produced by means of radiation or the influence of chemicals were either not viable or led to infertility and were thus lost to classical genetics as either an object of research or a tool of analysis.

Cytology also contributed to the reification of the gene as a locatable and measurable entity. In parallel with the reinterpretation of Mendel's rules, the assumption that the underlying units of heredity had to be, in one way or another, components of the cell nucleus and located on the chromosomes gained currency. As we recounted in chapter 4, cytology became engaged in the discourse of heredity around the middle of the 1880s. Theodor Boveri then laid the foundations, in parallel with Walter Sutton, of what came to be known as the "chromosome theory of heredity" in the first years of the twentieth century. Boveri also introduced experimental approaches to cytology, a discipline that until then had been more or less limited to microscopic observation, albeit of an increasingly sophisticated nature. His experiments on the double fertilization of sea urchin eggs, conducted with his wife Marcella Boveri at the Zoological Station in Naples, showed that homologous chromosomes were equivalent but qualitatively different and that they became randomly distributed among the germ cells.[80]

Experimental findings in cytology thus paralleled the results of breeding experiments. Even if arguments about the microscopic material constitution of the hereditary units, the elementary units of the new science, remained hypothetical, it proved possible nonetheless to diagnose macroscopic chromosomal changes that went hand in hand with changes in the phenotype. Chromosome anomalies developed into an essential tool of experimental cytogenetics, as illustrated, for example, by Blakeslee's early experiments on the radioactively treated jimson weed *Datura*[81] as well as by Emmy Stein's radiation experiments on the snapdragon in Baur's laboratory in Berlin. In addition, the phenomenon of polyploidy—that is, the multiplication of whole chromosome sets encountered in many cultured plants—invited a genetic interpretation of phenotypic effects.

As mentioned earlier, in the late 1920s, Morgan's student Muller succeeded in producing X-ray–induced mutations in *Drosophila* that were passed on to progeny according to Mendel's rules. In the wake of this achievement, radiation genetics established itself as a specialty; in the course of time, its activities concentrated increasingly on the determination of threshold values and dose effects. Yet, although these results strengthened the case for believing in the materiality of the hereditary factors—they could be transformed by physical influences and transmitted in their altered form—they offered no insight into the nature of the hereditary substances that underwent these changes. In 1950, on the occasion of the fiftieth anniversary of the "rediscovery" of Mendel's work, Muller confessed,

> The real core of gene theory still appears to lie in the deep un-
> known. That is, we have as yet no actual knowledge of the mech-
> anism underlying that unique property which makes a gene a
> gene—its ability to cause the synthesis of another structure like
> itself, a daughter gene in which even the mutations of the origi-
> nal gene are copied. . . . We do not know of such things yet in
> chemistry.[82]

This analysis notwithstanding, cytogenetics and formal genetics had joined forces in the meantime. During the 1930s, the zoologist Theophi-lus Painter succeeded in correlating changes in the banding pattern on the giant chromosomes of salivary gland cells of *Drosophila* with gene lo-cus changes on Morgan's formal gene maps. With her light microscope, Barbara McClintock identified chromosomal translocations, inversions, and deletions induced by radiation and related them to corresponding character changes in her maize plants. Morgan and Sturtevant described mutations such as "bar," in which changes in the eyes of the fruit fly could be ascribed to the chromosomal duplication of a gene. The occur-rence of what came to be called "position effect" also attracted consider-able attention: eye color, and with it the expression of a gene, could vary according to the position the gene occupied on the chromosome. But if the function of a gene depended on its position on the chromosome, then one could conjecture that its physiological and developmental func-tion might depend on the organization of the genetic material at large, as Richard Goldschmidt, among others, argued, and not simply on the properties of the hereditary units, which were presumed to be of a par-ticulate nature.[83] As we have seen, Johannsen had allowed for, and even claimed, such a possibility.

*Classical Genetics: Limits and Extensions*

Muller distinguished two basic properties of genes: "autocatalysis" al-lowed them to replicate, and "heterocatalysis" was responsible for their phenotypic effect—that is, the expression of a character. However, it was clear that neither crossing experiments nor cytogenetic investiga-tions with the microscope would yield decisive insights into these two properties. Having reflected on the nature of gene mutation and gene structure while he was still in Berlin, Max Delbrück had an inkling in the late 1930s that the problem of autocatalysis, and with it gene duplication, could profitably be studied in bacterial viruses known as bacteriophages.[84] Delbrück had been disappointed with the options presented by research on *Drosophila* during a previous visit to Morgan's laboratory in Pasadena. In phages, Delbrück saw the simplest naturally occurring model for the genetic material. They could induce infected bacteria to devote themselves completely to the multiplication of the phages that had infected them, and the infected bacteria finally perished as a consequence of this reprogramming. But initially, the phage system, which we will examine in more detail in the next chapter because it ushered in the era of molecular biology, remained as formal as the classical *Drosophila* genetics it was intended to replace.

Also in the late 1920s and the early 1930s, Alfred Kühn, Ernst Caspari, and their group in Göttingen turned to the other problem, that of heterocatalysis. They wanted to shed light on gene action, and Kühn aspired to reorient genetics toward physiological and developmental studies. Together with Caspari, Ernst Plagge, and Erich Becker, he showed by means of organ transplants between wild-type individuals and eye-color-pigment mutants of the flour moth *Ephestia kühniella*, and an analysis of the diffusible substance involved, that a distinction needed to be drawn between "gene action chains" and "substrate chains." Genes, Kühn concluded, acted through a "primary reaction" via ferments, which in turn catalyzed particular steps in a metabolic cascade (fig. 6.3). He saw the future of genetics in elucidating this connection:

> We stand at only the beginning of a vast research domain; its disclosure will still require a lot of biological-biochemical collaboration. But general characteristics of the action of hereditary dispositions are already appearing on the horizon. Our apprehension of the expression of hereditary traits is changing from a more or less static and preformationist conception to a dynamic and epigenetic one. The formal correlation of individual genes mapped to specific loci on the chromosomes with certain characters has only a limited meaning. Every step in the realization of characters is, so to speak, a knot in a network of reaction chains into which many gene actions radiate. One trait appears to have a simple correlation to one gene only as long as the other genes of the same action chain, and of other action chains that are part of the same knot, remain the same. Only a methodically

Über eine Gen-Wirkkette der Pigmentbildung bei Insekten.

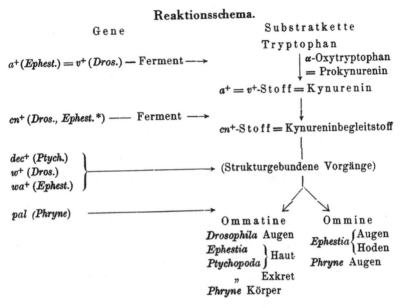

FIGURE 6.3 Diagram illustrating the relationship between a gene-action chain and a substrate chain. From Alfred Kühn, *Über eine Gen-Wirkkette*, 1941.

conducted genetic, developmental, and physiological analysis of a great number of single mutations can gradually disclose the hustle and bustle of effects of hereditary dispositions [*Wirkgetriebe der Erbanlagen*].[85]

Again, for a research program as envisioned by Kühn, a genetically homogeneous model organism system was a prerequisite for success. Kühn's experimental system was based on the flour moth *Ephestia*. Compared with *Drosophila*, with which George Beadle and Boris Ephrussi had embarked on similar work, the flour moth had the advantage that it allowed for relatively easy transplantation experiments. The disadvantage was that its chromosome pattern was cytologically complex and difficult to analyze.

Kühn saw his experiments as a first step away from the "preformationist" tendency and atomism that he deemed to be at work in the transmission genetics of his time. Instead, he pleaded for the establishment of an "epigenetics"—in the sense of the study of heterocatalysis—that would combine formal genetic, developmental, and physiological analyses in order to understand gene action. To cite an expression from the passage above, the "hustle and bustle of [genetic] effects" emanated for Kühn from the interweaving of both gene action chains *and* substrate chains. But again, Kühn's *Ephestia* system did not allow him to study the molecular mechanisms of the hypothesized gene-ferment relations that were the starting point of gene action. The "primary reactions" identified in his system remained beyond the reach of his investigations.

The practices of transmission genetics de facto remained limited to interpreting identifiable aspects of the phenotype as clues hinting at the formal structure of the genotype. The question of how the genes acted on the phenotype was thus set aside, as was the problem of their material constitution. Lenny Moss therefore rightly distinguishes the "gene-p" perspective of classical genetics (p standing for "preformationist" as well as for "predictive") from the "gene-d" perspective (d standing for "developmental") that became characteristic of molecular genetics in the decades that followed. In the latter perspective, genes are basically seen as developmental resources for the organism.[86] It is worth mentioning that Morgan himself had never lost sight of the formal character of his program. As late as in 1939, he declared on the occasion of his Nobel lecture, "At the level at which the genetic experiments lie, it does not make the slightest difference whether the gene is a hypothetical unit, or whether the gene is a material particle."[87]

The fundamental attraction of classical, formal genetics lay in its numerical nature and in the apparent precision of its results. It is therefore no coincidence that population geneticists like Ronald Fisher, J. B. S. Haldane, and Sewall Wright developed elaborate mathematical models of the effects of mutation and selection on the genetic constitution of populations with recourse to the abstract gene of classical genetics.[88] For some time, particularly in the aftermath of de Vries's mutation theory, the Mendelian inheritance of mutations had even been propagated as an alternative theory of evolution, a theory that would render obsolete Darwin's selection hypothesis as well as the gradualism postulated by the biometricians, be it from a Darwinian or an anti-Darwinian perspective. Now, evolution was redefined genetically, as the change of gene frequencies in the gene pool of populations. This redefinition rested on the formula derived by Weinberg and Hardy in 1908 that described the genetic equilibrium of an ideal population free from evolutionary factors such as mutation, environmental pressures, or selective mating. With regard to this formula, Jean Gayon has spoken of a principle of inertia as the "starting point for a mechanics of evolution."[89] The Hardy-Weinberg law became the foundation of neo-Darwinism and of the so-called evolutionary synthesis of the 1930s.[90] Darwin's evolutionary theory, which was a crystallization point for the discourse on heredity in the nineteenth century and was then brushed aside by classical genetics in the early twentieth century, now returned in the form of population genetics.

These developments at the same time demonstrate that the gene concept continued to cover several operational definitions. Depending on experimental context, the object of genetics was defined as a unit of transmission, a unit of recombination, a unit of mutation, or a unit of function, and combining these different units in experimental investigations rarely resulted in the one-to-one relations that had originally been presupposed to exist between them. Autocatalysis and heterocatalysis, one of the most important legacies of classical genetics, thus largely remained formal and operational concepts, despite their chemical connotations. One could also say that classical genetics was a phenomenological science: inferences were drawn from the phenotype, which was accessible to inspection, to the genotype, which was amenable to modification but could not be visualized as such. Yet the whole thrust of classical genetics remained directed toward that which lay beyond the phenotype, continuously referring to the underlying "real" target of its efforts, even though neither the crossing technique nor experimental cytology provided tools for condensing it *materially* into an epistemic object. We will show in the next chapter how molecular biology succeeded in this endeavor.

Before we do that, a few final observations on institutional developments are appropriate. Chairs, societies, and specialized journals are normally good indicators of the degree of institutionalization of a scientific discipline. With respect to genetics in Germany, which we take as an example here,[91] a detailed analysis of this kind has been provided by Jonathan Harwood. His book *Styles of Scientific Thought: The German Genetics Community, 1900–1933* documents a clash between an academic and a pragmatic research style during the formation of a community of German geneticists, a clash that turned out to be rather productive.[92] Erwin Baur, for instance, who originally trained in medicine and subsequently played a major role in the establishment of botanical genetics, became the founder of the *Zeitschrift für inductive Abstammungs- und Vererbungslehre* ("Journal for the Inductive Science of Descent and Inheritance") in 1908—the first genetics journal worldwide. Baur also gave the first lectures in genetics at the University of Berlin in 1907. In 1914, he founded the Institute for the Science of Heredity at the Agricultural College in Berlin and became its first director. Together with Carl Correns and Richard Goldschmidt, he founded the Deutsche Gesellschaft für Vererbungswissenschaft in 1921. In 1928, he became the first director of the newly established Kaiser Wilhelm Institute for Plant Breeding Research in Müncheberg, a position he occupied until his early death in 1933.

The title of Baur's journal explicitly brings together genetics—although the term had not yet found its way into the German language—and evolution. Baur maintained the option of transforming evolutionary biology into an inductive or experimental science by way of genetics. The architects of the evolutionary synthesis in the late 1930s and early 1940s,[93] and in their wake, a number of historians of biology,[94] have portrayed the situation largely as one in which genetics, evolutionary biology, and developmental biology began to drift apart after the turn of the twentieth century, a process that was supposedly reversed only with the rise to prominence of population genetics in evolutionary biology and with the emergence of molecular genetics in developmental biology. There are reasons to suspect that this view is an artifact of historiography, at least in general, but it has nevertheless heavily influenced the way in which the history of genetics has been written.[95] If it existed at all, this separation took different forms in different countries, such as the United States, Great Britain, France, and Germany. In fact, a significant portion of the research that we think of as being genetic today was carried out at chairs and institutes of zoology and botany during the first decades of the twentieth century. In Germany, one focus was on experimental cytogenetics, as represented by Boveri. On the other hand, Goldschmidt, Correns, and Hartmann established a physiologically oriented genetics that found its institutional home at the Kaiser Wilhelm Institute for Biology in Berlin-Dahlem from 1913 onward. In a different national context and five years earlier, William Bateson was endowed with a "Professorship of Biology" that was founded "with the understanding that the holder shall apply himself to a particular class of physiological problems, the study of which is denoted by the term Genetics."[96]

The fact that Bateson's professorship at Cambridge and the Kaiser Wilhelm Institute in Berlin carried the term "biology" in their names indicates that genetics was conceived as a kind of experimental "general biology" at the time. On the whole, however, the classical division of biology at the universities into zoology and botany remained in place. In contrast, zoological (Goldschmidt), botanical (Correns), and proto-zoological (Hartmann) research on questions of genetics was brought together at the Kaiser Wilhelm Institute for Biology. Interestingly, in one way or another, the work in this institute revolved around a physiological and developmental genetics of sexuality.[97] In 1927, the Kaiser Wilhelm Society, which saw itself as an institution at the intersection between basic and applied research, also founded an institute for human genetics, the Kaiser Wilhelm Institute for Anthropology, Human Genetics, and Eugenics. Baur, as well as Correns and Goldschmidt, had pressed for its establishment. Its first director, Eugen Fischer, is generally seen as an early representative of a eugenically motivated human genetics. As outlined in chapter 4, this institute, as well as Ernst Rüdin's Genealogical-Demographical Department at the German Research Institute (later Kaiser Wilhelm Institute) for Psychiatry in Munich—its emphasis being on human population genetics and on sibling research—played a central role in comparative hereditarian pathology and racial hygiene in the Weimar Republic as well as during National Socialism in Germany.[98] The Kaiser Wilhelm Institute for Brain Research in Berlin-Buch also had a department that focused on evolutionary population genetics: Nikolai Timofeeff-Ressovsky's department, which became the hub of *Drosophila* genetics in Germany in the 1920s. In summary, it appears that the biological institutes and departments of the Kaiser Wilhelm Society were among the key drivers behind the establishment of genetics in Germany in the first decades of the twentieth century, with physiological genetics, population genetics, and human genetics at its core.

Classical genetics took root in a social and political context in which eugenic thinking formed a self-evident horizon for research in Germany, elsewhere in Europe, and in the United States and Canada. The experimental procedures of the new genetics, however, had little relevance to eugenics as practiced on humans. As far as medical genetics is concerned, Ilana Löwy and Jean-Paul Gaudillière have in this respect spoken of an "elusive 'Mendelization' of the clinic." They argue that as far as epidemiology is concerned, questions of heredity, infection, and immunization remained confounded in such a way that the vertical and the horizontal dimensions of the transmission of diseases were impossible to separate.[99] A similar argument can be made about other areas of application. Mendelism by no means immediately came to dominate areas like plant breeding,[100] let alone animal breeding,[101] with the exception perhaps of the plant breeding stations in Groß-Enzersdorf near Vienna, Svalöf in Sweden, and the Agricultural Experiment Station of the State of New York, which became part of Cornell University in 1923. Essentially the same is true for the investigations into heredity that were carried out in the realm of microbiology with bacteria and in protozoology with eukaryotic unicellular organisms during the first half of the twentieth century. More often than not, forms of non-Mendelian heredity in these organisms were investigated as alternative generalizable models for decades.[102] As we have shown, all these areas of research and practice had been swayed by hereditary considerations well before the end of the nineteenth century, and this also prepared the ground for the reception of Mendelian genetics. But the complex phenomena observed here often escaped the Mendelian grasp even after classical genetics had established itself as a discipline. This situation was bound to change only after 1945.

The message of the new genetics was without doubt most immediately effective as what Georges Canguilhem once called a "scientific ideology"[103]—a form of thought that is hyperbolic with respect to the areas of investigation that it seeks to bring under its control. The solutions held out by this mode of thought therefore tend to be incommensurable with the objects under consideration, but nevertheless act as points of orientation and as action models through their very exorbitance. Neither should one underestimate discursive differentiation within the new discipline. Genetics soon split into a more theoretically oriented population genetics and a more practically oriented transmission genetics. At the time, the latter branch assumed a position of rhetorical distance from evolutionary biology, developmental biology, and physiology. However, the very fact that this position was largely rhetorical points to the fact that, in principle at least, genetics, evolutionary biology, and developmental biology remained compatible. The work of scientists like Kühn, Beadle, and Ephrussi, after all, addressed the connection between these disciplines directly. James Griesemer, for instance, insists that most of the early geneticists did not want to separate questions of development, evolution, and hereditary transmission, but had developmental and evolutionary "invariants" in view right from the beginning when they investigated how genes were transmitted.[104]

The establishment of classical genetics as a discipline therefore cannot be reduced to the relatively narrow realm of pure transmission genetics in the style of Morgan, nor did the whole realm of biological heredity at once succumb to the dictate of Mendelian heredity. As is well known, at the height of Stalin's regime in the Soviet Union, "Mendelism" in general was suspected of being a "capitalist science" and was combated in the name of a practical biology of adaptation and acclimatization that was based on the Russian tradition of plant breeding, from Ivan Vladimirovitch Mitchurin to Trofim Lysenko. This tale is usually revived to warn, in one way or another, against politically motivated manipulations of the autonomy of scientific research.[105] But with that, the social constitution and shaping of knowledge is turned into a deplorable but occasional exception. In contrast, in this and the foregoing chapter, we have seen that the crystallization of hereditary knowledge into a biological discipline not only required a new regime of experimentation—although it did require that too—but also occurred in a social and cultural context that was saturated with, if not obsessed by, the idea that biological solutions would have to be found for central questions of social and economic life and, moreover, that these solutions would be the ultimate ones to be desired.

# 7

### Heredity and Molecular Biology

The twentieth century has rightly been called the century of the gene.[1] Genetics not only left its mark on the life sciences and the medical sciences of the century; its organizing object, the gene, in all its variants, formed the epistemic core of a hereditary discourse that radiated into social, cultural, and political spheres in more ways than one. The development of genetics engendered its own historical breaks, but it also remained nested and entrenched in nation-state, world war, and Cold War history of that century. In the two preceding chapters we have traced these mutual relations at many levels with respect to classical genetics, which emerged during the first decade of the twentieth century and dominated the tone of genetic discourse until well into the 1940s.

The ensuing history can by no means be told as a linear continuation of the story into the present. The last two chapters of this book try to sketch, in broad strokes and in an effort to identify the relevant patches of a multidimensional patchwork, the shift to the molecular genetic discourse of heredity. It proceeded, as we will see, neither continuously nor in a consciously planned manner, but was rather configured by developments at the fringes of the discourse of genetics. Nevertheless, this transformation left little ground untouched in the end. By the middle of the 1960s, it dawned on even those who had spent their scientific lives in the tradition of classical genetics that times were changing. For example, near the end of his *History of Genetics* of 1965, the distinguished German geneticist Hans Stubbe drew attention to what he called "new problems which now mark a new, dynamic era in genetics."[2] In this era, the discourse of heredity was molecularized and embedded in a new experimental culture, whose main elements we will describe in more detail later in this chapter.[3] In the next and final chapter, we will follow up this compressed history of molecular genetics with a sketch of the third phase of genetics in the twentieth century—the catchwords here being gene technology and genomics. As we will see, this last transition again occurred under specific cultural and disciplinary constellations.

The molecularization of genetics did not arise directly from the experimental regime of classical genetics, which combined the construction of new strains, crossing experiments, and the statistical analysis of hybrid offspring with microscopic cytology. It was rather part of an all-encompassing molecularization of biology that occurred in parallel with the development of classical genetics but had largely independent roots. Accounts of the history of molecular biology from the perspective of molecular genetics, as it had crystallized in the late 1960s, have often been genocentric.[4] However, the molecularization of biology must be seen as a broader movement. To start with, molecular biology appropriated a whole arsenal of new physical and chemical procedures for the analysis of living processes.[5] Phenomena beyond the reach of traditional analytic chemistry and physical mechanics were thus brought into the realm of biological investigation. Moreover, in many cases, the questions that motivated the development of these techniques and their application to biological phenomena had nothing to do with genetics. Instead, the aim was to understand the conformation and function of biologically relevant molecules beyond their mere atomic composition. In this way, an assemblage of new research technologies and experimental systems took shape that made it possible to characterize a whole range of recently identified substances that were biologically active—vitamins, hormones, antibodies, enzymes, and antibiotics among them. Genes were added to this list not before the late 1930s, and it took until the 1950s for their molecular analysis to be brought on its way by Watson and Crick's discovery of the double-helical structure of DNA.

Early on in the context of this analytic enterprise, a new concept of biological specificity emerged, which is often described as the "lock and key" principle. Emil Fischer proposed this model in 1894 to describe the relation of ferments to their substrates.[6] Initially, enzymes secreted by glands into certain organs and body cavities were characterized. Subsequently, enzymes that functioned as biocatalysts within the metabolism of cells became the target of the new biological-chemical disciplinary specialty called biochemistry.[7] The lock and key vision of biological specificity proved helpful in a number of experimental domains, not only in enzymology but also in immunology and endocrinology. However, in the core area of molecular genetics, the lock and key model would later be abandoned in favor of a different vision of biological specificity, that of information. The lock and key principle therefore catalyzed the molecularization of biology, but at the same time acted as an "epistemological obstacle" to the molecularization of genetics, to use an expression of Gaston Bachelard, as its vivid imagery made it hard to adopt a different and new perspective.[8]

In these developments, taken together, we thus witness the establishment of a new biological research culture alongside classical genetics. The conjuncture of these two research cultures finally gave rise to molecular genetics.[9] Some of the characteristic features of this field, as it emerged in the first half of the twentieth century from practical contexts that are characteristic of the development of the sciences as a whole, will be pinpointed here. First, close cooperation between basic research and industry shaped the chemical as well as the physical aspects of the molecularization process, albeit in different ways.[10] In Germany, research into vitamins and their therapeutic uses was carried out by academic research laboratories, such as that of Adolf Windaus at the universities of Göttingen, in close cooperation with pharmaceutical companies, such as Merck and Bayer. Another example is the collaboration of Windaus's student Adolf Butenandt with research-intensive companies such as Schering or I. G. Farben. Investigating the structure and synthesis of sexual hormones, Butenandt first worked in Göttingen, then in Danzig, and finally at the Kaiser Wilhelm Institute for Biochemistry in Berlin.[11] A similar cooperation existed in Switzerland between Leopold Ruzicka from the Federal Polytechnical Institute in Zürich and CIBA in Basel.[12] Further comparable alliances existed in other European countries, such as Great Britain, France, and the Netherlands.[13] In Germany, several of the Kaiser Wilhelm Institutes founded in the 1920s specifically took up the molecular analysis of certain organic raw materials such as fibers and leather. These institutes too cooperated closely with relevant industries. The physical aspects of this molecularization process are exemplified by the development of the electron microscope by Ernst Ruska at Siemens & Halske in Berlin. Biological applications of the instrument were explored by physicians at the Charité in Berlin, who in turn cooperated with biologists from the Biologische Reichsanstalt in Berlin-Dahlem. Theodor Svedberg, who in the late 1920s and early 1930s developed his analytical ultracentrifuge in Uppsala, Sweden, was a staunch promoter of science-industry collaborations.[14] In the United States, the collaboration of the laboratories of the Radio Corporation of America (RCA) in Camden, New Jersey, with television pioneer Vladimir Zworykin and with the biologist Thomas Anderson adds to this overall picture.[15] And in Great Britain, the textile industry funded William Astbury's X-ray analyses of fibers such as keratin and collagen in Leeds. In the late 1930s, Astbury extended his work to deoxyribonucleic acid (DNA), samples of which he had obtained from Torbjörn Caspersson in Stockholm.[16] Other examples of such collaborations in which the molecularization of biology took place could be added.

The second pivotal context for the emergence of molecular genetics was the advancement of a molecular perspective in biology and medicine, taking the form of "biomedical platforms," to use an expression coined by Peter Keating and Alberto Cambrosio.[17] Keating and Cambrosio define scientific platforms as locally networked groups of experts being rooted in different working cultures and yet focusing on a common research agenda. At research clinics such as the Rockefeller Institute in New York, the molecularization of biology proceeded in different disciplinary domains that were situated between research and clinic, particularly microbiology and immunology, through the work of researchers such as the pathologist and virologist Peyton Rous, the oncologist and cytologist Albert Claude, and the physician and bacteriologist Oswald Avery and his group.[18] The same is true of the Massachusetts General Hospital in Boston, where oncologists Joseph Aub and Paul Zamecnik, as well as biochemist Fritz Lipmann, were active;[19] of the Pasteur Institute in Paris, where, among others, the physician and biologist Pierre Lépine practiced;[20] and of the Pathological Laboratories of the Ministry of Health in London, with its bacteriologist, Frederick Griffith.

The most important practical context in which molecular genetics emerged was probably cancer research.[21] After hygiene, bacteriology, and immunology had helped to curtail the impact of infectious diseases and had even provided some specific treatments, the investigation and the treatment of cancer attained new priority from the late nineteenth century onward. As far as the etiology of various malignant tumors was concerned, oncology at the beginning of the twentieth century recognized a plurality of potential causes, ranging from infection to heredity to environmental stress. In this fast-growing field, which attracted more and more funding from private organizations as well as publicly funded bodies, a variety of research avenues were opened. Soon it became clear that X-rays, radiotherapies, and chemotherapies, all of which were employed to supplement surgery, could be used efficiently in the long run only if more was learned about the cellular and molecular basis of malignant growth. In our description of the emerging laboratory culture of molecular biology, we will encounter cancer research repeatedly. While the discourse of heredity barely played a role during the emergence of the new research technologies—and it will temporarily recede into the background of our narrative for this reason—molecular genetics would have been unthinkable without the later appropriation of this discourse and the corresponding transformation of these new research environments.

## A Landscape of New Research Technologies

The emergence and development of new research technologies was one of the cardinal prerequisites for the molecularization of biology. Terry Shinn and Bernward Joerges have drawn attention to the fact that the research technologies that emerged in the nineteenth century—such as cytological microscopy and graphic recording devices in physiology— and that increasingly dominated research in the course of the twentieth century came to be located in an interstitial space between scientific research and industry from the beginning. In this interstitial space, transverse application profiles that cut across research and industry developed, resulting in a close linkage of science and technology and an acceleration of their respective processes of development.[22] Scientific instrument building received new impetus and became a branch of industry on its own. In short, the interlacing of science and technology created a new dynamic. In this context, research technologies may be characterized by the fact that they embody an excess: working with them generates questions that could not even have been posed without them. Insofar, research technologies are always research-*enabling* technologies as well.[23]

New research technologies also played a central role during the emergence of molecular biology, a point that has been stressed by a number of historians of biology. For Richard Burian, the core of molecular biology consists of a "battery of techniques."[24] Lily Kay has similarly spoken of a new "technological landscape."[25] In her concept of landscape, technology is thought of not as external to science, but as an embedding that empowers scientific thinking. Below, we consider a series of these biophysical techniques. They were a prerequisite for a new, molecular genetics and yet, as mentioned earlier, had little, if anything, to do with genetics or questions of heredity when they first appeared on the scene.

These points can be nicely brought home by considering the case of analytical ultracentrifugation, a technique that emerged in the 1920s and whose development became instrumental in the emergence of the concept of a biological macromolecule. It was not until the 1920s that the concept of macromolecules, introduced by Hermann Staudinger, replaced the concept of colloids or colloidal protoplasm.[26] Previously, the notion of protoplasm held sway in the hereditary theories of the late nineteenth century as well as in early-twentieth-century biochemistry. The key to this idea was the notion from colloid chemistry that the cellular plasm had a gel-like consistency and contained smaller molecules of changing proportions, compositions, and densities. The technology of the analytical ultracentrifuge—developed and constructed by the colloid chemist Theodor Svedberg in collaboration with the hematologist Robin Fåhraeus—put an end to this conception.[27] Although the ultracentrifuge was developed for the analysis of colloids, it yielded a surprising finding when it was applied to proteins like hemoglobin: in the gravitational field of the centrifuge, proteins behaved like discrete particles with relatively sharply defined, high molecular masses. Ultracentrifugation thus made available a level of the organization of living beings for quantitative dissection that so far had resisted experimental analysis. Moreover, it became a key research technology in the delineation of a field that was occasionally called "molecular biology" as early as the 1930s—a disciplinary label that became widespread, however, only around the middle of the century. Prior to this time, hybrid labels such as "physical biology," "biophysics," or "biophysical chemistry," which reflected the coming together of different areas of expertise, had been common.

This example shows how technology formation and concept formation can go hand in hand. Their intertwinement also appears to be characteristic in bringing about the "excess" associated with research technologies: their capacity to generate conceptual surplus. No less important for macromolecular analysis was the development of an electrophoresis apparatus by the Swedish biochemist Arne Tiselius. Here, separation was effected by virtue of the different electrostatic charges of a mixture of different macromolecules.[28] A student of Svedberg's, Tiselius, who also worked at the University of Uppsala, tried out and calibrated his electrophoresis apparatus on serum proteins. Preparative high-speed centrifugation, which was significantly easier to handle than analytical ultracentrifugation with its sophisticated optics, was soon applied to many separation problems. In particular, this research technology made it possible to isolate a number of cell components, such as mitochondria—which had been identified by cytological means for a

long time—and the newly described microsomes. Following this innovation, a cellular biochemistry that was based on the differential fractionation of components of the cytoplasm and was carried out in test-tube
environments emerged. Albert Claude and his colleagues Rollin Hotchkiss, Walter Schneider, and George Hogeboom from the Rockefeller Institute in New York did pioneering work in this area.[29]

The development of X-ray structure analysis by Henry and Lawrence Bragg in Cambridge, first applied to the crystals of small molecules
and later also used in industrial fiber research, raised the possibility of
a structural analysis of biological macromolecules. This technology was
turned into a biological research instrument at several international centers: Linus Pauling at the California Institute of Technology identified
the alpha-helical structure of proteins with it.[30] Lawrence Bragg, Max
Perutz, and John Kendrew at the Cavendish Laboratory of Cambridge
University sought to elucidate the molecular configuration of hemoglobin and myoglobin.[31] William Astbury at Leeds University specialized in
keratin and DNA analysis. John Desmond Bernal at Birkbek College in
London focused on the tobacco mosaic virus (TMV). Michael Polanyi
and Hermann Mark at the Kaiser Wilhelm Institute for Fiber Research in
Berlin-Dahlem were busy with the characterization of cellulose, silk, and
wool.[32] Torbjörn Caspersson in Stockholm and Jack Schultz at Caltech
developed a special kind of microscopy based on ultraviolet radiation,
which made it possible to visualize and localize DNA and RNA in cells
and their compartments. At the beginning of the 1940s, the first applications of electron microscopy to biological specimens—by Helmut
Ruska, Gustav Kausche, and Edgar Pfankuch in Germany (TMV) and
by Thomas Anderson and Salvador Luria in the United States (bacteriophages)—helped to consolidate the view that viruses such as TMV and
bacteriophages such as the T-phages of the bacterium *Escherichia coli*
possessed a particulate nature.

These examples give us an impression of how intensely the disclosure
of the world of biological macromolecules depended on new biophysical
research technologies and how the development of these technologies,
in turn, took place at the cutting edge of scientific research. It was not
only that the emerging picture of the structures and cellular locations of
macromolecules was based on the use of these techniques. These molecules—as objects of research—also tested the techniques as well as their
power of resolution, provoking further technical developments, particularly the development of preparation procedures, without which the
orientation toward biological entities would not have been possible. Crucially, the main thrust of this research was not simply to give the physical

principles that underlay a particular technology a robust expression. The focus was on the variable interface between instruments and biological objects, given the aim of determining biologically relevant properties that reflected the native state of those objects, not simply their physical or chemical parameters. A mutual "calibration" between research instruments and research objects therefore took place, first in terms of basic feasibility, then in ongoing processes of mutual adjustment.[33] This is also the reason why collaborations between physicists, chemists, biologists, instrument makers, and engineers, as well as the creation of interdisciplinary areas of research, played an increasing role in the context of a biology oriented toward macromolecules that was virtually unprecedented in the life sciences. More often than not, such research—together with the constitutive technologies—did not fit into existing disciplinary and national boundaries. This also explains why many of the key publications in these areas carry the names of two or more researchers with different disciplinary and national backgrounds.[34]

In the United States, work on transdisciplinary objects of research was promoted by the Rockefeller Foundation. Warren Weaver, the director of its Natural Science Division from 1932 to 1955, encouraged physicists and chemists in particular to acquaint themselves with biological problems.[35] The influence of the Rockefeller Foundation was, however, not restricted to the United States, it extended over all of Europe, in particular to Great Britain, France, and Germany. Between 1932 and 1959, its Natural Science Division provided over ninety million dollars to support scientific research, and a substantial part of this funding went to the biological—biophysical and biochemical—sciences.[36] Of the eighteen Nobel Prizes that highlighted molecular biological achievements between the identification of the structure of the DNA double helix in 1953 and the elucidation of the genetic code in 1965, no less than seventeen were awarded to scientists that in the decades before had been funded by the Rockefeller Foundation under the aegis of Warren Weaver.[37]

As Lily Kay has stressed, the biologically oriented funding policy of the Rockefeller Foundation during the interwar period was sustained by the conviction that control over the biological constitution and the biological functioning of human beings would in the end also provide the means for efficient social control. Weaver, who was one of those who coined the term "molecular biology," shared this conviction, albeit not the hope of its quick fulfillment. As he implemented his research policy, he took for granted that for the time being, all the physical and chemical preconditions to realize this vision were wanting.[38] The mobilization of the life sciences in the interwar period nonetheless led to irreversible

changes in the biological research landscape. This mobilization continued during World War II, practically without interruption, and during
the Cold War, public funding of molecular biological research reached a
level that was hitherto unheard of.[39]

*Experimental Systems*

Given the work of "calibration" described above, the instruments that
represent the core of particular research technologies are usually complex to operate and labor-intensive to maintain. However, an instrument is generally a central object of research interest only while it is being developed. In order to gain acceptance, research technologies need
to become part of different experimental systems. The epistemic interest
in running experimental systems is thus usually not focused on instruments, which tend to assume a generic character. As we show below, it
was experimental systems—not instruments per se, although they provided the boundary conditions—that came to define molecular biological research.[40] Here we want to describe in some detail the ramifications
of particular experimental systems that accompanied the molecularization of biology and, from the middle of the twentieth century onward,
the molecularization of genetics and heredity.

In vitro or test-tube systems in particular made it possible to extend
the analysis of the cell to the molecular level. The experimental opening
up of the subcellular level can be understood as an extension of Boveri's
early shaking experiments, which had the objective of separating and recombining nucleus and enucleated cytoplasm, as described in chapter 4.
In vitro experimental systems went a decisive step further. The aim of
these experiments was to fractionate the cytoplasm itself. Many of these
attempts commenced in cancer research and thus in a medical-clinical
context. At the Rockefeller Institute in New York, for example, the Bel-
gian physician Albert Claude was investigating the particulate nature of
Rous's chicken sarcoma virus in the 1920s. However, his research led
him away from the virus and toward a characterization of the granular
contents of healthy cells by way of differential high-speed centrifugation of homogenized tissue.[41] As a result of this work, isolated and well-
characterized mitochondria, as well as a sediment termed "microsomes,"
became available for enzymatic analysis. Immediately after World War II,
the extraction of cells and the separation of their components by ultracentrifugation were taken up by several research groups that devoted
themselves to the problem of protein biosynthesis, again in the context
of cancer research. Here, the work of the group around Paul Zamecnik

and Mahlon Hoagland at the Massachusetts General Hospital in Boston was of particular importance. In the course of a decade or so, the research trajectory of this group led from tissue slices to intact organelles to the macromolecules contained in the cytoplasm, and from there, as we will see, to the core of molecular genetics.[42]

The work of Zamecnik and Hoagland thus shared in a general trend toward physiological in vitro research in the first half of the twentieth century. The distinction between in vitro and in vivo systems was established around 1900, following the demonstration by chemists such as Eduard Buchner that not only glandular but also intracellular enzymes, or "ferments," were able to function outside the cell, in the test tube, at least under particular buffer conditions. Once again, one could characterize these developments as the emergence of an experimental "landscape" or "environment" in Kay's sense. Of course, work with parts of organisms, and with preparations of all sorts, had a long tradition in biology as well as in medicine. But research with in vitro systems in the early twentieth century was framed differently. It rested on assumptions that would have been anathema to a nineteenth-century physiologist such as Claude Bernard—namely, that the cell extract "repeats or mirrors the living system."[43] Artificial environments were created for the extracts, thus facilitating the characterization of processes outside the living body that normally occurred inside it.

Test-tube systems were thus crucial for the transition from an organismic and cellular to a subcellular and finally to a molecular knowledge regime. It is important to note in this context, however, that in vitro systems are typically simplified systems that enhance some characteristics or elements of a network of metabolic reactions and eliminate others. This simplification entails, necessarily, the tendency of in vitro systems to generate experimental artifacts. The results obtained in test-tube environments therefore need to be checked continuously by relating them back to the situation in vivo. However, the situation in vivo—the living system, in other words—can be represented only by means of further, perhaps less simplified, experimental systems—hence the proliferation of experimental systems. A good part of the history of biology in the twentieth century, and of molecular biology in particular, can be captured by this expanding spiral, a game of checks and balances that resulted in the recursive expansion of experimental systems.

The success of in vitro biochemistry hinged on the introduction of an additional technique that revolutionized biologically motivated test-tube chemistry around the middle of the twentieth century: the radioactive labeling of particular molecules and building blocks of macromolecules and the use of these molecular entities as tracers in the analysis of metabolic processes. We look at this technique here, rather than in the previous section, because its effects were conspicuously different from those of a tangible macroscopic instrument. One might say that radioactively labeled molecules became one with the molecular processes under investigation, or that they permeated those processes like an analytic capillary system. Although molecular in nature, the labeling technique was nevertheless reliant on big technological systems, as the production of radioactive phosphorus, sulfur, hydrogen, and finally carbon became feasible only when cyclotrons were built in the 1930s. Larger amounts of isotopes, as required for biological experiments, then became available as by-products of the first atomic piles and reactors in the United States after 1945. New measuring technologies soon followed.[44]

After World War II, the U.S. Atomic Energy Commission promoted the use of radioactive isotopes in biology and medicine as part of its "Atoms for Peace" campaign.[45] The pervasive uptake of labeling technology is perhaps best captured with the following statistic: between 1945 and 1956—that is, roughly in the time span of a decade—the percentage of papers in the *American Journal of Biological Chemistry* reporting the use of radioactive isotopes rose from 1 percent to circa 40 percent.[46] Complex molecules that incorporated radioactive labels soon became available for use as tracers in biological systems. Initially, these molecules were synthesized in the research laboratories themselves, but a newly forming chemical isotope industry soon seized on this opportunity. Radioactively labeled molecules such as the building blocks of proteins (amino acids) and nucleic acids (nucleotides) now became commercially available. Until they underwent radioactive decay, the chemical and physiological properties of these molecules were for all practical purposes identical to those of their unlabeled counterparts. They could thus be introduced as probes into metabolic processes, where they participated in chemical transformations, and the isotopes could then be detected experimentally in order to trace where in the system the labeled molecules had ended up.

In a way, radioactive tracing acted like a biochemical electron microscope. On the one hand, the technique made it possible to follow particular reactions, such as the incorporation of amino acids into proteins or of nucleotides into nucleic acids, not only within the cell but also in relatively crude test-tube environments still containing a number of various cellular components. At the end of the experiment, one could simply remove the free, unincorporated radioactivity from the sample. And since measurement did not involve microchemical methods, but rather the physical counting of radioactive decay events, one could largely dispense with the procedures of preparative and analytic chemistry. On the other hand, the technique boosted measurement sensitivity by several orders of magnitude compared with the measuring range covered by analytic chemistry.

The combination of ultracentrifuge-assisted differential cell fractionation and radioactive tracing brought the cellular synthesis of macromolecules within the scope of in vitro biochemistry. The experimental systems of biochemistry thus advanced to one of the centers of the burgeoning molecular genetics after World War II.[47] The techniques of biochemistry offered a unique advantage: Test-tube treatment usually dampened the biological functions of interest by several orders of magnitude. Radioactive tracing made it possible to offset this loss of activity. Whatever the cell sap lost in terms of biological activity was restored to it, so to speak, in terms of "radioactivity." The introduction of radioactivity into the laboratory cultures of biology also had other effects, not only on laboratory architecture but also on laboratory life as a whole. For instance, radioactive traces of vanishing strength in biological samples could be detected only in uncontaminated environments. A spread of "negative traces" of radioactivity through the laboratory therefore had to be avoided at all costs; otherwise, potential signals would disappear in high background radiation. Thus these techniques not only called for a completely new laboratory regime, but also had consequences for the organization and the conduct of the experiments themselves.

The technology of radioactive tracing in biological systems cannot be reduced to a single instrument or substance. It permeated a whole landscape of experimental systems and shaped its contours at the same time.[48] Tracer technology kindled the development of new measuring instruments, such as liquid scintillation counters, whose integration into experimental systems in turn changed their scope and character.[49] That it materially mediated the knowledge of biologists, chemists, physicists, and engineers—combining the physics of isotope production, the chemistry of radioactive labeling, the chemistry of liquid scintillation measurement, the physics of photomultiplying, electronics, and biological sample preparation—contributed to the historical fascination that emanated from molecular biology's notorious interdisciplinarity. Moreover, tracer technology called for a form of biological experimentation—a biological laboratory culture as a "life form"—that had been foreign to classical genetics, perhaps with the exception of that part of classical genetics that was concerned with the production of mutations via radiation. And the fascination associated with this technology was sustained by a scientific community that, besides being interdisciplinary in nature, also practiced a form of internationality that was based on an unprecedented circulation of people and materials between laboratories.[50]

*New Model Organisms*

The molecularization of biology and of the knowledge of heredity also went hand in hand with the introduction of new model organisms. Hardly any of the decisive experimental findings or concepts of the molecular era are tied to *Drosophila* or, for that matter, to any of the other model organisms associated with classical genetics—perhaps with the exception of maize, the system in which "jumping genes" were identified by Barbara McClintock.[51] In vitro biology strengthened the importance of other model organisms, such as guinea pigs, as well as that of mice and rats, which had been employed to study malignant growth since the 1920s. The rat in particular can be understood as a legacy of the biological-clinical contexts in which cancer research was undertaken before, during, and immediately after World War II—for example, at the Rockefeller Institute, at Massachusetts General Hospital, and at several American and European universities. Systems for the analysis of cell components and of the mechanism of protein biosynthesis were also developed in these contexts. The experiments that finally led to the creation of an in vitro system for the analysis of protein biosynthesis were originally carried out to find out whether there was a difference in the rate of protein production between normal and malignant rat liver. Subsequently, the rat liver system was widely used for a decade and a half until it was replaced by bacterial extracts, particularly from *Escherichia coli*, at the beginning of the 1960s.[52] In vitro systems based on *E. coli*—and on yeast—then became the foremost experimental systems for molecular genetics until yet another turning point in genetic research in the late 1970s led to a renewed emphasis on higher organisms, as we will discuss in chapter 8.

Like research technologies, model organisms are critical components of experimental systems. They often come into use under contingent circumstances, but then develop their own dynamics and temporal remanence. The knowledge they embody—together with the experimental systems in which they are used—is a precious holding that researchers will abandon only reluctantly once they have become acquainted with it and dependent on it. The protracted use of particular model organisms can therefore also turn into an epistemological obstacle. Which organisms are suitable models depends on the epistemic objects that stand in

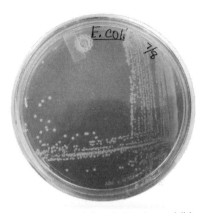

**FIGURE 7.1** *Escherichia coli* culture in a petri dish.
Courtesy American Society for Microbiology.

the center of scientific attention at any particular time and on the technologies that are available to work with them. Like research technologies, the organisms themselves need to be standardized to a certain degree in order to function as models. The production of homogeneous and, in particular, genetically pure lines has been at least as important in twentieth-century medical research and genetics as the development of new research technologies, and has accordingly become a focus of historical reflection.[53]

During the time span we consider here, the late 1930s to the 1960s, the molecularization of genetics developed on the basis of choosing viruses, bacteria, and lower fungi, especially tobacco mosaic virus and T-phages, the red bread mold *Neurospora crassa* and pneumococcus as well as *Escherichia coli* bacteria as model organisms (fig. 7.1). The characterization of viruses was of particular relevance for the discourse of heredity because they were assumed to represent something like "naked genes." More generally, the use of these model organisms facilitated the exploitation of new techniques of investigation that could not have been realized with higher organisms, among them new forms of crossing analysis.

It was nothing less than a sensation when Wendell M. Stanley of the Rockefeller Institute in Princeton announced the isolation and crystallization of a protein with the properties of the tobacco mosaic virus in 1935.[54] The *contagium vivum fluidum*, once described as a phytopathological ultra-bacterium of undefined nature by Martinus Beijerinck, had been transformed into a crystalline particle. Viruses—which were invisible under the light microscope, not retained by bacterial filters, and uncultivable on sterile growth media—became prototypical models for hereditary particles during the 1930s. They were assumed to represent something like bits and pieces of genetic material that reproduced auto-catalytically within infected bacterial cells. The crystallization of a virus was taken to symbolize the imminent and definitive breakdown of the boundaries between physics and biology, chemistry and genetics. In addition, the reported TMV protein consorted perfectly with the prevailing idea that the nature of genes was proteinaceous, and that prevailing belief was not at all shaken when the agriculturalist Frederick Bawden from the English Rothamsted Agricultural Station—one of the few early molecular biologists with a background in agriculture—and the chemist Norman Pirie in Cambridge demonstrated soon after Stanley's feat that TMV also had a RNA component. After all, that component made up less than 5 percent of its weight.

More than with any other model organism, all the physical research technologies that existed in the late 1930s were tried out on TMV, from ultracentrifugation (Stanley in Princeton) through X-ray structure analysis (Bernal in London) to electron microscopy (Ruska in Berlin). Research by the group of Gerhard Schramm and Georg Melchers at the Kaiser Wilhelm Institutes for Biochemistry and Biology in Berlin-Dahlem led to the first attempts at disassembling and reassembling the components of the TMV particle in the late 1930s and early 1940s. After World War II, the center of biochemical TMV research shifted to Stanley's newly founded virus laboratory at Berkeley, where Heinz Fraenkel-Conrat worked, while Schramm and Melchers continued their work at the Max Planck Institute for Virus Research in Tübingen.[55] Around the middle of the 1950s, both groups identified the RNA component of the TMV particle as the infectious principle and, with that, as its hereditary material.

With the red bread mold *Neurospora crassa*, George Beadle in Stanford adopted a model organism in 1937 that became the center of one of the most productive experimental systems in biochemically oriented genetics owing to its easily controllable metabolism and its high reproduction rate.[56] As described in the previous chapter, Kühn's physiological-genetic studies on the formation of eye pigments in the flour moth *Ephestia* in Göttingen, and similar work on *Drosophila* by Beadle and Ephrussi in Paris and in Pasadena, had already yielded remarkable results. However, the decisive biochemical turning point came with the transition to the *Neurospora* system, and it hinged on the fact that the "uncommon farmer" Beadle—as Paul Berg and Maxine Singer call him in their biography, referring to his background in the agricultural sciences[57]—teamed up with the chemist Edward Tatum and was therefore able to effectively combine classical genetic crossing experiments with biochemical analyses. Using X-rays, the team created a whole battery of mold mutants, each of which was no longer able to synthesize an essential amino acid. And by systematically recombining their deficient mutants, they arrived at postulating the existence of chains of enzymes that catalyzed the synthesis of the respective amino acids and whose intermediate products could be characterized chemically. This work was the basis for the far-reaching conclusion that each gene was responsible for the production of one enzyme, which in turn catalyzed one intermediate step in a metabolic cascade, a view that became known as the "one-gene–one-enzyme" hypothesis.

Toward the end of World War I, the bacteriologists Frederick Twort in Great Britain and Félix d'Hérelle in France had for the first time characterized viruslike entities that infested bacteria, multiplied within them, and finally induced the breakdown (lysis) of these bacteria.[58] However, not many other scientists followed up on these studies during the 1920s. Emory Ellis at Caltech did. When Max Delbrück arrived in California on a Rockefeller fellowship in 1937, he spurned Morgan's *Drosophila* lab and was enthralled by Ellis's work on phages. He was impressed by the extremely short multiplication cycle of the phages, the amazingly simple-looking techniques with which they could be visualized, and the possibilities for quantification this system offered via plaque counting and serial dilution. As mentioned in the previous chapter, Delbrück, a physicist by training, had already been thinking about the nature of genes and had contributed to theoretical work on gene mutation while he was still in Berlin. In the 1940s at Caltech, and in collaboration with the Italian physician Salvador Luria, he developed the phage technologies into quantitative procedures. Like virtually everybody at the time, Delbrück assumed that genes were proteins with autocatalytic properties and that phages presented the simplest models of such genes. Delbrück and Luria then demonstrated that phages, like higher organisms, could undergo mutations, and that these mutations were not induced by environmental changes, but rather arose spontaneously in cultures and could be "selected" for by changing the culture conditions. These results linked phage work with classical genetics. Moreover, phage research nurtured the hope that the use of bacterial viruses would lead genetics into the molecular age.

Much has been written about the "phage group" that Delbrück assembled around himself, and the emphasis has often been on the influence Niels Bohr and Erwin Schrödinger exerted on him[59]—Bohr with his ideas about complementarity, and Schrödinger by writing about a heritable "code-script" in the form of an "aperiodic crystal" in his book *What is Life?*[60] Gunther Stent has described the group around Delbrück as the cradle of "informational thinking" in molecular genetics,[61] a point that we will dwell on in the following section. Delbrück was one of the leading theoretical figures of the emerging molecular biology. In retrospect, however, it appears that his technical and organizational innovations, which secured phage its place in the history of molecular biology, were at least as important as his theoretical visions. He and his collaborators brought quantitative techniques to the analysis of virus replication; they standardized the phage system, particularly by means of a voluntary commitment by all members of the informal group to work only with T-phages; and they established a network of long-lasting international collaborations. From 1945, the legendary phage course at Cold Spring Harbor was instrumental in securing the ongoing information exchange among members of the group and the recruitment of newcomers to the field. Delbrück knew how to implement work on the phage system as community building. And yet, the expected breakthrough, the elucidation of the nature of the gene, did not result directly from these efforts.

All the roads taken by genetics in the first half of the twentieth century seem to have crossed at Cold Spring Harbor.[62] In 1904, the philanthropic Carnegie Institution, itself in existence since 1890, established a Station for Experimental Evolution in Cold Spring Harbor. The name reflects, as we have seen in the case of Germany, the idea of a tight connection of genetics with evolution, genetics being seen as the field in which evolutionary processes could be approached experimentally. In 1910, Charles Davenport became director of the Cold Spring Harbor Laboratory. Formerly a biometrician, then a Mendelian, Davenport studied aspects of human heredity and became one of the leading representatives of eugenics in North America.[63] The botanist Blakeslee, whose experiments with radium-induced mutations in *Datura* were mentioned in the previous chapter, came to Cold Spring Harbor in 1915. Milislav Demerec—who had studied agriculture in Austrian Hungary, then trained as a maize geneticist at Cornell, and began experiments with *Drosophila* on Long Island in the early 1920s—followed Blakeslee as director in 1941. Influenced by Delbrück, he then turned to bacterial and phage genetics. Yet he also supported the continuation of Barbara McClintock's protracted work on the cytogenetics of corn at the laboratory. Under his leadership, the Symposia on Quantitative Biology—to which we will come back later—were transformed into an intellectual hot spot of molecular research on heredity in the 1940s. Cold Spring Harbor's changing scientific staff thus reflects what one might call a transmutation of model organisms during the first half of the twentieth century: from man to agriculturally important plants, to *Drosophila*, and finally to bacteria and phage.

The conviction that genes were made of proteins was challenged neither in the field of biochemical *Neurospora* genetics, nor as a consequence of research with TMV, nor in phage genetics. The shift away from what one might call, following Kay, the "protein paradigm" of the gene was initiated by research with bacteria.[64] Only later did phage genetics as well as TMV research come to contribute to the fortification of the ensuing nucleic acid paradigm. Oswald Avery of the Rockefeller Institute is normally credited with the identification of DNA as the hereditary material in experiments with virulent and nonvirulent pneumococci (*Streptococcus pneumoniae*). The work of the group of physicians around him—including Colin MacLeod and Maclyn McCarty—had a medical background: the production of efficient immune sera.[65] All enzymological and biochemical methods of analysis then available at the institute were deployed to this effect. Following earlier work by Frederick Griffith in London, the group distinguished a rough (R) and a smooth (S) form of the pneumococci, which could be converted into each other. They took for granted the typological classification of the bacteria by Fred Neufeld from the Robert Koch Institute in Berlin. At the time, the types were regarded as different species, and the S and R forms as variants within a type. In 1928, Griffith had reported the transformation of a living R form of type I by a killed S form of type II. Avery, however, was convinced of the constancy of the types at that time and was reluctant to accept these results. These findings were not initially discussed in genetic terms, nor was it at all clear by then whether bacteria (with no identifiable nucleus) exhibited patterns of heredity similar to those of higher organisms. The language of bacteriology of the day stood in the tradition of ideas about the metabolism of foodstuffs and its adaptation to external conditions on the one hand and about immunospecificity on the other. These ideas also determined the search for the substance responsible for the bacterial transformation event, and his experiments had caused Griffith to suggest that this substance was sensitive to heat.

Again, it was an in vitro system that was at the heart of Avery's characterization of the transforming substance. He succeeded in breaking up pneumococcal cells in a gentle manner and in separating the cell sap from the cell walls with the help of a filter. A substance that continued to transform intact pneumococci of the nonvirulent form into virulent ones was then precipitated from the filtrate with alcohol. Leaving Griffith behind, Avery now introduced a whole battery of enzymological, chemical, and physical characterization techniques into his bacterial transformation system. Step by step, he narrowed down what kind of substance it might be that produced the transformation. All the indications pointed to deoxyribonucleic acid as the transforming agent, and this finding was finally published in 1944. At this point, Avery was aware of the possible genetic implications of his result, although up to then he had never connected his work with bold genetic hypotheses. The conviction that had guided him in his work was rather that *biological* specificity—in this case, the disease-causing property of a bacterium—ultimately had to rest on *chemical* specificity.

Leading geneticists, biochemists, and biophysicists such as Delbrück, Luria, Beadle, and Pauling did not accord much importance to Avery's report. The idea that DNA, still assumed to have a relatively simple, repetitive structure, was no candidate for the genetic material was deeply entrenched. There were exceptions. But if Avery's experiments heralded a "revolution," as his French colleague André Boivin held early on, then it was a remarkably silent revolution.[66] So, of the nineteen speakers at the 1947 Cold Spring Harbor symposium on nucleic acids, only three mentioned Avery's paper at all. Columbia University's biochemist Erwin Chargaff was one of the few who took up Avery's result, reasoning that if DNA was the genetic material, it had to be specific, and if it was specific, it had to vary in base composition from species to species. Indeed, toward the end of the 1940s, Chargaff showed that different kinds of microbes and lower fungi differed in their DNA composition: the ratio of guanine/cytosine to adenine/thymine varied between species, but the ratio of guanine to cytosine and of adenine to thymine always remained approximately one to one.[67] Chargaff had incidentally used a very simple technique to attain molecular resolution: paper chromatography. In contrast to most of the other new molecular research techniques, paper chromatography required only the toolkit of a modestly equipped chemical laboratory. Nevertheless, it was introduced in the analysis of organic molecules and their components, particularly amino acids, only around 1940.

With the analysis of *Neurospora* mutants, phage mutants, and the transformation of pneumococci, the outlines of a genetics of lower organisms emerged.[68] The field was definitively established toward the end of the 1940s by Beadle's collaborator Edward Tatum, who had moved from Stanford to Yale, and his student Joshua Lederberg. In 1946, they reported experiments that provided evidence for an exchange of genetic material between bacteria. From his time with Beadle at Stanford, Tatum had a collection of mutants of *E. coli* strain K12, each of which was defective in several metabolic reactions. Lederberg cultured pairs of the mutants together in a rich growth medium, harvested them, and then sowed them in a minimal medium in which the mutants were unable to survive. He found that some bacterial colonies nevertheless grew, so the genetic defects had obviously been reversed. Lederberg concluded that recombination had taken place, as he had hypothesized, during "conjugation" events—a mating phenomenon that was known to occur in protozoa. Half a decade later, at the beginning of the 1950s, he and Norton Zinder found, in experiments with salmonellae, that bacteriophages could carry bits and pieces of the bacterial genome from one bacterium to another—the effect was the same as that observed upon conjugation. In the combination of phage and bacterial genetics, classical genetic crossing assumed completely new forms. The first bacterial and phage gene maps were produced over the following decade.[69] Whereas *Drosophila* took about two weeks to reproduce, bacteria could do so within half an hour under optimal conditions. In addition, it was easy to change the composition of the medium and to grow lawns of bacteria on agar in a petri dish (see fig. 7.1). With these techniques, extremely sensitive and at the same time very simple instruments were at hand that could be used to find and identify mutants among millions of cells via phenotypic selection and to create genetically pure colonies—clones, in other words, that could then be used again as genetic tools. The bacterium *Escherichia coli* took pride of place in this context. Between 1945 and 1955, it became the most frequently used model organism in molecular biology and molecular genetics.[70]

*Assemblage*

To recapitulate briefly, the research technologies and the model organisms that were brought together in new experimental systems between 1930 and 1950 contained three different kinds of test-tube environments. First, enzymology, as well as cancer research, gave rise to biochemical in vitro systems that derived their components from homogenized cells. Second, a miniaturization of genetics took place: the focus on unicellular model organisms such as yeast and bacteria, particularly *E. coli*, and on viruses and phages such as TMV, T-phages, and later phage lambda, led to what one might characterize as in vitro genetics. Third, the tube and vessel systems of analytic instruments like the ultracentrifuge provided yet another form of "vitreous" environment. The resulting intersections with biological research materials gave rise to new, more or less molecular representations of biological phenomena and also brought forth a new language to talk about these phenomena.

Many of these developments had taken place independently of one another, even if workers in one area at times were aware of developments in other areas. Genetically motivated questions and high reproduction rates played a role in the choice of smaller and more quickly reproducing model organisms, but they did not dominate all research decisions in all areas of molecular biology. Moreover, the disciplinary expertise of the participating researchers was utterly diverse, even though they increasingly operated as parts of networks. The above-mentioned Symposia on Quantitative Biology at Cold Spring Harbor between 1940 and the early 1950s illustrate these points. Genetically oriented themes were discussed in the following years: 1941 ("Genes and Chromosomes"); 1946 ("Heredity and Variation in Microorganisms"); 1947 ("Nucleic Acids and Nucleoproteins"); 1951 ("Genes and Mutations"); and 1953 ("Viruses"). In the years in between, the symposia dealt, for instance, with "The Relation of Hormones to Development" (1942); "Biological Applications of Tracer Elements" (1948); "Origin and Evolution of Man" (1950); and "The Neuron" (1952). This list gives an impression of the broader context of the molecular, quantitatively oriented biology of which molecular genetics became a component part but which it did not dominate.

As a new language of molecular genetics emerged between 1953 and 1965, so did a new discourse of heredity. The beginning of this epoch is marked by the elucidation of the double-helical structure of DNA by Francis Crick and James Watson, its end by the definitive elucidation of the genetic code by the groups around Marshall Nirenberg and Heinrich Matthaei, Severo Ochoa, and Gobind Khorana. The double-helical model of DNA (fig. 7.2) is paradigmatic of a structurally oriented molecular biology drawing on biophysical research technologies, especially X-ray structure analysis (fig. 7.3). But molecular model building and results from nucleic acid chemistry, such as Chargaff's base-pair rules, also played a role in working out the structure of DNA.[71] The deciphering of the genetic code, on the other hand, is paradigmatic of a functionally oriented, biochemical molecular biology, with its predilection for experimental systems based on isolated cell components. Scientists in this field, which emerged between 1950 and 1960, characterized three classes of ribonucleic acid molecules that played a role in protein synthesis: ribosomal RNA, transfer RNA, and messenger RNA.[72] These nucleic acid types bridged the gap that had until then existed between genes and

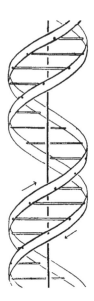

FIGURE 7.2 Model of the molecular structure of deoxyribonucleic acid (DNA). James Watson and Francis Crick, "Molecular Structures of Nucleic Acids: A Structure for Deoxyribose Nucleic Acid," *Nature* 171 (1953), p. 737.

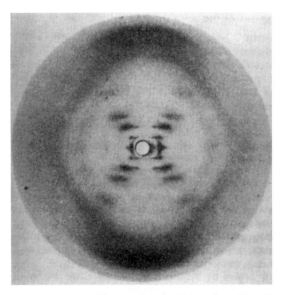

FIGURE 7.3 X-ray diffraction diagram of deoxyribonucleic acid (B-form) Rosalind Franklin and Raymond Gosling, "Molecular Configuration in Sodium Thymonucleate," *Nature* 171 (1953), p. 740.

enzymes, even in the most advanced molecular experimental systems, and they helped to demarcate, within ten years or so, the domain of molecular genetics in the narrow sense of the term.[73]

Crucially, basic genetic events increasingly became defined as a flow of information. The short sentence with which Watson and Crick ended their first paper on the structure of DNA initiated this consequential development: "It has not escaped our notice that the specific pairing we have postulated immediately suggests a possible copying mechanism for the genetic material."[74] A few years later, Crick formulated his "sequence hypothesis" and what came to be known as the "central dogma."[75] According to the sequence hypothesis, the specificity of the genetic material lies in the sequence of its nucleic acid bases, which in turn unambiguously determine the amino acid sequence of a protein synthesized under the auspices of DNA. The relation between the linear sequences of these two classes of macromolecules, one a carrier of information, the other a carrier of function, was conceived of as a "code."[76] At the beginning of the 1960s, it became clear that this code consisted of nonoverlapping triplets of nucleotides that formed a contiguous DNA sequence. Crick's central dogma stated that molecular information could flow from DNA through RNA to proteins; the reverse flow, from proteins to the genetic material, was deemed to be impossible. In the mid-1960s, the "central dogma" became widely seen as the molecular form of Weismann's "legacy" of the self-perpetuation of the germ line.

With that, the inheritance of acquired characters, as propagated by the plant breeder Trofim Lysenko in the Soviet Union during the founding decade of Western molecular genetics—which was decried by Lysenko as "Mendelism-Morganism"—was definitively ruled out of court. It was no coincidence that the Fifth International Congress of Biochemistry took place in Moscow, and when Nirenberg presented his results, amounting to the identification of the first "words" of the genetic code at the congress in August 1961, he dealt the final blow to Lysenkoism in socialist countries. Consequently, the "rehabilitation" of Gregor Mendel in Brno took place a few years later, on the occasion of the centenary of Mendel's lecture on plant hybrids in 1865.

Following the anthropologist of science Paul Rabinow, the events of the late 1950s and early 1960s could be characterized as the formation of a new "assemblage" as far as genetics is concerned. In his *Essays on the Anthropology of Reason*, Rabinow uses Gilles Deleuze's concept of "assemblage" in order to describe those special historical situations that occur from time to time in which knowledge enters a new configuration: constellations of actors, things, and institutions in which something unfolds that, according to Rabinow, "emerges out of a lot of small decisions; decisions that, to be sure, are all conditioned, but not completely predetermined,"[77] or in which, according to another passage, "new forms emerge that have something significant about them, something that catalyzes previously present actors, things, institutions into a new mode of existence, a new assemblage, an assemblage that made things work in a different manner. A manner that made many other things more or less suddenly possible."[78] Using our terminology of experimental cultures, we could reformulate these ideas as follows: The contemporary empirical sciences, especially the life sciences, are founded on experimental systems, a special kind of assemblage for the production of knowledge. From time to time, and usually through the coincidence of incremental decisions rather than on the basis of deliberate revisions, conjunctures occur in or between such systems that from then on not only let things appear in a new light, but also let them happen in a different way.

Following the shift to molecular genetics in Rabinow's sense of "assemblage," nothing in genetics remained quite what it had been in the first half of the twentieth century. The experimental systems changed. And the discourse and key concepts changed. In the context of molecular genetics, a new discourse of biological specificity emerged from which the irreducible biological forces and animate matter of the late eighteenth century had been definitively banished. Attempts to describe hereditary matter in atomistic or energetic terms borrowed from contemporary physics had likewise been dropped. They were replaced by notions such as "genetic program," "storage of information," "processing of information," "replication," "transcription," "translation," and "genetic code." The gene of classical genetics had been materialized, but at the same time its formal function as invisible placeholder for a visible effect was filled with a different interpretation: the gene now functioned as carrier of genetic information. The physical and chemical apprehension of living systems thus paradoxically brought about a conceptualization and terminology that situated itself beyond mechanics and energetics. This happened in an underhand fashion, for the informational view of genetics was prefigured neither in the functional thinking of the Delbrück school, nor in the structural thinking of biophysics, nor in the regulatory thinking of the group around Jacques Monod and François Jacob. Nevertheless, it inspired a new understanding of the organism. Heredity was no longer conceived of as the transmission of bodily characters, but as an information system, as a semiotic universe in its own right.

Molecular biologists of the first generation, such as Monod and Jacob, have stressed the influence that cybernetics, modern communication technology, and the computer sciences had on this new interpretation of the organism.[79] As early as 1948, Norbert Wiener had identified three ages in the conceptualization of the organism in his widely read book *Cybernetics; or, Control and Communication in the Animal and the Machine*: "If the seventeenth and early eighteenth centuries are the age of clocks, and the later eighteenth and the nineteenth centuries constitute the age of steam-engines, the present time is the age of communication and control."[80] To be sure, the early network of cyberneticians was more closely connected with neurobiological research than with genetics.[81] However, in the course of the 1950s, models of genetic regulatory circuits were clearly affected by cybernetic ideas. Efforts to decipher the genetic "code" with the help of the newest available computer technology and cryptologic procedures, which George Gamow, among others, used after he had become acquainted with the DNA double helix, were unsuccessful in the end. Nevertheless, they testify to the influence of digital information processing and control and feedback engineering on biology, and vice versa, that was typical of the first two decades after World War II. How far this influence went and what good it did for molecular biology has for quite some time been a matter of debate among historians and philosophers of biology.[82]

Classical genetics developed in the context of a larger discourse that essentially centered around questions of eugenics, racial identity, and sexuality—in short, a biopolitics of what came to be called the "racial body" (*Volkskörper*) in German racial ideology. This biopolitics culminated in World War II and the Holocaust. Although the early promoters of molecular genetics still had one foot in the discourse of eugenics, the new questions they raised appeared rather marginal with respect to that discourse or were even located outside it. A whole postwar generation of biologists understood molecular genetics as a new, international and interdisciplinary, outsider discourse that was as far away from a society-wide implementation of human genetics for eugenic ends as it was from the entanglement of physics with atomic politics. In the end, the DNA double helix gained acceptance as an icon for the heritage of the history of humankind,[83] and the old racial discourse tended to be replaced by narratives on the common origin of mankind in Africa.[84]

This is not at all to say that the discourse associated with molecular genetics was devoid of political salience. It incorporated—often explicitly—a biological determinism that Western scientists directed not only against Soviet Lysenkoist biology, but also against social ideals in their own countries, such as, for instance, the establishment of equal opportunity in education. So, during a CIBA symposium on "Man and His Future" in London in 1962, geneticists of the classical period, as represented by Hermann J. Muller, exchanged ideas with molecular geneticists of the younger generation, represented by Joshua Lederberg, about how mankind might in the future take control of its own evolution by means of modern genetics.[85] The symposium caused a public stir. A greater part of public attention and force of persuasion, however, accrued to the hereditary discourse of molecular genetics as a result of medium- and long-term hopes for the detection of the molecular bases of diseases. In 1949, Linus Pauling had already articulated a vision of "molecular medicine" with his research into sickle-cell anemia.[86] The identification of the underlying amino acid change in hemoglobin meant that sickle-cell anemia could be regarded as a "molecular disease" caused by a condition that was long known to be inherited in a Mendelian fashion. Future remedy would come through genetic repair, so went the hope. We will take up these hopes and promises in the next chapter. Visions of this kind also motivated contemporary American and European science policies to help molecular biology in general, and molecular genetics in particular, to establish themselves as academic specialties—if only for the span of a generation, as it would turn out.

# 8 Gene Technology, Genomics, Postgenomics: Attempt at an Outlook

It is always difficult for historians to judge contemporary trends. Yet in the last three decades of the twentieth century, molecular biology has patently undergone another transformation that has again deeply affected the practice of molecular genetics and how its participants and their customers relate to questions of heredity. In this chapter we trace the contours of the development of hereditarian thinking and intervention in the age of genomics, postgenomics, and molecular medicine. We also characterize concomitant reconfigurations in the relationship between academic research and industrial biomedical technology. To use the words of historian Eric Vettel, our aim is to investigate "how something so complex as the biotechnology industry was born, and how it became both a vanguard for contemporary world capitalism and a focal point for polemic ethical debate."[1] This massive cultural transformation stands as visibly before our eyes as it appears to be indeterminate in its future development.

We begin this chapter with a short description of what, following Gaston Bachelard, we could call a "phenomenotechnical" gestalt switch: a rapid, but lasting, change in the understanding of heredity as an object defined by technologies allowing for its direct manipulation.[2] As a result of this switch, the process of heredity came to be understood basically as a copying procedure that could be technically manipulated. The ground for this unprecedented shift, which took place between 1970 and 1980, was prepared by the purification of two classes of cellular enzymes: those that guide the duplication of the hereditary material and its transcription in the living cell, and those that cut and splice nucleic acids. Specifically, DNA polymerases—purified in the late 1950s by Arthur Kornberg at Washington University and later at Stanford University—copy DNA by synthesizing a strand complementary to each of the two strands of a double helix. RNA polymerases transcribe DNA into RNA and provide the cell with various RNA molecules, including messenger RNAs for protein synthesis. Some RNA polymerases were identified as early as 1960, but the purification of these enzymes in an active form proved to be a laborious and demanding task. In 1970, Howard Temin in Madison and David Baltimore in Boston additionally described a viral polymerase that went the other way round: it could transcribe RNA molecules into DNA molecules. As a consequence, the central molecular genetic dogma—"DNA makes RNA, RNA makes protein"—had to be revised.[3] Not only was DNA transcribed into RNA, but the reverse was also possible, and the enzyme that facilitated this process became known as "reverse transcriptase." Toward the end of the 1960s, Daniel Nathans in New York, Hamilton Smith in Baltimore, and Werner Arber in Basel characterized the first so-called restriction endonucleases. Proteins in this class of bacterial enzymes recognize short, well-defined double-stranded sequences of foreign DNA, and they cut DNA strands specifically at these recognition sites. In parallel, other bacterial enzymes were identified that could join together fragments of DNA: the DNA ligases.

These various enzymes could also be employed as molecular tools. They made it possible to create hybrid or, as they came to be known, "recombinant" nucleic acid molecules. Relatively small, circular DNA molecules, or plasmids, that were isolated from bacterial cells proved to be particularly suitable as vectors for transporting any arbitrary piece of genetic information into a bacterium, where it multiplied.[4] The aforementioned enzymes now provided a basis for the manipulation of hereditary material in the test tube as well as in bacterial cells, and the techniques were also soon adapted for DNA manipulation in the cells of higher organisms. Taken together, this molecular panoply raised the possibility of developing a gene *technology* in the strict sense of the term, a molecular *engineering* science that used molecular entities like polymerases, restriction enzymes, ligases, and plasmids as instruments.[5] This second generation of molecular biological techniques no longer relied on heavy analytic equipment built on physical principles. The "scissors" and the "needles" for cutting and splicing genes, as well as the vectors in which they were incorporated and with which they were transported, now themselves had macromolecular dimensions. The molecular entities represented what one might call a "soft" or "wet" technology, a molecular toolbox that made it possible to synthesize nucleic acids in the test tube and to incorporate the products into living cells.[6]

In the course of the 1970s, gene technology advanced in both these directions. In 1972, Paul Berg at Stanford University succeeded in creating a hybrid DNA molecule from two viruses: phage lambda and the infectious ape virus SV40.[7] A year later, Stanley Cohen, also at Stanford, and Herbert Boyer in San Francisco combined their work on plasmids and restriction enzymes, respectively, and found a way of inserting a piece of viral DNA into a plasmid from a bacterium in vitro. They then channeled the plasmid back into bacterial cells, where it replicated "naturally." Following these experiments, it became possible to clone whole genes in this way, a feat that provided the starting point for genetic bioengineering and for the industrialization of gene technology. These experimental results also unleashed a fierce debate about the possible risks of creating harmful recombinant organisms and the concomitant necessity for regulation. We will come back to these topics in more detail in the following section. Here we note that, given the possibility of deploying the molecular genetic reproduction machinery of the cell for the multiplication of deliberately constructed nucleic acid templates, molecular biologists left behind the working paradigm of classical biophysicists, biochemists, and geneticists and became gene technologists or genetic engineers. They no longer created test-tube conditions under which the molecules of the organism and their reaction sequences assumed the status of objects of scientific investigation. They turned the tables: molecular technologists now assembled molecules that carried genetic information and used the milieu of the cell as an appropriate technical medium for the reproduction of these molecules, for expressing them, and for investigating the effects of their products. Thus the organism itself was transformed into a laboratory, a *locus technicus*. Henceforth, the question was no longer how to represent intracellular structures and processes for studying them outside the cell, but how to realize extracellular projects within the cell. As Waclaw Szybalski, a contemporary observer and himself an oncologist at the McArdle Laboratory in Madison, wrote on the occasion of the Nobel Prize awarded to Werner Arber, Hamilton Smith, and Daniel Nathans, "The work on restriction nucleases not only permits us easily to construct recombinant DNA molecules and to analyze individual genes but also has led us into the new era of 'synthetic biology' where not only existing genes are described and analyzed but also new gene arrangements can be constructed and evaluated."[8] Sheila Jasanoff has drawn attention to the consequences of this avenue of research by claiming that this form of biotechnology "brings new entities into the world and through that process reorders our sense of rightness in both nature and society."[9]

Reverse transcriptase was of central importance for the fabrication of proteins from higher organisms—including those of humans—by means of gene technology. While attempting to express the genes of eukaryotes in bacterial cells in the late 1970s, Philip Allen Sharp and Richard Roberts surprised the community of molecular geneticists by reporting that the genes they studied were composed of stretches of DNA that coded for amino acid sequences in the corresponding protein (exons), but that these segments of DNA were interrupted by noncoding DNA sequences (introns). The primary RNA transcript was split and spliced to yield a functional messenger RNA composed of exons alone.[10] As this feature of gene expression does not exist in bacteria, the genes of higher organisms could not be expressed in bacterial hosts. The genes of higher organisms were therefore unsuitable as a starting point for genetic engineering. Reverse transcriptase made it possible to transcribe spliced eukaryotic messenger RNA into a complementary DNA (cDNA), which could then be inserted into a plasmid, proliferated, and expressed in bacteria—the "reactors" of molecular biotechnology. Subsequently, whole cDNA libraries of eukaryotic genes were created in this way.

Let us mention a few more enzymatic techniques developed in the late 1970s and early 1980s that proved to be crucial additions to the arsenal of gene technologies. In 1977, Frederick Sanger from Cambridge reported a procedure for sequencing DNA that relied on DNA polymerase. In contrast to the step-by-step degrading procedure developed concurrently by Allan Maxam and Walter Gilbert at Harvard, Sanger's procedure involved the stepwise, randomly interrupted prolongation of a DNA sequence. The resulting DNA fragments were then separated electrophoretically, and the final base in each fragment was identified (fig. 8.1). With this technique one could, in principle, sequence the DNA genome of a virus such as bacteriophage phiX-174, which spans more than five thousand nucleotides, in a few days. When Robert Holley embarked on a project to determine the sequence of the first transfer RNA in 1958, it took him eight years and the work of a whole research team to elucidate the primary structure of a molecule of some eighty nucleotides in length. Now the sequencing of gene fragments, whole genes, and

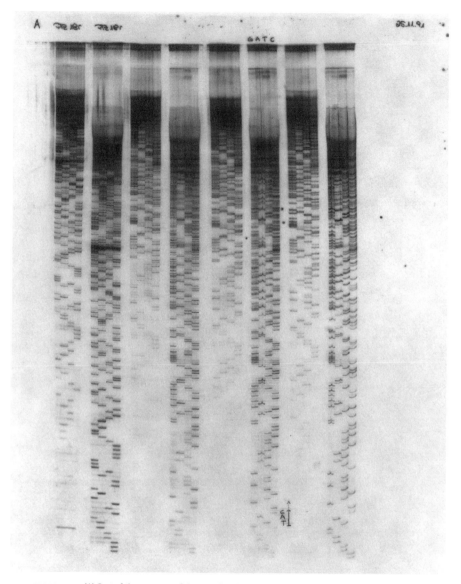

FIGURE 8.1 (A) Part of the sequence of the gene for ribosomal protein L1 in *Halobacterium halobium*.

even gene complexes became possible and was soon routine in molecular genetic laboratories.

Further, the construction of physical maps of chromosomes through a combination of fragmentation by restriction enzymes, amplification of the fragments in plasmids, and partial sequence analysis also became

```
POLYLINKER
SELECTION
   #resistance  Ap
   #indicator beta-galactosidase
SUMMARY  pTZ18R    #length 2871    #checksum 6457
SEQUENCE

Ptz18h11.Seq  Length: 3507  November 29, 1991  17:22 Check: 7065
..

     1 CCCATTCGCC ATTCAGGCTG CGCAACTGTT GGGAAGGGCG ATCGGTGCGG
    51 GCCTCTTCGC TATTACGCCA GCTGGCGAAA GGGGGATGTG CTGCAAGGCG
   101 ATTAAGTTGG GTAACGCCAG GGTTTTCCCA GTCACGACGT TGTAAAACGA
   151 CGGCCAGTGC CAAGCTTGCA TGCCTGCAGG TCGACTCTAG AG
                                                     >SEQED
           (include) of: L1HAHAECOBAM.REV check: 8386 from: 1
           to: 657>
                                                     GATCCTGC
   201 AGCTAGCAGT AGGCAACCTC CACGGCAGGC CCCATCGTTG TCTTCACGTA
   251 GACGGAGTCC ACGTTCAGCG GGCCTTTTTC GAGGTTCGCG TGCAGCCGAC
   301 GCATGATGAC GTCGATGTTG CTGGCGATGT CCTCGGCGGA CATGTCCTCC
   351 GCGCCGACGC GCGTGTGGAA CGTGCGGCGG TCGCGGCTGC GGATCTGCAC
   401 GGTGTTTTTC ATGCGGTTGA CTGTGTCGAC GACGTCGTCG TCGGGCTGGA
   451 GCGGGGTCGG CATTTTCCCG CGCGGACCAA GCACTTGACC GAGCGCACCC
   501 GCGATGTCCT GCATCATGGG TGCTTCCGCC ACGAAGAAGT CCGTCTCGTC
   551 TGCGAGATCC TTCGCGGCGT CGGTGTCGTC TGCGAGGTCG CTGAGGTCGT
   601 CCTCGTCGAG GACGTCGTCA GCGACGTCGT CCGCGCGAAC CGCGGTTTCG
   651 CCGTCTGCGA AAACCACGAT CTGCGTCTCC TGTCCGGTGC CCGACGGCAG
   701 CACGACGCCC TCGTCGACTC GTTGCGACGG GTCGTTGAGG TCGAGGTCGC
   751 GCAGGTTGAC TGCGAGGTCT ACCGTCTCAC GGAAGTTCCG CTGTGGGGCA
   801 TCCTCAAGTG CGCGAGCTAC GGCCTCTTCT ATATCGTTGT CTGCCA
                                                    <...►>SEQ
           ED : start codon for HL1 going to the left>
                                                     TGG
                                                       <
           SEQED (include) of: L1HAHAECOBAM.REV check: 8386
           from: 1 to: 657<
                                                     A
   851 ATTCCCTATA GTGAGTCGTA TTAAATTCGT AATCATGGTC ATAGCTGTTT
   901 CCTGTGTGAA ATTGTTATCC GCTCACAATT CCACACAACA TACGAGCCGG
   951 AAGCATAAAG TGTAAAGCCT GGGGTGCCTA ATGAGTGAGC TAACTCACAT
  1001 TAATTGCGTT GCGCTCACTG CCCGCTTTCC AGTCGGGAAA CCTGTCGTGC
  1051 CAGCTGCATT AATGAATCGG CCAACGCGCG GGGAGAGGCG GTTTGCGTAT
  1101 TGGGCGCTCT TCCGCTTCCT CGCTCACTGA CTCGCTGCGC TCGGTCGTTC
```

**FIGURE 8.1** (B) Computer printout of the letter sequence corresponding to fig. 8.1 (A). (A) and (B) are with kind permission from François Franceschi, Max Planck Institute for Molecular Genetics, Berlin.

feasible. In 1978, David Botstein from the Massachusetts Institute of Technology introduced the use of restriction enzymes to characterize human DNA in terms of restriction fragment length polymorphism (RFLP). Finally, in the early 1980s, Kary Mullis, a scientist at Cetus Corporation, developed the polymerase chain reaction, or PCR for short. With this procedure, the tiniest amounts of DNA could be amplified in a cyclical fashion in vitro with the help of a heat-stable DNA polymerase. Cetus Corporation, a biotech firm avant la lettre founded in 1971, patented the PCR technology, which became emblematic of the new molecular biotechnological turn of the 1980s, and sold the patent to Hoffman-La Roche in 1991 for no less than \$300 million.[11]

Further methods were added to this instrumentarium of enzymes and vectors during the late 1970s and 1980s, particularly techniques for producing point mutations in genes and for making large molecular vectors such as yeast artificial chromosomes (YACs) as well as elegant procedures for the technical synthesis of well-defined nucleic acid sequences and an electrophoretic technique for the separation of large DNA fragments. The textual metaphors that accompanied the rise of molecular biology now came to be materialized in these technologies: reading as DNA analysis; writing as DNA synthesis; copying as the polymerase chain reaction; and editing as procedures for mutating genes. As Kaushik Sunder Rajan has remarked, "Genomics allows the metaphor of life-as-information to become material reality that can be commodified."[12] We could say that on the one hand, bioinformation processing molecules became technical tools, and on the other hand, technical instruments came to embody those very processes of handling bioinformation.

Within the span of a decade, these procedures were adapted for automation. In June 1986, Leroy Hood and Lloyd Smith from the California Institute of Technology announced the development of the first automated DNA sequencing machine. Soon these machines populated most molecular biological and biomedical laboratories, together with the reagents for gene technology, which could now be obtained in handy packages; standardized "kits" of these materials were increasingly supplied by the biochemical and pharmaceutical industries.[13] On the occasion of the fortieth anniversary of the elucidation of the double-helical structure of DNA, David Jackson, a student of James Watson who worked for DuPont Merck and can be counted as a representative of the new generation of industrial gene technologists, characterized the situation as follows:

> I would argue that the ability to read, to write, and edit DNA is functionally unprecedented in human history. All we have ever been able to do before is to select among the various combinations of genes that the mechanisms of genetics have presented to us. And, while we have developed very powerful and very sophisticated selection procedures, selecting from among a set of

> alternatives over which one has almost no control is fundamentally different from being able to write and edit one's own text.

And fundamentally different from making an unlimited number of copies of one's own text, one feels tempted to add. Jackson continued, "The ability to write and edit DNA is the basis for a synthetic and a creative capability in biology that has not previously existed."[14]

A century of in vitro research had thus created the preconditions for biology to reenter the cell and the organism. Rather than extracting cell components and representing cellular processes in test-tube environments, researchers took components from outside the cell and incorporated and expressed them inside the cell—"synthetically" as well as "creatively." On the one hand, the mechanism of heredity was thus converted into a biotechnological copying process that allowed one to make medically relevant products; on the other hand, directed interventions in the gene inventory of an organism could be used not only to alter that particular organism, but also to reprogram its descendants.

*Political and Economic Reconfigurations*

The situation of the field of molecular genetics changed decisively as a consequence of the genetic engineering options that showed itself on the horizon at the beginning of the 1970s. Initially, the prospect of altering the hereditary stock of organisms by means of gene technology appeared to be restricted to viruses and microorganisms. However, as early as 1974, Rudolf Jaenisch at the Salk Institute in La Jolla managed to stably integrate "foreign" DNA into the genome of mouse embryos. The targeted modification of the "hereditary repository"of higher organisms, including humans, and a new manipulative genetics appeared on the horizon. (Interestingly, in German, the corresponding hereditarian conception of species is expressed in the general notion of *Erbgut*, a term referring both to a specific genome and to an inherited property, in particular, a farm that was traditionally bequeathed to the eldest son in agricultural families.)

This state of affairs stimulated lively debate about the potential risks of gene technology.[15] At first, the debate was conducted by scientists who were themselves involved in establishing the new field. Their manifestos of warning continued a tradition established by physicists following World War II. Foremost among the activists was Paul Berg, who, together with ten colleagues from the field of molecular genetics, called for a voluntary worldwide moratorium on gene technology until binding rules to govern this area of research were in place. Berg declared that he was ready to voluntarily suspend his own work on the recombination of viral and bacterial DNA for the time of such a moratorium. At a conference convened by Berg and his colleagues at Asilomar in 1975, more than a hundred scientists working in related areas then called for the public regulation of potential biohazards. Following this appeal, the National Institutes of Health (NIH) convened a Recombinant DNA Advisory Committee, and a year later its first rules were implemented. These rules became a model for regulation in many other countries.[16] On the one hand, the new rules emphasized the creation and provision of bacterial strains that were not viable outside a laboratory environment. On the other hand, the rules established graded prescriptions for laboratory security. They concerned the handling of genetically modified microbes, which, it was feared, could escape from the laboratory, spread epidemically as new pathogens, and cause unforeseeable damage. In their efforts to regulate a technology *in statu nascendi*, the scientists involved moved uneasily between a sense of social responsibility, their own scientific and potentially economic interest in the unimpeded search for and exploitation of results, and the management of fears circulating in the population.[17] In Germany, corresponding regulations were implemented in 1978, then replaced by a gene technology law in 1990 following a protracted public and parliamentary debate. Legal regulations were also created in other European countries, but not in the United States.

In 1976, the very year in which NIH issued its rules for handling gene technology, Herbert Boyer and the venture capitalist Robert Swanson founded Genentech, the first gene technology firm proper. This set in motion a development that unsettled the biological research regime of the postwar period and shifted it from a molecular biology that understood itself as a pure and basic science to applied research.[18] Actually, the distinction between basic and applied research was itself called into question in the course of this development. Here it must suffice to mention only a few milestones in the growing market orientation and commercialization of gene technology: Genentech reported the successful cloning of the first human protein gene, for somatostatin, one year after the formation of the company. In 1978, the firm announced that it had successfully expressed the human gene for insulin in bacteria. Five years later, insulin produced in this manner was on the market. Genentech went public in 1980. The so-called Bayh-Dole Act subsequently facilitated exchange of the products of gene technology between academic research sites and industry. That same year, the Supreme Court ruled in favor of Ananda Chakrabarty, a biochemist from General Electric, and allowed him to patent a genetically altered, oil-degrading microorganism from the *Pseudomonas* group of bacteria.[19] With this ruling, a taboo was overturned that had so far precluded the patenting of life forms in the United States. Three years later, Stanford University applied for a patent for a stable expression vector that could be used for multiple cloning purposes and that was suitable for use with *E. coli* systems.[20] Now American universities more generally, above all in California, joined the patenting bandwagon. Also in 1983, the first U.S. patents for genetically altered plants were granted. One year later, in 1984, Stanford received the first *product* patent for prokaryotic DNA—again a step beyond traditional patenting practice, which was oriented toward protecting invented *processes*. In 1985, outdoor experiments with so-called transgenic plants began; genes that conferred resistance against insects, viruses, and bacteria had been inserted into their genomes. Ten years after the founding of Genentech, the number of comparable start-up businesses in the United States had risen to about four hundred. In 1988, Harvard scientists Philip Leder and Timothy Stewart received the first patent for a genetically altered mammal, the so-called oncomouse. By 1990, Stanford and the University of California in San Francisco—a private and a publicly funded institution, respectively—held on the order of a hundred recombinant DNA patents, and the licensing of patents itself started to become a profitable business.

During the 1980s, the debate about the risks associated with genetically modified microbes, particularly as a result of uncontrolled proliferation, faded into the background. It gave way to a broader discussion encompassing ethical, legal, social, and cultural perspectives on the consequences of what was increasingly perceived as a sea change in the technologies for manipulating genetically relevant molecules, and with that, living beings, including human beings.[21] Topics under discussion included the prospect of a molecularized human medicine guided by the genetic makeup of the individual; the possibility of somatic and even germ line gene therapy; the diagnosis of a growing number of genetically determined or genetically influenced diseases, and concomitantly, the right to know or not to know about such risks; changes in reproductive behavior as a consequence of the availability of new reproductive technologies such as cloning, including the production of tailor-made embryonic stem cells for therapeutic or even reproductive purposes; and finally, controversies around genetically manipulated foodstuffs and renewable biological raw materials, including the ecological, economic, and potential health risks related to their cultivation and consumption. Since then, debate in these areas has become perpetual because none of the implicated factors, regardless of whether they are political, economic, social, cultural, technical, or scientific, are unambiguous. They are constantly rearticulated and reconnected in context-dependent ways. Herbert Gott-weis therefore speaks of a "micropolitics of boundary drawing."[22] With Sunder Rajan, we could also summarize thus: "The beginning of the bio-technology industry in the late 1970s and early 1980s was . . . marked by a coproduction of new types of science and technology and changes in the legal, regulatory, and market structures that organized the conduct of that technoscience."[23]

The politics of boundary drawing involved several objects of negotia-tion and several negotiating parties at the same time, and it was, on the whole, innovation-driven. At stake was, on the one hand, a reconfigura-tion of the relationship between basic and application-oriented research. This reconfiguration found expression in the forging of new alliances be-tween commercial and academic interests, not only between the conflict-ing priorities of science and industry at large, but also, within industry, between established pharmaceutical corporations and the new biotech firms, and within universities and academic institutions, between estab-lished disciplines that sustained and were sustained by particular moral codes on the one hand and emerging areas of research on the other. In addition, questions were raised about how to assess the potential dan-gers and risks connected with medical and agricultural applications of gene technology as well as how to draw social and political boundaries for such applications—for example, by regulating access to individual genetic information or regulating the ownership of gene-derived prod-ucts and genes per se. Finally, ethical boundaries came into play. These boundaries become particularly conspicuous in places where the ma-nipulation of human cells eclipsed the use of model organisms altogether and thus subjected human-derived biological material to genetic manip-ulation. Within discursive networks of this complexity, relatively small epistemic events can trigger major shifts; emergent nodes can reorganize the network, but prominent issues can also collapse and dissipate. A good example of the latter process is the debate about the risk of biohaz-ards associated with bacterial gene manipulation in the 1970s.

In the span of a single decade, gene technology and its discussion as a key technology of the future thus generated a powerful economic, politi-cal, and cultural dynamic.[24] The Asilomar conference had already given expression to a massive reconfiguration of academic and economic in-terests in an area of research that, until then, had not lent itself to profit-able exploitation. But biotechnological possibilities and options also cast new light on the hereditary process. With the increasing production of enzymes and other active substances by means of gene technology, the cellular mechanisms of heredity—the duplication and multiplication of the hereditary substance in particular—were transformed into a biological production method under technical control. Bacterial cultures were the reactors for this production process. Like phages, which redirect bacterial cells to produce viral components, genetic engineers now used bacterial cells as machinery for copying genes and as factories for making proteins.

*Genome Research*

In 1985, ten years after the first Asilomar conference, Robert Sinsheimer—then president of the University of California, Santa Cruz—organized a meeting at which leading molecular biologists considered the feasibility of sequencing the whole human genome. Concurrently, Charles DeLisi and David Smith from the U.S. Department of Energy made plans for a similar meeting that would bring together relevant experts at the Los Alamos National Laboratory. These activities unleashed a controversy that split the community of molecular biologists in two.[25] Sydney Brenner at the University of Cambridge, for example, seized on the American initiative and called for European involvement in sequencing the human genome. Others, among them David Botstein, then at MIT, were skeptical of such initiatives for three reasons. First, many biologists, regardless of whether they worked in academic or in industrial laboratories, felt uneasy about the involvement of the Department of Energy (DOE), which had previously conducted large-scale military projects staffed by physicists. Botstein reportedly referred to the human genome initiative as a "program for unemployed bomb-makers."[26] Second, many molecular biologists were concerned about the sheer size of the project. They felt that the work carried out in biological laboratories was incompatible with such a mammoth global enterprise and that the bureaucracy required for its coordination would be too cumbersome and time-consuming. Third, many experimental molecular biologists could not imagine subordinating their genuine research agendas to a large-scale project that would not, in all likelihood, generate research questions that were scientifically interesting. The debate reached a first peak during the Cold Spring Harbor symposium of June 1986, whose theme was "The Molecular Biology of *Homo sapiens*."

Despite these points of contention, the human genome initiative gained momentum. In 1987, the DOE began to support a pilot project. The National Institutes of Health countered by founding its own Office of Human Genome Research in 1988, and Noble laureate and enfant terrible of molecular biology James Watson, then director of the Cold Spring Harbor Laboratory, was enlisted as leader of what soon became known as the Human Genome Project. One of his first undertakings was to devote 3 percent of the total funding for the initiative to the investigation of the ethical, legal, and social implications (ELSI) arising from the mapping and sequencing of the human genome. In that same year, the international Human Genome Organization (HUGO) was brought into being. Not incidentally, Victor McKusick, founder of the Mendelian Inheritance in Man database, became its first president. In 1989, NIH's Office of Human Genome Research was promoted to become the National Center for Human Genome Research (NCHGR). On October 1, 1990, NIH and DOE jointly announced the official start of a sequencing project, which was envisioned to last fifteen years and was to be implemented in five-year phases. At the same time, sequencing efforts for four model organisms, *Mycoplasma capricolum* (one of the smallest existing microbes), the intestinal bacterium *Escherichia coli*, the roundworm *Caenorhabditis elegans*, and baker's yeast *Saccharomyces cerevisiae*, were included as additional strands in the project.[27]

Hence the Human Genome Project had gotten under way.[28] It soon grew into a broad international scientific collaboration. National institutes with the objective of pursuing human genome research were founded in several countries. In France, for example, Généthon was founded in 1990 with support from a private patient organization, the "French Association against Myopathies" (AFM). Under the directorship of Daniel Cohen, Généthon constructed a physical map that spanned the human genome in 1994.[29] In the United States, Craig Venter left NIH after a controversy about the patenting of DNA fragments derived from brain tissue in 1992 and established a nonprofit enterprise, The Institute for Genomic Research (TIGR), which published the first complete sequence of a free-living organism—*Haemophilus influenzae*—in 1995.[30] James Watson resigned from his post at the NCHGR following the patenting dispute in 1992; Francis Collins became its new director—and Venter's rival for the years to come. In 1993, the Wellcome Trust and the Medical Research Council in Great Britain founded the Sanger Centre near Cambridge.[31] Consortia for sequencing the genomes of further model organisms were created and appended to the project. In Japan, a rice genome project took off in 1991, and the European Community launched a project for sequencing the genome of yeast in 1992.

During the 1990s, technical advances accelerated genome research to an extent nobody had foreseen: new computer algorithms for DNA sequence comparison were developed; new strategies for physical mapping and sequencing, such as "sequence tagged sites" (STS) and "expressed sequence tags" (ESTs), were quickly adopted; the first DNA chips came on the market in 1996; and capillary DNA sequencing machines started to become available a year later. In 1996, the European yeast consortium completed the genome sequence of *Saccharomyces cerevisiae*. The sequence of *E. coli* followed in 1997. In 1998, Venter founded the commercial firm Celera Genomics and signaled his intention to sequence the complete human genome for only $300 million within three years by means of a strategy that had never been used on this scale—the "whole-genome shotgun" method.[32] John Sulston, director of the Sanger Centre, reported an essentially complete genome sequence of *C. elegans* at the end of 1998, Venter that of *Drosophila melanogaster* in March 2000, and in the same year, an international consortium published a genome sequence for the plant model organism *Arabidopsis thaliana*. In June 2000, Collins and Venter, representing NIH and Celera Genomics, and thus publicly funded and private research respectively, announced what was called a "draft" sequence of the human genome during a ceremony at the White House in Washington. In 2001, the preliminary results of the international Human Genome Project and Celera Genomics were published in the journals *Nature* and *Science*. Competition with a genomics company provided such a boost to the Human Genome Project that the latter could be declared complete, at least in preliminary form, prematurely yet auspiciously in 2003, on the occasion of the fiftieth anniversary of the publication of Watson and Crick's paper on the DNA double-helical structure.[33] At the time of this writing, the database of the European Bioinformatics Institute at the European Molecular Biology Laboratory displays the complete sequences of 796 phages, 2,601 viruses, 119 archaebacteria, 1,705 bacteria, and 144 eukaryotes.[34] All of these lists grow on a weekly basis.

The initiative to sequence the human genome provided the backdrop for a multiplicity of sequencing projects, and genomics now forms an integral part of molecular genetics. Apart from unprecedented forms of cooperation, genome research has also given rise to new forms of competition between universities, publicly financed research institutes, commercial biotechnology companies, and privately financed laboratories. In the research landscape thus shaped, scientific and commercial interests have increasingly become indistinct. Moreover, novel research organizations have come onstage, such as laboratories financed by patient interest groups. The creation of Généthon in France illustrates the intervention of a nongovernmental organization in the international research process.[35] With genome research, a new form of life science emerged that could be characterized as "big science," even though it was in no way modeled on the forms of big science in physics, which were typically organized around a very large scientific instrument. The patchwork character of this "distributed" big science generated new challenges, but the innovation-friendly nature of networks also created opportunities. Without being predictable in any particular performance dimension, new technologies could emerge anywhere in the network, prove themselves locally, and spread in capillary fashion. This was a decisive feature, since the technologies necessary for the completion of the Human Genome Project simply did not exist at the outset; they were developed en route. Genome research thus generated not only scientific products, but also the means for their production. In this respect, Charles Cantor, then a molecular geneticist at DOE and one of the inventors of pulse-field gel electrophoresis, remarked at the beginning of the 1990s, "None of these methods existed ten years ago"—and added immediately, "It would be extremely presumptuous to assume that any of them will necessarily be in common use ten years from now."[36]

The most significant challenges arose in the scientific coordination and in the social organization of genome research, on the one hand, and in the coordination and synchronized evaluation of the generated DNA sequence data on the other.[37] The latter feat of coordination depended on the development of bioinformatic databases and algorithmic instruments that made it possible to feed newly generated data into databases continuously, that rendered genomic data retrievable and manipulable, and that made data comparisons feasible. Databases—first for protein sequences, then for DNA sequences—had been part and parcel of the development of molecular biology,[38] but the explosion of molecular data associated with genomics necessitated the development of novel ordering structures for data.[39] In genomics, the development of new instruments and technical procedures was intimately coupled to the development of electronic data processing.

The data-driven nature of genome research also amplified questions about data delivery and access. The basic conflict arose where noncommercial and commercial institutions jointly carried out research projects. The latter had an interest in the commercial protection and exploitation of data; these interests clashed with the disclosure of data in a continuous and ad hoc fashion, as was typical for noncommercial research. But questions of scientific priority, of authorship, of control, of liability, and about the possibility of localizing sources of error were equally pertinent. Which data should be accessible, when, and for whom? These issues connected scientific communication in genomics with a broader debate about media technologies in the second half of the twentieth century.[40] While that debate extends far beyond the realm of science, genomics has been, and continues to be, one of the central scientific arenas in which questions of this kind are being raised.

Possibly the most remarkable surprise associated with genome research concerns the number of human genes. While it had long been taken for granted that humans harbor on the order of a hundred thousand genes, in 2001 this number was revised downward to something between twenty thousand and thirty thousand. Hence the number of genes in humans appeared to be more in line with that in other organisms—particularly the higher organisms used as model organisms—than had commonly been assumed. Not only the figure was surprising, however. In parallel, since at least the late 1970s, evidence had accumulated bit by bit that gene expression in higher organisms was much more complex than previously thought. The parts of many coding DNA sequences were observably processed and recombined in the cell to yield a variety of functional products. The French molecular biologist François Gros had recognized this trend already at the end of the 1980s by noting that the gene concept had "exploded."[41] Coding sequences were no longer understood as determinants of defined characters—"genes for"—but as resources deployed according to the developmental and metabolic situation of the organism.[42] The contours of the key concept and object of twentieth-century biology, the gene, which had been firmly established as a formal entity with the rise of classical genetics at the beginning of the century, and which was so neatly materialized in early molecular biology, became more and more fuzzy.[43] While "the gene" had been conjured up again and again at the outset of the Human Genome Project, a language of postgenomics and epigenetics took over as human genome sequencing reached its high point. Since then, the gene has lost pride of place in contemporary biology, a point we will come back to at the end of the chapter.

Finally, the discourse associated with genome research raised challenging questions about individual and ethnic identity. Comparative human genome research—facilitated by ever more powerful DNA sequencing techniques—has made it plain that human genetics has always been a hybrid discourse. Apart from the inheritance of individual characteristics, human genetics has always concerned itself with the ascription of individuals to groups and with the origin of those groups as well. Let us illustrate these claims with three examples. In 1998, the Icelandic parliament passed a law that authorized the company deCODE Genetics, founded in 1996 by geneticist Kári Stefánsson, to develop a database that would bring together established national registers of disease data—in other words, data about the whole Icelandic population—with genome sequencing data.[44] The project set off intense debates about the legitimacy of creating such a database containing private health information—particularly under a commercial umbrella.[45] Anthropologist Gísli Pálsson and historian of science Mike Fortun have documented and critically assessed how genomic and more conventional medical data here became amalgamated in a way that implicates the identity politics of the whole Icelandic population, with all the associated political as well as economic, ethnic as well as epidemiological, public as well as private connotations.[46] After a protracted controversy, the Supreme Court of Iceland stopped the planned data bank project at the end of 2003. However, in the view of its promoter Stefánsson, the kind of computerized population genetics that deCODE had set out to develop remains indispensable for a modern, up-to-date system of disease diagnosis and prevention; he predicts that, in ten to fifteen years, every society will have such a database at its command. Meanwhile, Estonia has launched a "Gene Bank," and so has Mexico; Great Britain has established a steadily growing "Biobank," and the list keeps growing.

Another project that touched on sensitive identity issues, perhaps even more than its scientific proponents ever imagined, was the Human Genome Diversity Project, which was conceived by the population geneticist Luca Cavalli-Sforza at Stanford University during the 1990s. Although the project merely aimed to explore the genetic diversity of human populations, several debates ensued—particularly about the propriety of potential commercial exploitation of data that had been collected for scientific purposes, but also about the danger of a renewed, molecular racism and about the origin and identity of ethnic groups and minorities.[47] On the other hand, we witness today an increasing demand for individual genome profiles that are requested in the context of searches for one's own "ancestors," which is met by a growing number of private genetic testing companies, demonstrating how precision can turn into arbitrariness.[48]

Identity is also at stake in a third area in which the methods of genomics have come to be used routinely—namely, in forensic medicine. Since the development of "genetic fingerprinting" by the British geneticist Alec Jeffreys in the mid-1980s, DNA samples can be systematically compared with one another. One of the main areas of application of this technique is paternity testing; another is the identification of individuals in the context of criminal investigations and prosecutions. The refinement of PCR now allows those who enforce the law to identify even the tiniest amounts of DNA—a hair or a skin particle will suffice—and to compare the resulting genetic profile with one derived from a blood sample. DNA fingerprinting illustrates how techno-scientific methods can intervene in and change the routine administration of justice, but also shows how dilemmas concerning scientific expertise in the courtroom ensue.[49]

*Genetically Modified Organisms*

Gene technology has been motivated from the outset—in addition to its medical promise—by the commercial exploitation of genetically modified microbes, plants, and animals. As we have seen, in a first application, bacteria carrying a genetically engineered plasmid were used as reactors for the production of gene products such as human insulin and erythropoietin. Since then, a multiplicity of transgenic microorganisms have been developed and employed in research, industrial production, nutrition, environmental care, and biomedicine. Comparative genomics of microorganisms—now that new complete genome sequences for bacteria are reported almost daily—has facilitated the construction of recombinant microorganisms in a previously unheard-of manner.

The development of transgenic plants—an area of application dubbed "green" gene technology—began in 1980 with Jeff Schell's successful use of the soil bacterium *Agrobacterium tumefaciens* as a vector for the introduction of foreign genes into plants. The first patents on transgenic plants were issued in the United States three years later, and in 1985 the first field trials commenced with plants that carried genes for resistance to various viruses, bacteria, and insects. The U.S. Department of Agriculture saw no need for fundamentally new regulations concerning the introduction of transgenic crops and dealt with them on the basis of existing regulations. In 1994, the FlavrSavr tomato came on the market; genetically altered soybeans, corn, and cotton followed in 1995 and 1996. By 1996, genetically engineered crops were being cultivated on almost 2 million hectares in the United States and five other countries; four years later, the number of countries growing transgenic crops had risen to thirteen, and the number of hectares cultivated to 44 million. However, in the recent past, alternative procedures for generating commercially promising plant varieties have gained in prominence, particularly new methods of reshuffling plant genomes (*cis*-genetic plants) and new methods of breeding (smart breeding).[50]

In Europe, plant gene technology has encountered considerable public resistance. In 1998, the European Union declared a de facto moratorium on the importation of plants modified by genetic engineering; however, this arrangement was revoked in 2004. Yet, until this day, the cultivation of such plants remains insignificant in Europe, and field trials are subject to strict regulation.[51] Public debate hinges on three issues. First, more than two-thirds of the population of the European Union simply prefers not to consume genetically altered foods. Second, there is an intense and ongoing discussion about the risks associated with large-scale cultivation of such plants. In this context, concerns center on the problem of outcrossing and other potentially detrimental effects on traditional crop plants and the surrounding ecosystems in general. The third argument is that if Southern countries were pushed to adopt such crops for agricultural production, the result would be only new dependencies and an increase, not a decrease, in the North-South divide between rich and poor countries.

The commercial production of transgenic animals started with Phil Leder's oncomouse, which was patented in 1988, four years after the submission of the initial patent application.[52] During the 1980s, the work of veterinary physician Ralph Brinster from the University of Pennsylvania and molecular biologist Richard Palmiter from the University of Washington also attracted considerable attention. By inserting a growth hormone gene into the genome of normal laboratory mice, they constructed "giant" mice, and they subsequently developed so-called super-pigs via the same procedure.[53] However, according to Daniel Kevles, a profitable transgenic livestock breeding industry has not been established in the United States to this day, neither for the production of improved meat or other agricultural products nor in the broader area of "molecular farming"; that is, the instrumentalization of animals, rather than bacteria, as breeding reactors for gene products like enzymes. Nevertheless, nearly five hundred patents for genetically engineered animals were issued in the United States in the years up to 2003, roughly half for agricultural applications and the other half in the area of biomedical research.[54] The scope of biomedical applications can be illustrated by considering so-called knockout mice, generated by Mario Capecchi, Martin Evans, and Oliver Smithies at the end of the 1980s. In these experimental animals, a particular gene has been knocked out, or can be knocked out, in a directed fashion. Knockout mice therefore facilitate the study of isolated gene actions. These mice have led, among other insights, to the surprising observation that the metabolism of higher animals possesses a considerable compensation potential, a robustness that is due to inherent genetic as well as metabolic redundancy. Animals are often able to compensate for the loss of particular genes that have been eliminated, even in cases in which those genes had previously been considered essential.[55]

The horizontal transfer of genetic material across species boundaries—in principle, all the way from archaebacteria to mammals—was explored and assumed increasing practical relevance during the construction of genetically modified organisms for the purpose of consumption or for further research. Such transfers had not been possible yet in classical genetics. The experimental practices of classical genetics were restricted to sexual reproduction and hence to the mixing and transfer of species-specific, or at least congeneric, hereditary characters. To be sure, we noted in chapter 2 that with the work of Darwin, heredity had already come to be conceived less as a vertical relation of descent than as a horizontal relation within populations, later expressed in the concept of the "gene pool" by population geneticists. The molecular techniques of genetic engineering have decisively radicalized this "horizontalization" of hereditarian thinking and genetic practice. Exchange of hereditary material is now conceivable across the whole spectrum of life forms. Molecular gene technology has widened the circulation of hereditary material from closely circumscribed sexually reproducing populations to the world of living beings as a whole. In a sense, it has globalized the circulation of hereditary material and rendered permeable not only species boundaries, but equally the boundaries between the kingdoms of plants, animals, and microbes. Thus, organic boundaries that were seen as largely fixed in the eighteenth century and as changeable only over evolutionary time spans in the nineteenth century have been subject to renegotiation. The entire gene pool brought about by evolution has become a universal toolbox, one might say. And potentially, each of these tools can now be subjected to further "tinkering," to take up an expression Jacob used to characterize evolution itself.[56] The implications of this transgression of boundaries, as adumbrated in adjectives like "recombinant," "transgenic," and "chimeric," for the future of biological evolution still lie beyond reckoning.

## Molecular Medicine and Medical Genetics

As mentioned already at the end of the last chapter, the world-renowned, eccentric protein chemist and eventual Nobel laureate Linus Pauling published a paper in 1949 in which he explained the medical phenomenon of sickle-cell anemia as being caused by an electrical charge difference in the oxygen-transporting blood pigment hemoglobin. The publication was accompanied by Pauling's triumphal statement that sickle-cell anemia was thus the first pinpointed "molecular disease." This statement went hand in hand with the claim that a molecularization of medicine would soon follow. The publication of this paper is often heralded as the "birth" of molecular medicine. The amino acid substitution that caused the difference in charge associated with sickle-cell anemia was soon identified. And it took another number of years until the emerging field of molecular genetics made it clear that a nucleotide point mutation in the hemoglobin gene was behind this amino acid substitution. Human medical genetics at the time was still confined to tracing pedigrees of familial diseases, identifying cytologically observable chromosome aberrations, and according genetic counseling.

Sickle-cell anemia also provides a good example underscoring the ambiguous difference between molecular medicine and "genetic medicine," the latter expression having gained currency more recently. Thomas Caskey of the National Institutes of Health has also spoken about "DNA-based medicine."[57] The fact that Pauling invoked the prospect of a "molecular medicine" immediately after World War II, but *not* the possibility of genetic therapy, reveals a change of perspective that took place only in the course of the second half of the twentieth century. With Pauling's findings, a molecular diagnosis became possible for a disease that was particularly common among the black population of the United States—one in twelve Americans of African descent is a carrier, and one out of five hundred manifests the recessive disease—but a genetic therapy for sickle-cell anemia remains elusive to this day. Following Keith Wailoo and Stephen Pemberton's history of sickle cell diseases, after the war, the increased infection risks associated with the condition could be effectively treated with antibiotics. From the vantage point of genetics, however, the only option was to identify carriers of the "sickle cell gene" and to make them aware of the transmission risks, particularly if they chose to have children with another carrier. When sickle-cell anemia came to symbolize the bodily suffering of the African-American population at the height of the civil rights movement in the United States, and when the National Sickle Cell Anemia Control Act was passed by Congress in 1972, recognition of the condition was heightened at roughly the same time the prospects of gene technology appeared on the horizon. Yet, at the beginning of the 1970s, urea therapy was promoted as the main way of alleviating the symptoms of the disease. Undesirable side effects soon became apparent, however, and this approach culminated in a medical debacle. During the 1980s, the spotlight moved onto drugs suspected to act as "gene switches." For example, 5-aza-cytidine, which was administered to cancer patients with some success, also raised hopes for treating sickle-cell disease, but those hopes were soon dashed. Then, in the 1990s, the use of hydroxy-urea took the place of the modified nucleotide, but again this approach disappeared because of long-term side effects. Only rather recently, the transplantation of bone marrow, as a kind of gene therapy based on the transplantation of tissue capable of forming healthy blood cells, has been discussed. But this approach, too, appears to be fraught with significant risks.[58]

Technically, three different hurdles have to be overcome to achieve successful gene therapy. First, defective genes and their healthy counterparts have to be identified. Second, appropriate and safe vectors for introducing these DNA sequences into diseased cells have to be constructed. Third, conditions need to be found under which the transferred genetic material becomes integrated into the target genome at the appropriate place, in a stable manner, and does not in turn hamper other gene expression processes. Despite the complexity of the situation, at the technical as well as at the scientific level, the ethics committee of the University of California, Santa Cruz had already approved the first gene-therapeutic intervention on a human being in 1984. However, the authorities were alarmed, the experiment was called off, and in 1985, the National Institutes of Health issued guidelines for gene-therapeutic experiments on humans. Five years later, French Anderson, then at NIH, commenced the first gene therapy, on a four-year-old girl suffering from an immune deficiency. The therapy was temporarily successful. In 1995, therapeutic trials on patients suffering from cystic fibrosis failed. In this case, Ron Crystal, a scientist at NIH and co-founder of the company GenVec, had utilized a modified adenovirus as a vector. Like sickle-cell anemia, cystic fibrosis is caused by an autosomal and recessive genetic defect, but in contrast to the former, it is a "white" condition: the gene is present in one out of every twenty-five to thirty Americans of Caucasian or Ashkenazi descent, and one out of twenty-five hundred to four thousand children develops the condition. Gene-therapeutic trials with the aim of curing muscular dystrophy also failed. In 1999, Jesse Gelsinger, an eighteen-year-old suffering from ornithine transcarbamylase deficiency, died following the administration of a gene therapy vector. Prior to this incident, after a dose-escalation trial involving six patient groups, the responsible researcher—James Wilson from the University of Pennsylvania—had abandoned gene therapy trials for cystic fibrosis that turned out to be ineffective.[59] Finally, a patient suffering from severe combined immunodeficiency linked to the X chromosome (X-SCID), who had been treated with genetically modified blood stem cells three years earlier, began to show cancerous growths in 2002 that were traced to the viral gene vector that had been used in the experiment. As a consequence, that same year, moratoriums on gene therapies in which retroviruses were used as vectors were announced in Germany, France, and the United States.

Keith Wailoo and Stephen Pemberton sum up the historical course of the "genetization" of Tay-Sachs disease, cystic fibrosis, and sickle-cell anemia and the development of genetic treatments for these conditions as follows: "The enterprise deflated precipitously beginning in the late 1990s, its promise sapped by experimental failures, by the deaths of patients, by new concerns about cancer risk, and by the bursting of the inflated biotechnology bubble."[60] In 1996, at the apex of the gene therapy wave, French Anderson was still able to boldly announce that medical practice would be revolutionized by gene therapy and that an appropriate gene therapy for almost every genetically based disease would be found within twenty years. Today, gene therapists have returned to their research laboratories, and their prognoses have become more cautious. The sequencing of the human genome was driven not least by broader hopes, stoked by biomedical experts, for efficacious gene-based therapies; it probably would not have gotten out of the starting gate in the absence of these promises. Yet the project has spawned not a single efficacious strategy for the treatment of a genetic disease. Sickle-cell anemia exemplifies the gap between the molecular biological diagnosis of genetically determined diseases on the one hand and the far more complex territory of genetic therapies on the other. Eric Lander has encapsulated the state of affairs in the following formula: "Genetics is a very good way to pin down the system that has gone awry—to figure out which genes and which biological pathways are behind disease. But it doesn't mean that you're going to treat it by genetics."[61]

Monogenetic diseases—that is, diseases that are triggered by a single defective gene and are generally recognizable as a consequence of their more or less unambiguous Mendelian inheritance patterns—are not only comparatively few, but most of them are also much less prevalent in human populations than "common" complex biological perturbations: heart and circulatory disorders, numerous cancers, diseases of the immune system, afflictions of the nervous system, and dispositions toward infections and inflammations. All these common disorders have genetic backgrounds, to be sure, but these are multifactorial and can vary considerably from individual to individual. This also means that the search for gene therapies, as conducted in the 1990s, was not likely to gain etiological significance. Today, scientists focus on the analytic, diagnostic assessment of individual genetic predispositions; the hope is to offer individualized drug therapies one day. This prospect of a genetically based, but not genetic, medicine is often referred to as pharmacogenetics.

For the time being, however, gene therapy, as well as pharmacogenetics, is still more of a promise than reality. In contrast, the scope of genetic diagnosis has vastly expanded since the introduction of the RFLP technology, but it is associated with its own set of social, legal, and ethical problems.[62] Tests for a hereditary disposition toward certain kinds of cancer furnish one example. As such, disease predictions are statistical in nature; what is at stake here is not only the right to genetic knowledge but also the right to genetic ignorance. In addition, we face the quandaries associated with irreversible preventive interventions—such as the removal of whole organs—that can have intrusive consequences for the life of the individual.[63] The gathering of genetic data by third parties—employers and health insurers—also fuels debate in this area. Another important area of genetic diagnosis associated with promise as well as ethical dilemmas is the screening of embryos, either through prenatal diagnosis (PND) or, in cases of in vitro fertilization, through preimplantation diagnosis before the embryo is transferred to the uterus (PID). In this context, one often encounters reference to "therapeutic" options that are and may one day become available. These practices, however, are a de facto form of genetic selection. Here the right of individuals to self-determination and self-determined choice of offspring, economic interests, the standards of public health, and moral convictions stand in an indissoluble tension.

These few examples amply show that individual and public health practices and expectations are currently undergoing a wide-ranging reconfiguration impelled by new genetic concepts.[64] This reconfiguration is driven by the prospects of a molecular medicine that increasingly gropes at the genetic level. However, the associated hopes and fears, further fueled by phantasms of a genetic optimization of man, are often more powerful than tangible therapeutic results. So far, only genetic diagnostics have attained broad coverage. Tests targeted at particular genes and the mutations that underlie certain diseases therefore represent a more and more important segment of the biotechnology market. Moreover, techniques for the analysis of nucleic acid sequences are increasingly deployed in the realm of infectious disease, for these techniques now allow one to identify pathogens that are otherwise difficult to detect. The present expansion of genetic analysis, however, also leads to the realization that although the genetic makeup of an organism reaches down to the particularities of metabolism and development, no genetic determinism can be founded on this observation—in other words, that the genetic level of DNA is but one among others in bodily processes. With that, phenomena of heredity will undoubtedly not disappear, but the discourse of heredity will no longer be controlled by a single discipline—genetics—that governs biology at large. Rather, one might expect that the discourse of heredity—at the level of individual bodies as well as at the level of their social interactions—will undergo a concomitant complication and differentiation. If Walter Bodmer, director of the Imperial Cancer Research Fund and president of the Human Genome Organization from 1991 to 1993, assumed less than twenty years ago that the total information about the human genome "will enable genetic analysis of essentially any human difference,"[65] today's molecular and human geneticists would certainly eschew such a bold confusion of *pars* and *totum*.

*Postgenomics*

At the beginning of the Human Genome Project, one often heard it said that the sequence would explain everything. Twenty years on, new catchwords like postgenomics, epigenetics, proteomics, and even "organomics" have gained currency, telling us that the sequence is not everything. They testify to the fact that in biology, the devil is in the details: in the manifold networks of development and metabolism. Ernst Ludwig Winnacker, then president of the German Research Organization, was one of the first bioscientists to conjecture, in 1997, that a "postgenomic" age was imminent. To paraphrase his ideas: although biologists had, until recently, focused on particular, isolated genes, the decisive research question in the future would be how individual genes were involved in the formation of whole cells, whole tissues, and finally, whole organisms. The time of fixation on parts of the organism would soon be over. The organism as a totality would then be the order of the day.[66]

For historians of biology, these claims are not entirely without precedent. Biologists have claimed to have overcome reductionist perspectives in favor of a holistic view of living beings more than once. The cry appears to ebb and flow in centennial waves. At the beginning of the nineteenth century, it was articulated in Romantic *Naturforschung*, whose supporters often mounted attacks against what they perceived as an undue "*Zergliederung*" (dismemberment) of biological phenomena.[67] Following a lengthy interlude, holistically motivated biologists at the beginning of the twentieth century again wanted to leave behind developmental mechanics and Mendelian transmission genetics, which they criticized as utterly reductionist and positivist. Their call to arms was in the name of systems theoretical visions of the organism.[68] Now, at the beginning of the twenty-first century, molecular biologists and molecular geneticists have set out to vanquish genetic reductionism in the name of and with the united forces of genome research itself. And herein may lie a new aspect of the present "holistic turn": For the first time in history, we are not dealing with a minority opposition within biology at large. Rather, the very pertinaciousness with which the reductionist strategies of molecular biology and gene technology were pursued seems itself to have led to a more systemic vision of the functioning, the development, and the evolution of organisms. To conclude this final chapter, we would like to draw together a few of the threads in this story in order to characterize—even if only tentatively and in outline—the place of heredity in the life sciences at the beginning of the twenty-first century.

DNA or gene chips, a novel technology for the analysis of nucleic acids, were developed in the mid-1990s. DNA chips essentially contain specific genes, or gene fragments, or fragments of whole chromosomes, or fragments of whole genomes that are immobilized on an appropriate support, normally a glass or silicon chip. The first DNA chip, with more than six thousand yeast genes, was presented in 1996 in California. DNA chips take advantage of the possibility of complementary base pairing between nucleic acids, and the applications of the technology range from genetic diagnosis to the investigation of cellular activity patterns. For example, it is possible to compile differential activity patterns of cells by hybridizing the messenger RNAs that the cells contain at a particular time with the corresponding background of gene sequences. Fluorescence labeling then makes the gene expression pattern visible. The rapid progress from the analysis of single genes in early molecular biology to the whole-genome analysis typical of genomics is not least due to the complementary base pairing properties of nucleic acids.[69] Almost all molecular technologies developed along the way exploit this property. Compared with those of proteins, which are far more difficult to handle, these molecular properties present invaluable advantages.

With the arrival of DNA chip technology, the acquisition of information about the systemic states of cellular metabolism and about developmental transitions from one such state to another has been put on a new basis. Such assessments have now become possible to an extent that was unthinkable when genome projects were first conceived. New physical methods for the identification of micro amounts of protein have raised the efficiency of acquiring systemic cellular information even further— far beyond what could be achieved with traditional two-dimensional procedures for the separation of proteins. This advance has opened up a field of research called proteomics. Many DNA sequences in higher organisms, including human beings, have been found to give rise to a multiplicity of spliced RNA products, and thus to an astounding number of proteins that are themselves subject to posttranslational modification. In addition, each state of differentiation and of metabolic activity of a cell is associated with a corresponding expression pattern. Taking all this into account, one can imagine that the structure of the proteome, and its successive interaction patterns in particular, present challenges of still another order of magnitude compared with the analysis of the human genome. This is where epigenetics and epigenomics come into their own, namely by investigating the stabilization as well as the transmission of cellular activity states. Both processes rest on cellular systems of short-term information storage that embed and contain the genomic structures that are kept stable over evolutionary time spans.[70]

For the further development and differentiation of such a systemic view of the cell, and for the integration of the vast amounts of data delivered by these technologies, concepts of interactive networks will be decisive. They can be supplied only by a biology based on theoretical modeling and bioinformatics tools. With that effort, a novel feature of today's systems biology, as compared with the speculative systems approaches of the past, is being identified: one of the decisive challenges of bioinformatics consists in designing appropriate formal languages and so-called ontologies that help to meaningfully nest together the genomic, proteomic, and epigenomic databases of the future.[71]

This survey would be essentially incomplete without a word on molecular developmental biology as one of the major inroads into postgenomics. Whereas many of the molecular biologists who unleashed the first wave of gene technology at the beginning of the 1970s moved into the development and commercialization of nucleic acid–based technologies and products, finally launching the Human Genome Project, the concomitant molecularization of developmental biology was much less spectacular, at least in its beginnings.[72] Molecularization here began with the identification of genes that manifestly intervene in processes of embryonic development and differentiation. At the beginning of the 1980s, a handful of molecular biologists—equipped with the new molecular tools—once again focused on *Drosophila melanogaster*, the pet of classical genetics and genetically the most extensively characterized higher organism at the time.[73] In doing so, they drew on the work of Ed Lewis at Caltech and Antonio García-Bellido in Madrid. A former student of Thomas Hunt Morgan, Lewis had continued to describe *Drosophila* genes with the methods of classical genetics at a time when molecular biologists had begun to focus on bacteria. He identified genes that influenced the segmentation or body plan of fruit flies (among them a four-winged mutant). García-Bellido, who was in turn a student of Lewis and of Ernst Hadorn in Zürich, had also investigated segmentation mutants and the morphogenesis of the wings by means of cytogenetic methods. Now, Walter Gehring in Basel started to elucidate the function of homeotic genes in the development of the insect eye with molecular genetic methods. This molecular approach, together with Eric Wieschaus's and Christiane Nüsslein-Volhard's exhaustive collections of mutants that affected the oogenesis and embryogenesis of the fly, led to a new understanding of developmental genes as a system of molecular "switches" that were to be activated and deactivated in an orderly manner during development. New model organisms representing vertebrates, such as zebrafish (*Danio rerio*), have since been introduced in developmental biology. At the other end of the complexity scale, the tiny worm *Caenorhabditis elegans* became established as a model of cellular and molecular development following a call for action by Sydney Brenner in Cambridge.[74]

Many of these molecular switches turned out to be highly conserved throughout the animal kingdom. They are reminiscent of the regulatory genes postulated by Jacob and Monod as part of their operon model, yet they function differently. While regulatory genes in bacteria adjust the metabolism reversibly to changes in the environment, mostly responding to the presence or absence of particular nutrients, developmental genes initiate and control irreversible differentiation processes. They code for so-called transcription factors that bind to the control regions of chromosomal DNA and determine the rate at which a gene or a group of genes is transcribed and expressed. Among them, one also finds second-order regulators that in turn control and modulate the activity of transcription factors. Developmental genes are important gatekeepers and belong to the evolutionarily most conserved genomic structures. For example, the *pax* gene family controls the whole process of eye formation, regardless of whether one considers insects or vertebrates. To the surprise of Walter Gehring and collaborators, a *pax* gene isolated from a mouse and inserted into the genome of *Drosophila* still triggered the formation of an eye—not of a vertebrate, but of an insect eye. Many of these genes or gene families, such as the *hox* genes of the homeobox family, play a role in spatial pattern formation and in the temporal succession of these patterns during embryogenesis.

Michel Morange has highlighted two points that are by now firmly established, even though this area of research is still in a state of flux.[75] First, without developmental genes, there is no orderly development.

Second, not only single genes, but whole gene complexes together with the associated genomic structures, appear to have been evolutionarily conserved. These findings thus support the view that genomes—of bacteria and of higher organisms—are not mere conglomerates of genes, but are themselves hierarchically structured ensembles. Another class of highly conserved genes controls intracellular and extracellular signal transmission processes, which are obviously of decisive importance in the development of multicellular organisms.

And yet, as mentioned earlier, one of the most surprising insights gleaned by means of gene knockout technology was that organisms often appear to be able to compensate via alternative pathways for the loss of certain genes that were deemed indispensable. With this observation, molecular developmental biology—a disciplinary label that is increasingly replacing "molecular developmental genetics"—was confronted with the fact that developmentally relevant processes function in a redundant fashion.[76] They appear to be massively buffered and remarkably robust with respect to changing inner and outer conditions. Gene products play an important role in this context, but the processes themselves are only partially determined by genes. Rather unexpectedly, studies of embryonic gene expression using DNA chip technology have also revealed that the same coding DNA sequences can be called up in different developmental stages and in different tissues and that the corresponding gene products can have different functions in different cellular processes. This multifunctionality further suggests that the relation between genetic determining factors and the life of the organism needs to be rethought and modeled in a more complex fashion.[77]

Another area that at least needs to be mentioned relates to recent findings about the reproductive biology of higher organisms, the manipulative control of their propagation and the establishment of cloning techniques in particular.[78] In the early 1960s, the British developmental biologist John Gurdon in Oxford reared tadpoles produced after the nuclei of intestinal cells had been transplanted into enucleated, fertilized egg cells. These experiments, which were done with the South African clawed frog *Xenopus laevis*, provided the first demonstration that the somatic cell nuclei of higher organisms can conserve the genetic potential for the differentiation of all cell types during development. Nonetheless, it took twenty-five years until Ian Wilmut and Keith Campbell at the Roslin Institute in Scotland successfully performed this procedure in a mammal, and the eventual birth of Dolly, the first cloned sheep, in 1996 was widely received as a spectacular feat.[79] In 1998, three generations of mice were produced from adult cumulus cells of the mouse ovary at the University of Hawaii by the same procedure, and that same year scientists at the Kinki University in Japan cloned eight genetically identical calves via nuclear transfer. The generation of genetically identical individuals through "embryo splitting" was successfully carried out for cows and mice in 1981 and applied to primates (rhesus macaques) at the Oregon Regional Primate Research Center in Beaverton. The first creation of a chimera from two mammals was achieved in 1984 with a sheep and a goat.

Human reproductive biology was not, and is not, exempt from this development. In 1978, the first human being generated in vitro, the "test-tube baby" Louise Brown, made headlines worldwide. Since then, in vitro fertilization has become routine in human medicine. In November 1998, two teams, one from Johns Hopkins University in Baltimore, and the other at the University of Wisconsin under the direction of John Gearhart, reported that they had isolated and propagated human embryonic stem cells in the laboratory for the first time. In 2001, scientists from Advanced Cell Technology, a biotech firm in Worcester, Massachusetts, carried out the first somatic nuclear transfers with human embryonic cells and thus laid the basis for human therapeutic cloning. In 2004, Hwang Woo Suk from the Seoul National University in South Korea claimed to have derived pluricellular human embryos from embryonic stem cells that had been cloned in this manner. A year later, however, these results were exposed as fraud.[80] Therapeutic cloning holds out the hope of growing immunocompatible tissues and organs in order to treat a myriad of diseases. However, in the minds of many, the development of this technology is associated with the possibilities of reproductive cloning of human beings, and with the fear of this eventuality. Human stem cell research is currently one of the most controversial research areas in human biomedicine, particularly in view of the generation of embryos for research purposes (also known as "consumptive human embryo research"). This area is being regulated differently in different countries. In the United States, public funding for stem cell research is restricted to stem cell lines that have been produced from surplus embryos stemming from in vitro fertilization treatment. Commercial research is not subject to such restrictions. In Great Britain, embryonic stem cell research has been permitted up to a certain developmental stage of the embryo under an act of Parliament since 2000. In Germany, such research remains prohibited owing to an embryo protection law passed in 1990; work with a number of imported stem cell lines is permitted subject to rigorous restrictions.[81]

Genetics is an integral part of both molecular developmental biology and reproductive biology. But in both areas, questions of heredity are also visibly and multiply embedded in broader biological, biosocial, and biopolitical contexts that are becoming more and more prominent as new interventions become feasible. Any problems associated with these interventions can therefore neither be defined in purely genetic terms nor solved by purely genetic means. In a way, in the life sciences, and in human biology and biomedicine in particular, we have come to the realization that the objects of study operate at several different and irreducible levels, and that determinations of these levels can overlap and intersect with one another in multiple ways.

To conclude, we agree with Ken Waters, who has argued that the widespread genocentrism in twentieth-century biology and the concomitant disciplinary exaltation of genetics were epistemologically and not ontologically founded, albeit genocentrism was often interpreted in an ontological fashion.[82] Classical genes offered preferential access to the makeup of organisms, and molecular genes, manipulable as they were through an array of nucleic acid technologies, increased the options for experimental access exponentially. There is, however, no reason to ontologize this historically conditioned episteme. On the contrary, if current moves toward systems biology are carried through and prove viable, it is conceivable that the discipline of genetics as conceived of in the twentieth century will recede into the background, if not dissolve entirely, in the life sciences of the future. For a long century, the genetic level of the organism was the most amenable to experimental manipulation. But not least as a consequence of accumulated gene-technological knowledge, it has now become clear that even the most complete genetic analysis can never reveal more than a fraction of the totality of biological phenomena. The analysis and manipulation of living beings at the genetic level and within the bounds of what is genetically possible will of course continue, but it will be increasingly embedded in transgenomic contexts.

The world picture of twentieth-century genetics colligated around two influential sets of images: genes as biological atoms, and genes as biological carriers of information. Both sets were nested in experimental technologies, or more precisely, they were materialized by those technologies and thus became efficacious. This observation implies that new biological technologies will historicize those older guiding ideas. For the late-nineteenth-century and early-twentieth-century image of discrete, independent elementary units that are linearly arranged along their respective chromosomes and are responsible for a particular bodily or mental character, that historicization has already taken place. And more recently, molecular, gene-technological "genes" have turned out to be

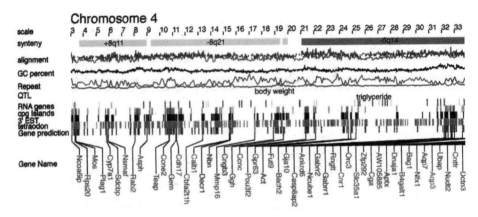

FIGURE 8.2 Detail from an annotated map of the mouse chromosome 4. *Nature*, 5 December 2002.

far more complicated as well. Their definition is, as Morange aptly observes, "impossible today."[83]

The second set of images, a legacy of what can now be called "classical" molecular genetics and its toolbox, is composed of metaphors related to information and writing. Its cultural reach and reverberations can hardly be overestimated. Molecular copying techniques, for some time the apparent climax of the molecular domination of life, have materialized these metaphors in a seemingly uncircumventable fashion and have transformed the encounter with the polysemic "book of life" into a concerted project of "deciphering." Lily Kay as well as Dorothy Nelkin and Susan Lindee have described this process, which also underwrites biology's inflated concept of isolatable information, in detail.[84]

Finally, a third, related image is emerging in the life sciences today: the map. To be sure, the roots of this image lie in nineteenth-century hereditary thought, and it has also been inscribed in mapping projects during the classical and molecular phases of genetics.[85] However, maps have recently begun to play a much more prominent role in the representation of genetic and postgenetic data.[86] Maps of chromosomes (fig. 8.2), developmental maps, metabolic maps, and evolutionary maps are becoming indispensable orientation devices in a landscape inundated by scientific data that no longer make sense unless ordered and compressed. As we have seen, the epistemic space of heredity was delineated in its outlines in the late eighteenth century and throughout the nineteenth century, and the concept of biological heredity was condensed into experimental epistemic objects in the twentieth century. Now, at the beginning of the twenty-first century, these epistemic objects appear to be spreading out again into spatial, reticular structures.

# Notes

EPIGRAPH

William James, "Two Reviews of *The Variation of Animals and Plants under Domestication* by Charles Darwin," in *The Works of William James: Essays, Comments, and Reviews*, ed. F. H. Burkhardt, Fredson Bowers, and I. K. Skrupskelis (Cambridge, MA: Harvard University Press, 1987): 234; see Louis Menand, *The Metaphysical Club* (New York: Farras, Straus and Giroux, 2002), 142, for a discussion of James's review.

PREFACE

1. Leslie C. Dunn, *A Short History of Genetics: The Development of Some of the Main Lines of Thought: 1864–1939* (New York: McGraw-Hill, 1965); Hans Stubbe, *History of Genetics: From Prehistoric Times to the Rediscovery of Mendel's Laws*, trans. T. R. W. Waters (Cambridge, MA: MIT Press, 1972); Elof A. Carlson, *The Gene: A Critical History* (Philadelphia: Saunders, 1966); Carlson, *Mendel's Legacy: The Origin of Classical Genetics* (Cold Spring Harbor: Cold Spring Harbor Laboratory Press, 2004); James Schwartz, *In Pursuit of the Gene: From Darwin to DNA* (Cambridge, MA: Harvard University Press, 2010).

2. François Jacob, *The Logic of Life: A History of Heredity*, trans. B. E. Spillman (Princeton, NJ: Princeton University Press, 1993), 10–17.

3. In this we are following Dominique Pestre, who coined the expression "knowledge regime" (*regime de savoir*) in order to move beyond disciplinary perspectives on science while continuing to focus on its long-term development; see Pestre, *Science,*

*argent et politique: Un essai d'interprétation* (Paris: INRA, 2003), 30–37; for a short discussion of the concept in English, see Pestre, "The Technosciences between Markets, Social Worries and the Political: How to Imagine a Better Future?" in *The Public Nature of Science under Assault: Politics, Markets, Science and the Law*, ed. H. Nowotny et al. (Berlin: Springer, 2005), 29–52.

4. Henning Schmidgen, "Fehlformen des Wissens," in Georges Canguilhem, *Die Herausbildung des Reflexbegriffs im 17. und 18. Jahrhundert*, trans. H. Schmidgen (Munich: Fink, 2008): ix. Translations, unless otherwise noted, are our own.

5. See Ana Barahona, Edna Suarez-Díaz, and Hans-Jörg Rheinberger, eds., *The Hereditary Hourglass: Genetics and Epigenetics, 1868–2000*, preprint 392 (Berlin: Max Planck Institute for the History of Science, 2010).

6. Evelyn Fox Keller, *The Century of the Gene* (Cambridge, MA: Harvard University Press, 2000).

7. See Hans-Jörg Rheinberger, *Experiment—Differenz—Schrift* (Marburg: Basilisken-Presse, 1992).

8. For a full documentation of the research project "A Cultural History of Heredity," see http://heredity.mpiwg-berlin.mpg.de/heredity/Heredity.html (accessed September 19, 2011). Some contributions have been published in Staffan Müller-Wille and Hans-Jörg Rheinberger, eds., *Heredity Produced: At the Crossroads of Biology, Politics and Culture, 1500–1870* (Cambridge, MA: MIT Press, 2007). A second volume is in preparation: Staffan Müller-Wille and Christina Brandt, eds., *Heredity Explored: Between Public Domain and Experimental Science, 1850–1930*.

CHAPTER ONE

1. Porphyry, *Introduction*, trans. J. Barnes (Oxford: Clarendon Press, 2003), 7.

2. Klaus Heinrich, *Tertium datur: Eine religionsphilosophische Einführung in die Logik* (Basel: Stroemfeld/Roter Stern, 1983), 99. Original emphasis. See also Jean Gayon and Jean-Jacques Wunenburger, "Présentation," in *Le paradigme de la filiation*, ed. J. Gayon and J. J. Wunenburger (Paris: L'Harmattan, 1995), 8.

3. Heinrich, *Tertium datur*, 123.

4. Eva Bäckstedt, "Rasbiologens barnbarn gör upp med arvet," *Svenska Dagbladet Kultur*, March 2, 2002, 4–5. Translations, unless otherwise noted, are our own.

5. Carlos López Beltrán, "Natural Things and Non-natural Things: The Boundaries of the Hereditary in the 18th Century" in *Conference: A Cultural History of Heredity I: 17th and 18th Centuries*, preprint 222 (Berlin: Max Planck Institute for the History of Science, 2003), 71–74.

6. The absence of ideas of transmission in premodern theories of generation has been noted before in Jacob, *Logic of Life*, and Peter Bowler, *The Mendelian Revolution: The Emergence of Hereditarian Concepts in Modern Science and Society* (Baltimore: Johns Hopkins University Press, 1989).

7. Alfred W. Crosby, *Ecological Imperialism: The Biological Expansion of Europe, 900–1900* (Cambridge: Cambridge University Press, 1986).

8. A similar argument is made by Phillip Thurtle concerning the formation

of genetics around 1900; see Thurtle, *The Emergence of Genetic Rationality: Space, Time, and Information in American Biological Science, 1870–1920* (Seattle: University of Washington Press, 2008), chap. 12.

9. See Ludmilla Jordanova, "Interrogating the Concept of Reproduction in the Eighteenth Century," in *Conceiving the New World Order*, ed. F. D. Ginsburg and R. Rapp (Berkeley: University of California Press, 1995), 375.

10. Staffan Müller-Wille and Hans-Jörg Rheinberger, *Das Gen im Zeitalter der Postgenomik: Eine wissenschaftshistorische Bestandsaufnahme* (Frankfurt/M.: Suhrkamp, 2009).

11. See Robert C. Olby, *Origins of Mendelism*, 2nd ed. (Chicago: University of Chicago Press, 1985), 55–63, and Jean Gayon, *Darwinism's Struggle for Survival: Heredity and the Hypothesis of Natural Selection* (Cambridge: Cambridge University Press, 1998), 105–106.

12. A detailed overview of ancient generation theories is provided by Erna Lesky, *Die Zeugungs- und Vererbungslehren der Antike und ihr Nachwirken*, Abhandlungen der Geistes- und Sozialwissenschaftlichen Klasse der Akademie der Wissenschaften und der Literatur in Mainz, no. 19 (1950) (Wiesbaden: Franz Steiner, 1951).

13. Francis Galton, "A Theory of Heredity," *Journal of the Anthropological Institute* 5 (1876): 330.

14. Ibid., 331.

15. Ibid., 336.

16. Ibid., 330.

17. Ibid., 331.

18. On the concept of epistemic things, see Hans-Jörg Rheinberger, *Toward a History of Epistemic Things: Synthesizing Proteins in the Test Tube* (Stanford, CA: Stanford University Press, 1997).

19. See Barry Barnes and John Dupré, *Genomes and What to Make of Them* (Chicago: University of Chicago Press, 2008).

20. Francis Galton, *Inquiries into Human Faculty and Its Development* (London: Macmillan, 1883), 24–25.

21. Galton, "Theory of Heredity," 336.

22. Ibid.

23. Francis Galton, "Hereditary Improvement," *Fraser's Magazine* 7 (1873): 127.

24. On Galton's political positions, see Allen R. Buss, "Galton and the Birth of Differential Psychology and Eugenics: Social, Political, and Economic Forces," *Journal of the History of the Behavioral Sciences* 12 (1976): 52–53, and Ruth Schwartz Cowan, "Nature and Nurture: The Interplay of Biology and Politics in the Work of Francis Galton," in *Studies in the History of Biology*, vol. 1, ed. W. Coleman and C. Limoges (Baltimore: Johns Hopkins University Press, 1977), 154–157.

25. See Donald A. MacKenzie, *Statistics in Britain, 1865–1930: The Social Construction of Scientific Knowledge* (Edinburgh: Edinburgh University Press, 1981), 51–72.

26. Francis Galton, *Hereditary Genius: An Inquiry into Its Laws and Consequences* (London: Macmillan, 1869), 1.

27. Cyril Dean Darlington, introduction to *Hereditary Genius: An Inquiry into Its Laws and Consequences*, by Francis Galton (London: Collins, 1962), 9.

28.Galton, *Hereditary Genius*, 350.

29. Michel Foucault, *The History of Sexuality: Volume 1: An Introduction*, trans. R. Hurley (New York: Vintage Books, 1990), 106–107.

30. Ibid., 137.

31. Ibid., 141–143; on the concept "bio-politics," see also Foucault, "Faire vivre et laisser mourir: La naissance du racisme," *Les Temps Modernes* 535 (1991): 37–61.

32. Galton, *Hereditary Genius*, 376.

CHAPTER TWO

1. Georges Louis Leclerc, Comte de Buffon, *Natural History, General and Particular, by the Count de Buffon*, 2nd ed., vol. 2, trans. W. Smellie (London: W. Strahan and T. Cadell, 1785), 29.

2. For the theological background, see Peter McLaughlin, *What Functions Explain: Functional Explanation and Self-Reproducing Systems* (Cambridge: Cambridge University Press, 2001), 174–179; on the history of the concept of reproduction in the life sciences, see Roselyne Rey, "Génération et hérédité au 18e siècle," in *L'ordre des caractères: Aspects de l'hérédité dans l'histoire des sciences de l'homme*, ed. J.-L. Fischer (Paris: Sciences en situation, 1989), 7–41; and Jordanova, "Interrogating the Concept of Reproduction."

3. Jacob, *Logic of Life*, 19–20.

4. See Nicholas Russell, *Like Engend'ring Like: Heredity and Animal Breeding in Early Modern England* (Cambridge: Cambridge University Press, 1986), 40; Garland E. Allen, "T. H. Morgan and the Split between Embryology and Genetics, 1910–1926," in *A History of Embryology*, ed. T. Horder, I. A. Witkowski, and C. C. Wylie (Cambridge: Cambridge University Press, 1986), 113–144; Frederick B. Churchill, "From Heredity Theory to 'Vererbung': The Transmission Problem, 1850–1915," *Isis* 78 (1987): 337–364; Jane Maienschein, "Heredity/Development in the United States, circa 1900," *History and Philosophy of the Life Sciences* 9 (1987): 79–93; Peter J. Bowler, *Mendelian Revolution*, 6; and Carlos López Beltrán, *El sesgo hereditario: Ámbitos históricos del concepto de herencia biológica* (Mexico City: Universidad Nacional Autónoma de México, 2004), 43–62.

5. See Ron Amundson, *The Changing Role of the Embryo in Evolutionary Thought* (Cambridge: Cambridge University Press, 2005), 34–39; for detailed historical evidence, see Conway Zirkle, *The Beginnings of Plant Hybridization*, Morris Arboretum Monographs, vol. 1 (Philadelphia: University of Pennsylvania Press, 1935); and Zirkle, "Species before Darwin," *Proceedings of the American Philosophical Society* 103 (1959): 636–644.

6. Aristotle, *Generation of Animals*, trans. A. L. Peck (Cambridge, MA: Harvard University Press, 1990), 157 (735a25); for similar references, see Lesky, *Zeugungs- und Vererbungslehren*, 139; and Bentley Glass, "The Germination of the Idea of Biological Species," in *Forerunners of Darwin, 1745–1859*, ed. B. Glass, O. Temkin, and W. L. Strauss (Baltimore: Johns Hopkins Press, 1968), 31.

7. Aristotle, *Generation of Animals*, 253 (747b33–36).

8. Aristotle, *History of Animals*, trans. A. L. Peck and D. M. Balme (Cambridge, MA: Harvard University Press, 1991–1993), 203 (606b19–607a8). This story provides the background to the well-known proverb that everything new comes from Africa; see Harvey M. Feinberg and Joseph B. Solodow, "Out of Africa," *Journal of African History* 43 (2002): 255–261.

9. See Zirkle, *Beginnings of Plant Hybridization*, 7–41.

10. See Mary Terrall, "Speculation and Experiment in Enlightenment Life Sciences," in Müller-Wille and Rheinberger, *Heredity Produced*, 267.

11. John Locke, *An Essay Concerning Human Understanding* (Oxford: Clarendon Press, 1979), 451.

12. John Ray, *Historia plantarum species hactenus editas aliasque insuper multas noviter inventas et descriptas complectens*, vol. 1 (London: Clark and Faithorne, 1686), 40–42.

13. On the definitions of spontaneous and equivocal procreation, see Peter McLaughlin, "Spontaneous versus Equivocal Generation in Early Modern Science," *Annals of the History and Philosophy of Biology* 10 (2005): 79–88. Edmund O. Lippmann, *Urzeugung und Lebenskraft: Zur Geschichte dieser Probleme von den ältesten Zeiten an bis zu den Anfängen des 20. Jahrhunderts* (Berlin: Julius Springer, 1933), and John Farley, *The Spontaneous Generation Controversy from Descartes to Oparin* (Baltimore: Johns Hopkins University Press, 1974) review the history of these ideas.

14. See Justin E. H. Smith, "Imagination and the Problem of Heredity in Mechanist Embryology," in *The Problem of Animal Generation in Early Modern Philosophy*, ed. J. E. H. Smith (Cambridge: Cambridge University Press, 2006), 80–99.

15. See Silvia De Renzi, "Resemblance, Paternity, and Imagination in Early Modern Courts," in Müller-Wille and Rheinberger, *Heredity Produced*, 61–83.

16. See Jean Céard, *La Nature et les Prodiges: L'insolite au XVIe siècle*, 2nd ed. (Geneva: Droz, 1996); and Lorraine Daston and Katharine Park, *Wonder and the Order of Nature, 1150–1750* (New York: Zone Books, 1998).

17. We use the adjective "specific" in the sense of "pertaining to . . . a distinct species of animals or plants," rather than in one of its many, more general senses; see *Oxford English Dictionary*, online edition, http://www.oed.com (accessed September 19, 2011), s.v. "specific."

18. Hippocrates, *Hippocratic Writings*, ed. G. E. R. Lloyd (Harmondsworth, UK: Penguin Books, 1950), 317; see Stubbe, *History of Genetics*, 27–30, on this Hippocratic tract.

19. Hippocrates, "De Genitura," in *Hippocratis Coi Opera qvae Graece et latine extant*, ed. and trans. G. Mercuriale (Venice: Industria ac sumptibus Iuntarum, 1588), 10 and 15; both translations are "right" insofar as the Greek κρατύνει means both "to govern" and "to strengthen."

20. Claude Lévi-Strauss, *Totemism*, trans. R. Needham (London: Merlin Press, 1991), 74. In *The Savage Mind*, Lévi-Strauss suggests that the "logic of totemic classification" permeated the works of "naturalists and alchemists of antiquity and the middle ages: Galen, Pliny, Hermes Trismegistos, Albertus Magnus"; see Lévi-Strauss, *The Savage Mind* (London: Weidenfeld and Nicolson, 1974), 42.

21. Aristotle, *Lectures on the Science of Nature*, trans. C. G. Wallis (Annapolis: The St. John's Bookstore, 1940), 33 (194b13); see Andrea Falcon, *Aristotle and the Science of Nature: Unity Without Uniformity* (Cambridge: Cambridge University Press, 2005), 9.

22. René Descartes, *Oeuvres de Descartes*, vol. 11, ed. C. Adam and P. Tannery (Paris: Vrin, 1986), 507.

23. René Descartes, *The Philosophical Writings of Descartes*, trans. J. Cottingham, R. Stoothoff, and D. Murdoch, vol. 1 (Cambridge: Cambridge University Press, 1998), 322. The comparison with beer suggested itself because the consistency of male sperm was often described as "foamy" at the time; see Jacques Roger, *The Life Sciences in Eighteenth-Century French Thought*, ed. K. R. Benson, trans. R. Ellrich (Stanford, CA: Stanford University Press, 1997), 40–42.

24. Laurence Sterne, *The Life and Opinions of Tristram Shandy, Gentleman—A Sentimental Journey Through France and Italy* (Munich: Günter Jürgensmeier, 2005), 5.

25. Lesky, *Zeugungs- und Vererbungslehren*; Roger, *Life Sciences*; see also Justin E. H. Smith, ed., *The Problem of Animal Generation in Early Modern Philosophy* (Cambridge: Cambridge University Press, 2006).

26. For what follows, see also Ohad Parnes, Ulriko Vedder, and Stefan Willer, *Das Konzept der Generation* (Frankfurt/M.: Suhrkamp, 2008), on the parallel historical change in the use of the term "generation."

27. Henry Peter Bayon, "William Harvey (1578–1657): His Application of Biological Experiment, Clinical Observation, and Comparative Anatomy to the Problems of Generation," *Journal of the History of Medicine and Allied Sciences* 2 (1947): 51–52.

28. Walter Pagel, "The Philosophy of Circles—Cesalpino—Harvey," *Journal of the History of Medicine and Allied Sciences* 12 (1957): 153; see also Peter M. Jucovy, "Circle and Circulation: The Language and Imagery of William Harvey's Discovery," *Perspectives in Biology and Medicine* 20 (1976): 92–107.

29. See Roger, *Life Sciences*, 89.

30. William Harvey, *The Works of William Harvey*, trans. R. Willis (London: Sydenham Society, 1847), 363; see also ibid., 513–514.

31. Roger, *Life Sciences*, and Smith, *Problem of Animal Generation* provide excellent overviews of early modern generation theories.

32. See Michael Wolff, *Geschichte der Impetustheorie: Untersuchungen zum Ursprung der klassischen Mechanik* (Frankfurt/M.: Suhrkamp, 1978), 184–191; Johannes Fritsche, "The Biological Precedents for Medieval Impetus Theory and Its Aristotelian Character," *British Journal for the History of Science* 44 (2010): 1–27.

33. Aristotle, *Generation of Animals*, 11 (716a14); see also Lesky, *Zeugungs- und Vererbungslehren*, 127, on this distinction between male and female.

34. Harvey, *Works of William Harvey*, 354.

35. Ibid., 372–373; translation slightly altered (the quoted translation renders *filius* as "child"; see William Harvey, Exercitationes de generatione animalium, Amsterdam: Elzevir, 1651, p. 299). The passage contains paraphrases from Aristotle, *De gen. anim.*, 734a17 and 740b35–a6.

36. See James G. Lennox, "The Comparative Study of Animal Development: William Harvey's Aristotelianism," in Smith, *Problem of Animal Generation*, 27–28.

37. For a more detailed discussion, see Erna Lesky, "Harvey und Aristoteles," *Sudhoffs Archiv* 41 (1957): 369–372.

38. See Roger, *Life Sciences*, 91–94.

39. Harvey, *Works of William Harvey*, 373.

40. Ibid., 281; see Eve Keller, "Making Up for Losses: The Workings of Gender in William Harvey's *De generatione animalium*," in *Inventing Maternity: Politics, Science, and Literature, 1650–1865*, ed. S. C. Greenfield and C. Barash (Lexington: University of Kentucky Press, 1999), 46–48, for a detailed discussion.

41. Harvey announced this method as a "new and untrodden path" in the preface to his *Exercitationes*; see Harvey, *Works of William Harvey*, 166. On Harvey's method and its reliance on, as well as departure from, Aristotle, see Lennox, "William Harvey's Aristotelianism."

42. Harvey, *Works of William Harvey*, 235; see John S. White, "William Harvey and the Primacy of Blood," *Annals of Science* 43 (1986): 239–255.

43. Dietlinde Goltz, "Der leere Uterus: Zum Einfluß von Harveys *De generatione animalium* auf die Lehren von der Konzeption," *Medizinhistorisches Journal* 21 (1986): 242–268.

44. See Bayon, "William Harvey," 58–72.

45. Harvey, *Works of William Harvey*, 465.

46. Ibid., 554.

47. Ibid., 321; on Harvey's speculation about spontaneous generation, see Edward T. Foote, "Harvey: Spontaneous Generation and the Egg," *Annals of Science* 25 (1969): 139–163.

48. Justin E. H. Smith, "Introduction," in Smith, *Problem of Animal Generation*, 9; on the significance of microscopic observations for the emergence of microsubstantialism, see Andrew Pyle, "Malebranche on Animal Generation: Preexistence and the Microscope," in Smith, *Problem of Animal Generation*, 194–214.

49. Foote, "Spontaneous generation."

50. See François Duchesneau, *Les modèles du vivant de Descartes à Leibniz* (Paris: Vrin, 1998), 218–220; I. Bernhard Cohen, "Harrington and Harvey: A Theory of the State Based on the New Philosophy," *Journal of the History of Ideas* 55 (1994): 187–210; Antonio Di Meo, "Il concetto di 'circolazione'": Storia di una rivoluzione transdisciplinare," in *Le rivoluzioni nelle scienze della vita*, ed. G. Cimino and B. Fantini (Firenze: Leo S. Olschki, 1995), 31–84.

51. Harvey, *Works of William Harvey*, 367, translation slightly adapted; the quoted passage stems from Aristotle, *On Coming-to-Be and Passing-Away*, trans. E. S. Forster (Cambridge, MA: Harvard University Press, 1992), 315 (336a32–b2).

52. William Harvey, *Exercitationes*, 291.

53. See Andrew Gregory, "Harvey, Aristotle and the Weather Cycle," *Studies in the History and Philosophy of the Biological and Biomedical Sciences* 32 (2001): 153–168.

54. See Thomas Fuchs, *Die Mechanisierung des Herzens: Harvey und Descartes; Der vitale und der mechanische Aspekt des Kreislaufs* (Frankfurt/M.: Suhrkamp, 1992), 95–109.

55. See Christopher Hill, "William Harvey and the Idea of Monarchy," *Past and Present* 27 (1964): 54–72, and Keller, "Making Up for Losses."

56. See Roger, *Life Sciences*, 99–104.

57. On the distinction between preexistence and preformation, see Roger, *Life Sciences*, 259ff., and Peter J. Bowler, "Preformation and Pre-Existence in the Seventeenth Century: A Brief Analysis," *Journal of the History of Biology* 4 (1971): 221–244.

58. Quoted from Francis Joseph Cole, *Early Theories of Sexual Generation* (Oxford: Clarendon Press, 1930), 12; on theories of preexistence in the seventeenth and eighteenth centuries in general, see also Elizabeth Gasking, *Investigations into Generation 1651–1828* (London: Hutchinson, 1967), and Clara Pinto-Correia, *The Ovary of Eve: Egg and Sperm and Preformation* (Chicago: University of Chicago Press, 1997).

59. Nicholas Malebranche, *Dialogues on Metaphysics and Religion*, trans. J. Bennett, http://www.earlymoderntexts.com/pdfbits/ml5.pdf, 2007 (accessed September 29, 2011), 121. Malebranche refers to bees and their larvae in the original.

60. See Howard B. Adelmann, *Marcello Malpighi and the Evolution of Embryology*, vol. 2 (Ithaca, NY: Cornell University Press, 1966), 872; Wolf Lepenies, *Das Ende der Naturgeschichte: Wandel kultureller Selbstverständlichkeiten in den Wissenschaften des 18. und 19. Jh.* (Munich: Hanser, 1976), 45; and Helmut Müller-Sievers, *Self-Generation: Biology, Philosophy, and Literature around 1800* (Stanford, CA: Stanford University Press, 1997), 26–30.

61. Carl Linnaeus, *Amoenitates academicae, seu Dissertationes variae Physicae, Medicae, Botanicae antehac seorsim editae*, vol. 1 (Stockholm: Godofredus Kiesewetter, 1749), 327.

62. Ibid., 347–349.

63. Ibid., 368.

64. Ibid., 359.

65. Caspar Friedrich Wolff, *Theorie von der Generation: In 2 Abh. erkl. u. bewiesen* (Berlin: Birnstiel, 1764), 25.

66. Linnaeus, *Amoenitates academicae*, vol. 1, 344, 351, 354, 358.

67. Carl Linnaeus, *Linnaeus' Philosophia Botanica*, trans. S. Freer (Oxford: Oxford University Press, 2003), 52.

68. See Peter F. Stevens and Sean P. Cullen, "Linnaeus, the Cortex-Medulla Theory, and the Key to His Understanding of Plant Form and Natural Relationships," *Journal of the Arnold Arboretum* 71 (1990): 179–220.

69. Carl Linnaeus, *Systema Naturae, sive Regna Tria Naturae systematice proposita per classes, ordines, genera, et species* (Leiden: De Groot, 1735), unpag. [p. 1].

70. Carl Linnaeus, *Select Dissertations from the Amoenitates Academicae*, trans. F. J. Brand (1781; repr. New York: Arno Press, 1977), 77.

71. Linnaeus, *Systema Naturae*, unpag. [p. 1]; see Staffan Müller-Wille, "Genealogie, Naturgeschichte und Naturgesetz bei Linné und Buffon," in *Genealogie als Denkform in Mittelalter und Früher Neuzeit*, ed. K. Heck and B. Jahn (Tübingen: Niemeyer, 2000), 109–119.

72. See Staffan Müller-Wille, *Botanik und weltweiter Handel: Zur Begrün-*

*dung eines Natürlichen Systems der Pflanzen durch Carl von Linné (1707–1778)* (Berlin: Verlag für Wissenschaft und Bildung, 1999), 267–284.

73. Carl Linnaeus, *Miscellaneous Tracts Relating to Natural History, Husbandry, and Physick*, trans. B. Stillingfleet (London: J. Dodsley; Leigh and Sotheby; T. Payne, 1791), 40.

74. Linnaeus, *Amoenitates academicae*, vol. 6, 18; see Camille Limoges, "Introduction," in C. *Linné: L' équilibre de la nature: L'histoire des sciences; textes et études*, trans. B. Jasmin (Paris: Vrin, 1972), 7–24; and Staffan Müller-Wille, "Nature as a Marketplace: The Political Economy of Linnaean Botany," in *Oeconomies in the Age of Newton*, ed. N. Di Marchi and M. Schabas, History of Political Economy, Supplement, vol. 35 (Durham, NC: Duke University Press, 2003), 155–173.

75. See Amundson, *Changing Role of the Embryo*, pp. 36–38, for a critical evaluation of this view.

76. Carl Friedrich Kielmeyer, *Über die Verhältnisse der organischen Kräfte unter einander in der Reihe der verschiedenen Organisationen*, with an introduction by K. T. Kanz (Marburg: Basilisken-Presse, 1993), 5; see Robert J. Richards, *The Romantic Conception of Life: Science and Philosophy in the Age of Goethe* (Chicago: University of Chicago Press, 2002), 238–251.

77. Michel Foucault, *The Order of Things: An Archaeology of the Human Sciences*, trans. A. M. Sheridan Smith (London: Routledge, 1997), chap. 7.3; Jacob, *Logic of Life*, chap. 2.

78. William Coleman, *Biology in the Nineteenth Century: Problems of Form, Function, and Transformation* (New York: John Wiley, 1971) misses this dimension in his otherwise compelling portrait of nineteenth-century biology.

79. See Philip C. Ritterbush, *Overtures to Biology: The Speculations of Eighteenth Century Naturalists* (New Haven, CT: Yale University Press, 1964), chap. 5; Roger, *Life Sciences*, 459–472; Jacob, *Logic of Life*, 76–80; Timothy Lenoir, *The Strategy of Life: Teleology and Mechanics in Nineteenth-Century Germany* (Chicago: University of Chicago Press, 1982), chap. 1; James L. Larson, *Interpreting Nature: The Science of Living Form from Linnaeus to Kant* (Baltimore: Johns Hopkins University Press, 1994), chap. 5; Emma C. Spary, *Utopia's Garden: French Natural History from Old Regime to Revolution* (Chicago: University of Chicago Press, 2000), chap. 3.

80. Quoted from John Lyon and Phillip Reid Sloan, eds., *From Natural History to the History of Nature: Readings from Buffon and His Critics* (Notre Dame, IN: University of Notre Dame Press, 1981), 111.

81. Ohad Parnes, "On the Shoulders of Generations: The New Epistemology of Heredity in the Nineteenth Century," in Müller-Wille and Rheinberger, *Heredity Produced*, 315–346.

82. Immanuel Kant, *Kant's Critique of Judgement*, 2nd ed., trans. J. H. Bernard (London: Macmillan, 1914), 278; see Jacob, *Logic of Life*, 88–92, and Peter McLaughlin, *Kant's Critique of Teleology in Biological Explanation* (Lewiston, NY: Edwin Mellen Press, 1990), 44–51; on the history of Kant's concept of self-organization in the late eighteenth and nineteenth centuries, see Lenoir, *Strategy of Life*, and Müller-Sievers, *Self-Generation*.

83. Charles Bonnet, *Considérations sur les corps organisés, où l'on traite de leur origine, de leur développement, de leur réproduction, &c*, vol. 1 (Amsterdam: Rey, 1762), 52; see François Duchesneau, "Charles Bonnet's Neo-Leibnizan Theory of Organic Bodies," in Smith, *Problem of Animal Generation*, 285–314.

84. Released upon the death of an individual, the organic molecules became reassociated according to the inner mould of the organism that took them up. On Blumenbach's concept of "formative drive," see Timothy Lenoir, "Kant, Blumenbach, and Vital Materialism in German Biology," *Isis* 71 (1980): 77–108; Peter McLaughlin, "Blumenbach und der Bildungstrieb: Zum Verhältnis von epigenetischer Embryologie und typologischem Artbegriff," *Medizinhistorisches Journal* 17 (1982): 357–372; and Robert J. Richards, "Kant and Blumenbach on the Bildungstrieb: A Historical Misunderstanding," *Studies in the History and Philosophy of Biology and the Biomedical Sciences* 31 (2000): 11–32; on Buffon's concept of an "inner mould," see Roger, *Life Sciences*, 439–452; and Annie Ibrahim, "La notion de moule intérieur dans les théories de la génération au XVIIIe siècle," *Archives de Philosophie* 50 (1987): 555–580.

85. Karen Detlefsen, "Explanation and Demonstration in the Haller-Wolff Debate," in Smith, *Problem of Animal Generation*, 235–261.

86. Charles Darwin, *On the Origin of Species: A Facsimile of the First Edition*, introduction by E. Mayr (Cambridge, MA: Harvard University Press, 1966), 422.

87. Ibid., 345.

88. Ibid., 117 and 317; on Darwin's diagram and earlier drafts thereof, see Julia Voss, *Darwin's Pictures: Views of Evolutionary Theory, 1837–1874* (New Haven, CT: Yale University Press, 2010), chap. 2.

89. Darwin, *On the Origin of Species*, 315.

90. See Jonathan Hodge, "Darwin as a Lifelong Generation Theorist," in *The Darwinian Heritage*, ed. D. Kohn (Princeton, NJ: Princeton University Press, 1985), 204–244.

91. Charles Darwin, *The Works of Charles Darwin*, vols. 19 and 20, *The Variation of Animals and Plants under Domestication*, ed. P. H. Barrett and R. B. Freeman, vol. 2 (New York: New York University Press, 1988), 30–31. Quatrefages applied his concept of a *tourbillon vital* ("vital whirlpool") to "the strange activity . . . in the semi-sarcodic substance which surrounds [the] siliceous or horny skeleton [of sponges]"; see Armand de Quatrefages de Bréau, *The Human Species* (London: Kegan Paul, Trench & Co, 1883), 4.

92. See Staffan Müller-Wille, "Hybrids, Pure Cultures, and Pure Lines: From Nineteenth-Century Biology to Twentieth Century Genetics," *Studies in History and Philosophy of the Biological and Biomedical Sciences* 38 (2007): 796–806.

93. Gaston Bachelard, *The New Scientific Spirit*, trans. A. Goldhammer (Boston: Beacon Press, 1984), 16.

94. Alexandre Koyré, "Galileo and the Scientific Revolution of the Seventeenth Century," *Philosophical Review* 52 (1943): 333–348.

## CHAPTER THREE

1. See Carlos López Beltrán, "In the Cradle of Heredity: French Physicians and *l'hérédité naturelle* in the Early Nineteenth Century," *Journal of the History of Biology* 37 (2004): 39–72.

2. *Oxford English Dictionary*, online edition, http://www.oed.com (accessed September 19, 2011), s.v. "heredity."

3. Bryan A. Gardener, *Black's Law Dictionary*, 7th ed. (St. Paul, MN: West Publishing, 1999), s.v. "heredity."

4. Francis Ludes, Arnold O. Ginnow, Lawrence J. Culligan, Robert J. Owens, and Matthew J. Canavan, eds., *Corpus Juris Secundum: A Contemporary Statement of American Law as Derived from Reported Cases and Legislation*, vol. 39A, s.v. "heredity" (Saint Paul, MN: Thomson-West, 2003).

5. See Maaike van der Lugt, "Les maladies héréditaires dans la pensée scolastique (XIIe–XVIe siècle)," in *L'hérédité entre Moyen Age et époque moderne*, ed. M. van der Lugt and C. de Miramon (Florence: SISMEL—Edizioni del Galluzzo, 2008), 273–322.

6. Jean François Fernel, *Medicina ad Henricum II Galliarum regem christianissimum*, vol. 2, *Pathologia* (Paris: Andreas Wechel, 1554), 15.

7. Robert Burton, *The Anatomy of Melancholy*, 1621, Projekt Gutenberg Edition, 61, http://www.gutenberg.org/files/10800/10800-h/10800-h.htm (accessed March 4, 2010).

8. Lesky, *Zeugungs- und Vererbungslehren*, 148–155.

9. Michel de Montaigne, *The Complete Essays of Montaigne*, trans. D. M. Frame (Stanford, CA: Stanford University Press, 1976), 578.

10. Johann Heinrich Zedler, *Großes vollständiges Universallexikon aller Wissenschaften und Künste*, vol. 47 (1747) (Graz: Akademische Druck- und Verlags-Anstalt, 1993), 510; see also Peter McLaughlin, "Kant on Heredity and Adaptation," in Müller-Wille and Rheinberger, *Heredity Produced*, 281. On the meaning of Latin *hereditas*, see Franck Roumy, "La naissance de la notion canonique de consanguinitas et sa réception dans le droit civil," in Lugt and Miramon, *L'hérédité entre Moyen Age*, 41–66.

11. Roumy, "La naissance de la notion canonique," emphasizes that *hereditas* in Roman law was used in that technical spirit. In legal practice, family backgrounds were of course considered, increasingly so from the late classical period.

12. See Jack Goody, *The Development of the Family and Marriage in Europe* (Cambridge: Cambridge University Press, 1983), 123.

13. See ibid., 136–144.

14. See David W. Sabean, *Kinship in Neckarhausen, 1700–1870* (Cambridge: Cambridge University Press, 1998), chap. 3.

15. See Kilian Heck and Bernhard Jahn, eds., *Genealogie als Denkform in Mittelalter und Früher Neuzeit* (Tübingen: Niemeyer, 2000).

16. Goody, *Development of the Family*, 142–146.

17. See David W. Sabean, "From Clan to Kindred: Kinship and the Circulation of Property in Premodern and Modern Europe," in Müller-Wille and Rheinberger, *Heredity Produced*, 37–59.

18. Henry de Bracton, *On the Laws and Customs of England*, vol. 2, ed. G. E. Woodbridge, trans. S. L. Thorne (Cambridge, MA: Belknap Press, 1968–1977), 184.

19. Charles de Miramon, "Aux origines de la noblesse et des princes du sang: France et Angleterre au XIVe siècle," in Lugt and Miramon, *L'hérédité entre Moyen Age*, 157–210.

20. David W. Sabean, *Power in the Blood: Popular Culture and Village Discourse in Early Modern Germany* (Cambridge: Cambridge University Press, 1984), chap. 3.

21. Miramon, "Aux origines de la noblesse," 210.

22. Karl Marx and Friedrich Engels, *Collected Works*, vol. 3 (London: Lawrence & Wishart, 1975), 106; original emphasis.

23. See Klaus Oschema, "Maison, noblesse et légitimité: Aspects de la notion d'hérédité dans le milieu de la cour bourguignonne (XVe siècle)," in Lugt and Miramon, *L'hérédité entre Moyen Age*, 211–244.

24. Marx and Engels, *Collected Works*, vol. 3, 99.

25. Eduard Gans, *Das Erbrecht in weltgeschichtlicher Entwickelung*, 4 vols. (Berlin: Maurer, 1824–1835).

26. See *Civil Code*, http://www.napoleon-series.org/research/government/c_code.html (accessed October 2, 2011), §§ 731, 732, 735, 745, 756.

27. Examples are E. T. A. Hoffmann's narration "Das Majorat" of 1817 or Achim von Arnim's narration "Die Majoratsherren" of 1819; see Ulrike Vedder, "Continuity and Death: Literature and the Law of Succession in the Nineteenth Century," in Müller-Wille and Rheinberger, *Heredity Produced*, 85–101.

28. Alexis de Tocqueville, *Democracy in America*, vol. 1 (New York: Alfred A. Knopf, 1948), 48.

29. See Janet Browne, *Charles Darwin: The Power of Place; Volume II of a Biography* (New York: Alfred A. Knopf, 2002), chap. 8.

30. Sabean, "From Clan to Kindred," 45 and 52.

31. See De Renzi, "Resemblance, Paternity, and Imagination."

32. See Anne Carol, *Histoire de l'eugénisme en France: Les médecins et la procréation, XIXe–XXe siècle* (Paris: Seuil, 1995), 17–26; and Lugt, "Les maladies héréditaires."

33. See Lugt, "Les maladies héréditaires."

34. On humoral pathology, see Erich Schöner, *Das Viererschema in der antiken Humoralpathologie* (Wiesbaden: Steiner, 1964).

35. On Dino del Garbo and John of Gaddesden, see Lugt, "Les maladies héréditaires."

36. Ibid., p. 300.

37. Quoted according to David F. Musto, "The Theory of Hereditary Disease of Luis Mercado, Chief Physician to the Spanish Habsburgs," *Bulletin of the History of Medicine* 35 (1961): 361.

38. See Lugt, "Les maladies héréditaires," p. 293.

39. See Carlos López Beltrán, "The Medical Origins of Heredity," in Müller-Wille and Rheinberger, *Heredity Produced*, 105–132.

40. See Robert C. Olby, "Constitutional and Hereditary Disorders," in *Com-*

*panion Encyclopedia of the History of Medicine,* ed. W. F. Bynum and R. Porter, vol. 1 (London: Routledge, 1993), 412–437.

41. See López-Beltrán, "Medical Origins of Heredity."

42. Gianna Pomata, "Comments on Session III: Heredity and Medicine," in *Conference: A Cultural History of Heredity II: 18th and 19th Centuries,* preprint 247 (Berlin: Max Planck Institute for the History of Science, 2003), 150.

43. See Phillip K. Wilson, "Erasmus Darwin and the 'Noble' Disease (Gout): Conceptualizing Heredity and Disease in Enlightenment England," in Müller-Wille and Rheinberger, *Heredity Produced,* 133–154; on gout as a "patrician disease," see also Roy Porter and George S. Rousseau, *Gout: The Patrician Malady* (New Haven, CT: Yale University Press, 1998).

44. See Laure Cartron, "Degeneration and 'Alienism' in Early Nineteenth-Century France," in Müller-Wille and Rheinberger, *Heredity Produced,* 155–174.

45. See John C. Waller, "'The Illusion of an Explanation': The Concept of Hereditary Disease, 1770–1870," *Journal of the History of Medicine* 57 (2002): 410–448.

46. See López Beltrán, "Medical Origins of Heredity."

47. See Ann F. La Berge, *Mission and Method: The Early Nineteenth-Century French Public Health Movement* (Cambridge: Cambridge University Press, 1992), and Elizabeth A. Williams, *The Physical and the Moral: Anthropology, Physiology, and Philosophical Medicine in France, 1750–1850* (Cambridge: Cambridge University Press, 1994).

48. See Maaike van der Lugt and Charles de Miramon, "Introduction," in Lugt and Miramon, *L'hérédité entre Moyen Age,* 3–40.

49. Immanuel Kant, "Determination of the Concept of a Human Race," trans. H. Wilson and Günter Zöller, in *The Cambridge Edition of the Works of Immanuel Kant,* vol. 7, *Anthropology, History, and Education,* ed. R. B. Loudon and G. Zöller (Cambridge: Cambridge University Press, 2007), 143–153; "On the Use of Teleological Principles in Philosophy," trans. G. Zöller, in ibid., 192–218. On Kant's racial anthropology, see Raphaël Lagier, *Les races humaines selon Kant* (Paris: Presses Universitaires de France, 2004), and McLaughlin, "Kant on Heredity."

50. Staffan Müller-Wille, "Ein Anfang ohne Ende: Das Archiv der Naturgeschichte und die Geburt der Biologie," in *Macht des Wissens: Die Entstehung der modernen Wissengesellschaft,* ed. R. v. Dülmen and S. Rauschenbach (Cologne: Böhlau, 2004), 587–605.

51. For Pisa, see Dietrich v. Engelhardt, "Luca Ghini (um 1490–1556) und die Botanik des 16. Jahrhunderts: Leben, Initiativen, Kontakte, Resonanz," *Medizinhistorisches Journal* 30 (1995): 3–49; for Padua, see Alessandro Minelli, ed., *L'Orto botanico di Padova 1545–1995* (Venice: Marsilio, 1995).

52. On the history of botanical gardens in the seventeenth century, see William T. Stearn, "Botanical Gardens and Botanical Literature in the Eighteenth Century," in *Catalogue of Botanical Books in the Collection of Rachel McMasters Miller Hunt,* ed. J. Quinby and A. Stevenson, vol. 2 (Pittsburgh, PA: Hunt Botanical Library, 1961), 41–140; and D. Onno Wijnands, "Hortus auriaci: The

Gardens of Orange and their Place in Late 17th-Century Botany and Horticulture," *Journal of Garden History* 8 (1988): 61–86, 271–304; on botany and colonialism, see Richard Drayton, *Nature's Government: Science, Imperial Britain, and the "Improvement" of the World* (New Haven, CT: Yale University Press, 2000).

53. Ray, *Historia Plantarum*, vol. 1, 40

54. Georges Louis Leclerc, Comte de Buffon, *Natural History, General and Particular, by the Count de Buffon*, trans. W. Smellie, vol. 3 (Edinburgh: Printed for William Creech, 1780), 405.

55. See Staffan Müller-Wille, "'Varietäten auf ihre Arten zurückführen'—Zu Carl von Linnés Stellung in der Vorgeschichte der Genetik," *Theory in Biosciences* 117 (1998): 346–376.

56. See Conway Zirkle, "The Early History of the Idea of the Inheritance of Acquired Characters and of Pangenesis," *Transactions of the American Philosophical Society*, new series, 35 (1946): 91–151; Peter Bowler, *The Eclipse of Darwinism* (Baltimore: Johns Hopkins University Press, 1992), chap. 4; and Wolfgang Lefèvre, "Inheritance of Acquired Characters: Heredity and Evolution in Late Nineteenth-Century Germany," in *A Cultural History of Heredity III: 19th and Early 20th Centuries*, preprint 294 (Berlin: Max Planck Institute for the History of Science, 2005), 53–66.

57. For a late synopsis, see Charles Naudin and Ferdinand von Müller, *Manuel de l'acclimateur ou choix des plantes recommandées pour l'agriculture, l'industrie et la médecine et adaptées aux divers climates de l'Europe et des pays tropicaux* (Paris: Société d'Acclimatation, 1887).

58. Michael A. Osborne, *Nature, the Exotic, and the Science of French Colonialism* (Bloomington: Indiana University Press, 1994).

59. Joseph Gottlieb Kölreuter, *Vorläufige Nachricht von einigen, das Geschlecht der Pflanzen betreffenden Versuchen und Beobachtungen nebst Fortsetzungen 1, 2 und 3 (1761–1766)* (Leipzig: Wilhelm Engelmann, 1893).

60. Carl Friedrich Gärtner, *Versuche und Beobachtungen über die Bastarderzeugung im Pflanzenreich* (Stuttgart: At the expense of the author, 1849), 234–235.

61. See Olby, *Origins of Mendelism*, chaps. 1 and 2, as well as Staffan Müller-Wille and Vítězslav Orel, "From Linnaean Species to Mendelian Factors: Elements of Hybridism, 1751–1870," *Annals of Science* 64 (2007): 171–215.

62. On Buffon's hybridization experiments, see Spary, *Utopia's Garden*, 112–114.

63. For details on animal husbandry in the medieval ages, see Lugt and Miramon, "Introduction."

64. See Roger J. Wood, "The Sheep Breeders' View of Heredity Before and After 1800," in Müller-Wille and Rheinberger, *Heredity Produced*, 230–232.

65. Peter Lauremberg, *Horticultura, libris II. comprehensa; huic nostro coelo et solo accomodata* (Frankfurt/M.: Merian, 1631), 8.

66. See Russell, *Like Engend'ring Like*, and Wood, "The Sheep Breeders' View."

67. John Saunders Sebright, "The Art of Improving the Breeds of Domestic Animals, in a Letter, addressed to the Right Hon. Sir Joseph Banks," in *Artificial Selection and the Development of Evolutionary Theory*, ed. C. J. Bajema (Stroudsburg, PA: Hutchinson Ross, 1982), 98; original emphasis.

68. Roger J. Wood and Vítězslav Orel, *Genetic Prehistory in Selective Breeding: A Prelude to Mendel* (Oxford: Oxford University Press, 2001), 125–131.

69. See Marc J. Ratcliff, "Duchesne's Strawberries: Between Growers' Practices and Academic Knowledge," in Müller-Wille and Rheinberger, *Heredity Produced*, 205–228.

70. See Robert C. Olby, "Mendel No Mendelian?" *History of Science* 17 (1979): 53–72; and Bowler, *Mendelian Revolution*, chap. 5.

71. See Vítězslav Orel and Roger J. Wood, "Essence and Origin of Mendel's Discovery," *Comptes rendus de l'Académie des Sciences Paris, Sciences de la vie* 323 (2000): 1037–1041.

72. Georges Louis Leclerc, Comte de Buffon, *De l'Homme*, ed. M. Duchet (Paris: L'Harmattan, 1971), 352.

73. See Staffan Müller-Wille, "Schwarz, Weiß, Gelb, Rot: Zur Darstellung menschlicher Vielfalt," in *Dingwelten: Das Museum als Erkenntnisort*, ed. A. te Heesen and P. Lutz (Cologne: Böhlau, 2005), 161–170.

74. See Benjamin Braude, "Sons of Noah," *William and Mary Quarterly*, 3rd series, 54 (1997): 103–142.

75. See Renato G. Mazzolini, "Il colore della pelle e l'origine dell'antropologia fisica (1492–1848)," in *L'epopea delle scoperte*, ed. R. Zorzi (Firenze: Leo S. Olschki, 1994), 227–239.

76. See Nicolas Pethes, "'Victor, l'enfant de la forêt': Experiments on Heredity in Savage Children," in Müller-Wille and Rheinberger, *Heredity Produced*, 399–418.

77. Philip R. Sloan, "The Gaze of Natural History," in *Inventing Human Science: Eighteenth-Century Domains*, ed. C. Fox, R. Porter, and R. Wokler (Berkeley: University of California Press, 1995), 114.

78. Renato G. Mazzolini, "Las Castas: Inter-Racial Crossing and Social Structure (1770–1835)," in Müller-Wille and Rheinberger, *Heredity Produced*, 365.

79. On "wolf children," see Pethes, "'Victor, l'enfant de la forêt'"; on the ideology of the self-made man, see Paul White, "Acquired Character: The Heredity Material of the 'Self-Made Man,'" in Müller-Wille and Rheinberger, *Heredity Produced*," 375–398; and for the genius controversy, see Stefan Willer, "Sui Generis: Heredity and Heritage of Genius at the Turn of the Eighteenth Century," in Müller-Wille and Rheinberger, *Heredity Produced*, 419–440.

80. See Helmut Müller-Sievers, "The Heredity of Poetics," in Müller-Wille and Rheinberger, *Heredity Produced*, 443–465.

81. Michel Foucault, *The Archeology of Knowledge and the Discourse of Language* (New York: Pantheon Books, 1972), 187.

CHAPTER FOUR

1. López Beltrán, "Medical Origins of Heredity," 125.

2. See Lesky, *Zeugungs- und Vererbungslehren*; Stubbe, *History of Genetics*, chaps. 1–3; Rey, "Génération et hérédité."

3. Pierre Louis Moreau de Maupertuis, *Oeuvres*, vol. 2, 115 and 118 (Hildesheim: Olms, 1965); see Mary Terrall, *The Man Who Flattened the Earth:*

*Maupertuis and the Sciences in the Enlightenment* (Chicago: University of Chicago Press, 2002), chap. 7.

4. *Forerunners of Darwin, 1745–1859* is the title of the classical collection of papers on eighteenth-century theories of generation edited by Bentley Glass, Owsei Temkin, and William L. Strauss).

5. Maupertuis, *Oeuvres*, vol. 2, 149; see Terrall, *Man Who Flattened the Earth*, 215–221.

6. See Terrall, "Speculation and Experiment."

7. See Lagier, *Les races humaines*.

8. See Harriet Ritvo, *The Animal Estate* (Cambridge, MA: Harvard University Press, 1987).

9. See Lisbet Koerner, *Linnaeus: Nature and Nation* (Cambridge, MA: Harvard University Press, 1999).

10. See White, "Acquired Character."

11. Darwin, *On the Origin of Species*, 12–13; for Darwin's theory of inheritance, see Rasmus G. Winther, "Darwin on Variation and Heredity," *Journal of the History of Heredity* 33 (2000): 425–455.

12. Darwin, *On the Origin of Species*, 331.

13. See Cartron, "Degeneration and 'Alienism.'"

14. Cf. Carl J. Bajema, *Artificial Selection and the Development of Evolutionary Theory* (Stroudsburg, PA: Hutchinson Ross, 1982).

15. Gayon, *Darwinism's Struggle for Survival*, 11.

16. Prosper Lucas, *Traité philosophique et physiologique de l'hérédité naturelle* (Paris: J. B. Baillière, 1847–1850), vol. 1, XVI.

17. Ibid., vol. 2, 509.

18. Ibid., 900.

19. Darwin, *On the Origin of Species*, chap. 1; see Peter J. Bowler, "Variation from Darwin to the Modern Synthesis," in *Variation: A Central Concept in Biology*, ed. B. Hallgrimsson and B. K. Hall (New York: Elsevier, 2005), 414–426.

20. Ibid., 13.

21. Darwin, *Variation of Animals and Plants*, vol. 2, 321.

22. See Voss, *Darwin's Pictures*.

23. Francis Galton, "Hereditary Talent and Character," *Macmillan's Magazine* (1865): 322.

24. Gayon, *Darwinism's Struggle for Survival*, 9; see for example Darwin, *On the Origin of Species*, 350.

25. Galton, "Hereditary Talent and Character," 322; original emphasis.

26. Galton, "Theory of Heredity," 331.

27. Ibid., 343.

28. Ibid., 346. For the reception of this conclusion in the twentieth century, see also chapter 6.

29. Francis Galton, "On Blood Relationship," *Proceedings of the Royal Society* 20 (1872): 400–401; original emphasis.

30. See Manfred D. Laubichler and Jane Maienschein, eds., *From Embryology to Evo-Devo: A History of Developmental Evolution* (Cambridge, MA: MIT Press, 2007).

31. Churchill, "From Heredity Theory to 'Vererbung,'" 337.

32. François Duchesneau, "The Delayed Linkage of Heredity with the Cell Theory," in Müller-Wille and Rheinberger, *Heredity Produced*, 310.

33. Rudolf Ludwig Virchow, *Cellular Pathology as Based upon Physiological and Pathological Histology*, trans. F. Chance (London: John Churchill, 1860), 10.

34. See Thomas Cremer, *Von der Zellenlehre zur Chromosomentheorie: Naturwissenschaftliche Erkenntnis und Theoriewechsel in der frühen Zell- und Vererbungsforschung* (Berlin: Springer, 1985).

35. Theodor Boveri, *Zellen-Studien*, vol. 2 (Jena: Gustav Fischer, 1888), 5.

36. Theodor Boveri, "An Organism Produced Sexually Without Characteristics of the Mother," *American Naturalist* 27 (1893): 222–232; see also Manfred D. Laubichler and Eric H. Davidson, "Boveri's Long Experiment: Sea Urchin Merogones and the Establishment of the Role of Nuclear Chromosomes in Development," *Developmental Biology* 314 (2007): 1–11.

37. Boveri, "Organism Produced Sexually."

38. Churchill, "From Heredity Theory to 'Vererbung,'" 355; see also Brian Bracegirdle, *A History of Microtechnique: The evolution of the Microtome and the Development of Tissue Preparation*, 2nd ed. (Lincolnwood: Science Heritage Ltd., 1986), and Frederick B. Churchill, "Hertwig, Weismann, and the Meaning of Reduction Division circa 1890," *Isis* 61 (1970): 429–457.

39. Carl Wilhelm von Nägeli, *Mechanisch-Physiologische Theorie der Abstammungslehre* (Munich: R. Oldenbourg, 1884), 46; see Hans-Jörg Rheinberger, "Heredity and Its Entities around 1900, *Studies in History and Philosophy of Science* 39 (2008): 370–374, on the theories discussed in this section.

40. Ibid., 23.

41. Ibid., 42 ff.

42. Ibid., 42.

43. Ibid., 26.

44. Ibid., 583.

45. Ibid., 583.

46. Ibid., 44.

47. "Carl Wilhelm von Nägeli, "Die Individualität in der Natur mit vorzüglicher Berücksichtigung des Pflanzenreiches," *Veröffentlichungen des Wissenschaftlichen Vereins in Zürich* 1 (1856): 203; cf. Parnes, "On the Shoulders of Generations."

48. Nägeli, *Mechanisch-Physiologische Theorie der Abstammungslehre*, 41.

49. Ibid., 275.

50. Ibid., 275.

51. Ibid., 25.

52. See Hans-Jörg Rheinberger, "Morphologie bei Claude Bernard," *Aufsätze und Reden der Senckenbergischen Naturforschenden Gesellschaft* 41 (1994): 137–150; Jacob, *Logic of Life*, chap. 4.

53. Claude Bernard, *Leçons sur les phénomènes de la vie communs aux animaux et aux végétaux* (Paris: Vrin, 1966), 313.

54. Ibid., 311.

55. Claude Bernard, *De la physiologie générale* (Bruxelles: Culture et Civilisation, 1965), 148.

56. Ibid., 177.

57. Bernard, *Leçons*, 315 and 336.

58. Ibid., 331; original emphasis.

59. Ibid., 342.

60. Edmund Beecher Wilson, *The Cell in Development and Heredity* (New York: MacMillan, 1896), 261–263; on this book see Ariane Dröscher, "Edmund B. Wilson's *The Cell* and Cell Theory between 1896 and 1925," *History and Philosophy of the Life Sciences* 24 (2002): 357–389.

61. Hugo de Vries, *Intracellular Pangenesis*, trans. C. S. Gager (Chicago: Open Court Publishing, 1910), 34.

62. Ibid., 43.

63. Ibid., 199.

64. Ibid., 195.

65. See, for example, de Vries, *Intracellular Pangenesis*, 222.

66. August Weismann, *The Germ-Plasm: A Theory of Heredity*, trans. W. Newton Parker (New York: Charles Scribner's Sons, 1893), 61.

67. Ibid., 37.

68. Ibid., 49.

69. Ibid., 44–45; original emphasis.

70. Ibid., 77.

71. August Weismann, "The Continuity of the Germ-Plasm as the Foundation of a Theory of Heredity," in *August Weismann: Essays upon Heredity and Kindred Biological Problems*, ed. E. B. Poulton, S. Schönland, and A. E. Shipley, vol. 1 (Oxford: Clarendon Press, 1891), 195.

72. See Rasmus G. Winther, "August Weismann on Germ-Plasm Variation," *Journal of the History of Biology* 34 (2001): 517–555.

73. Weismann, *The Germ-Plasm*, 415.

74. See Ernst Mayr, "The Recent Historiography of Genetics," *Journal for the History of Biology* 6 (1973): 125–154, and Ernst Mayr, *The Growth of Biological Thought: Diversity, Evolution and Inheritance* (Cambridge, MA: Belknap Press, 1982), 793–797.

75. See Olby, *Origins of Mendelism*, chap. 3.

76. See Stephen Snelders, Frans J. Meijman, and Toine Pieters, "Bismarck the Tomcat and Other Tales: Heredity and Alcoholism in the Medical Sphere, the Netherlands 1850–1900," in *A Cultural History of Heredity III*, 193–211.

77. See George W. Stocking, *Race, Culture, and Evolution: Essays in the History of Anthropology* (Chicago: Chicago University Press, 1982), chap. 10.

78. See Theodore M. Porter, "The Biometric Sense of Heredity: Statistics, Pangenesis and Positivism," in *A Cultural History of Heredity III*, 31–42; Michael Bulmer, *Francis Galton: Pioneer of Heredity and Biometry* (Baltimore: Johns Hopkins University Press, 2003).

79. See Renato G. Mazzolini, *Politisch-Biologische Analogien im Frühwerk Rudolf Virchows* (Marburg: Basilisken-Presse, 1988).

80. Virchow, *Cellular Pathology*, 14.

81. Rudolf Ludwig Virchow, "Der Staat und die Ärzte," in *Rudolf Virchow: Sämtliche Werke*, ed. C. Andree, vol. 28.1 (Hildesheim: Georg Olms Verlag, 2006), 60.

82. Rudolf Ludwig Virchow, "Über die Reform der pathologischen und therapeutischen Anschauungen durch die mikroskopischen Untersuchungen," in *Ar-*

*chiv für pathologische Anatomie und Physiologie und für klinische Medicin,* ed. R. L. Virchow and B. Reinhardt (Berlin: G. Reimer, 1847), 216–217.

83. See Jean Gayon, "Entre force et structure: Genèse du concept naturaliste de l'hérédité," in *Le paradigme de la filiation,* 61–75.

84. Darwin, *Variation of Animals and Plants,* vol. 2, 17.

85. See Jean Gayon and Doris T. Zallen, "The Role of the Vilmorin Company in the Promotion and Diffusion of the Experimental Science of Heredity in France, 1840–1920," *Journal of the History of Biology* 31 (1998): 241–262.

86. See Frederick B. Churchill, "Living with the Biogenetic Law: A Reappraisal," in Laubichler and Maienschein, *From Embryology to Evo-Devo,* 37–81.

87. See Sara Paulson Eigen, "A Mother's Love, a Father's Line: Law, Medicine and the 18th-Century Fictions of Patrilineal Genealogy," in Heck and Jahn, *Genealogie als Denkform,* 87–107.

88. Diane B. Paul, *Controlling Human Heredity: 1865 to the Present* (Amherst, NY: Prometheus Books, 1995), 41.

89. See Ursula Mittwoch, "Sex Determination in Mythology and History," *Arquivos Brasileiros de Endocrinologia e Metabologia* 49 (2005): 7–13.

90. See Helga Satzinger, "The Chromosomal Theory of Heredity and the Problem of Gender Equality in the Work of Theodor and Marcella Boveri," in *A Cultural History of Heredity III,* 101–114.

91. Jonathan Harwood, *Styles of Scientific Thought: The German Genetics Community, 1900–1933* (Chicago: University of Chicago Press, 1993), 315–350; see also Jan Sapp, *Beyond the Gene: Cytoplasmatic Inheritance and the Struggle for Authority in Genetics* (Oxford: Oxford University Press, 1987).

92. See Stefan Willer, "Heritage—Appropriation—Interpretation: The Debate on the Schiller Legacy in 1905," in *A Cultural History of Heredity III,* 167–178.

93. Raphael Falk, "Mendel's Impact," *Science in Context* 19 (2006): 229.

CHAPTER FIVE

1. Warren Thompson, "Population," *American Journal of Sociology* 34 (1929): 959–975.

2. Demographic figures from Otto Andersen, "Denmark," in *European Demography and Economic Growth,* ed. W. R. Lee (London: Croom Helm, 1979), 79–122.

3. Jürgen Osterhammel, *Die Verwandlung der Welt: Eine Geschichte des 19. Jahrhunderts* (Munich: Beck, 2009), 191.

4. Paul Weindling, *Health, Race and German Politics between National Unification and Nazism, 1870–1945* (Cambridge: Cambridge University Press, 1989), 261.

5. See Philipp Sarasin, *Reizbare Maschinen: Eine Geschichte des Körpers 1765–1914* (Frankfurt/M.: Suhrkamp, 2001).

6. See Richard A. Soloway, *Demography and Degeneration: Eugenics and the Declining Birthrate in Twentieth-Century Britain* (Chapel Hill: University of North Carolina Press, 1995).

7. Weindling, *Health, Race and German Politics,* 5.

8. See ibid., chaps. 3 and 4.

9. On eugenic movements in Germany, see Peter Weingart, Kurt Bayertz, and Jürgen Kroll, *Rasse, Blut und Gene: Geschichte der Eugenik und Rassenhygiene in Deutschland* (Frankfurt/M.: Suhrkamp, 1992); in Great Britain and the United States, Daniel J. Kevles, *In the Name of Eugenics: Genetics and the Use of Human Heredity* (Cambridge, MA: Harvard University Press, 1985); in France, William H. Schneider, *Quality and Quantity: The Quest for Biological Regeneration in Twentieth-Century France* (Cambridge: Cambridge University Press, 1990), and Carol, *Histoire de l'eugénisme en France*; in Scandinavia, Gunnar Broberg and Nils Roll-Hansen, eds., *Eugenics and the Welfare State* (East Lansing: Michigan State University Press, 1996); in Latin America, Nancy Stepan, *The Hour of Eugenics: Race, Gender, and Nation in Latin America* (Ithaca, NY: Cornell University Press, 1991); in colonial Kenya, Chloe Campbell, *Race and Empire: Eugenics in Colonial Kenya* (Manchester: Manchester University Press, 2007).

10. See Stefan Kühl, *Die Internationale der Rassisten: Aufstieg und Niedergang der internationalen Bewegung für Eugenik und Rassenhygiene im 20. Jahrhundert* (Frankfurt/M.: Campus, 1997).

11. Quoted from Weingart, Bayertz, and Kroll, *Rasse, Blut und Gene*, 110.

12. See Diane B. Paul, "Eugenics and the Left," *Journal of the History of Ideas* 45 (1984): 567–590; on eugenics and the labor movement in Germany, see Reinhard Mocek, *Biologie und soziale Befreiung: Zur Geschichte des Biologismus und der Rassenhygiene in der Arbeiterbewegung* (Frankfurt/M.: Peter Lang, 2002); and R. Mocek, "Biology of Liberation: Some Historical Aspects of 'Proletarian Race Hygienics,'" in *From Physico-Theology to Bio-Technology*, ed. M. Teich and K. Bayertz (Amsterdam: Rodopi, 1998), 224–231.

13. Paul, *Controlling Human Heredity*, 10.

14. Kevles, *In the Name of Eugenics*, 59.

15. See statistical data in Weindling, *Health, Race and German Politics*, 145–146 and 499.

16. See Donald A. MacKenzie, "Eugenics in Britain," *Social Studies of Science* 6 (1976): 523–527, and Weingart, Bayertz und Kroll, *Rasse, Blut und Gene*, chap. 3.

17. See Kevles, *In the Name of Eugenics*, 24–27, and Weindling, *Health, Race and German Politics*, 146.

18. See Angelique Richardson, *Love and Eugenics in the Late Nineteenth Century: Rational Reproduction and the New Woman* (Oxford: Oxford University Press, 2003).

19. Quoted from Paul, *Controlling Human Heredity*, 20.

20. See John M. Efron, *Defenders of the Race: Jewish Doctors and Race Science in Fin-de-Siècle Europe* (New Haven, CT: Yale University Press, 1994); Gregory Michael Dorr, *Segregation's Science: Eugenics and Society in Virginia* (Charlottesville: University of Virginia Press, 2008).

21. On Nazi eugenics, see especially Robert N. Proctor, *Racial Hygiene: Medicine Under the Nazis* (Cambridge, MA: Harvard University Press, 1988).

22. See Lene Koch, "The Meaning of Eugenics: Reflections on the Govern-

ment of Genetic Knowledge in the Past and the Present," *Science in Context* 17 (2004): 315–331.

23. See Mark B. Adams, ed., *The Wellborn Science: Eugenics in Germany, France, Brazil, and Russia* (New York: Oxford University Press, 1990).

24. Paul, *Controlling Human Heredity*, 44–45.

25. See Götz Aly and Susanne Heim, *Architects of Annihilation: Auschwitz and the Logic of Destruction*, trans. A. G. Blunden (London: Weidenfeld & Nicolson, 2002).

26. "Protokoll über die Verhandlungen der Schafzüchter-Versammlung in Brünn am 1. und 2. Mai 1837," *Mittheilungen der k. k. Mährisch-Schlesischen Gesellschaft zur Beförderung des Ackerbaues, der Natur- und Landeskunde in Brünn*, 1837, 227.

27. See Hannah Arendt, *The Origins of Totalitarianism* (New York: Schocken Books, 2004), chap. 6 and 7, as well as Foucault, "Faire vivre et laisser mourir."

28. See Jenny Reardon, *Race to the Finish: Identity and Governance in an Age of Genomics* (Princeton, NJ: Princeton University Press, 2005), 18.

29. See Michel Foucault, *Society Must Be Defended: Lectures at the Collège de France, 1975–1976*, trans. David Macey (New York: Picador, 2003), 76–81 and 99–110.

30. Translated from Loïc Rignol, "Augustin Thierry et la politique de l'histoire: Genèse et principes d'un système de pensée," *Revue d'histoire du XIXe siècle* 25 (2002): 87–88.

31. Marx and Engels, *Collected Works*, vol. 3.7, 130.

32. George W. Stocking, "The Turn-of-the-Century Concept of Race," *Modernism/Modernity* 1 (1994): 6.

33. Quoted from Jacob W. Gruber, "Ethnographic Salvage and the Shaping of Anthropology," *American Anthropologist* 72 (1970): 1293; see also George W. Stocking, *Victorian Anthropology* (London: Macmillan, 1987), 240–245.

34. See Patrick Brantlinger, *Dark Vanishings: Discourse on the Extinction of Primitive Races, 1800–1930* (Ithaca, NY: Cornell University Press, 2003), and Stocking, *Victorian Anthropology*, chap. 6. This train of thought is already present in Kant's "Idea for a Universal History with a Cosmopolitan Purpose," in *Kant: Political Writings*, 2nd ed., (Cambridge: Cambridge University Press, 1991), 41–53; see Thomas A. McCarthy, "On the Way to a World Republic? Kant on Race and Development," in *Politik, Moral und Religion: Gegensätze und Ergänzungen*, ed. L. Waas (Berlin: Duncker & Humblot, 2004), 223–242.

35. Charles Lyell, *Principles of Geology, Being an Attempt to Explain the Former Changes of the Earth's Surface, by Reference to Causes Now in Operation*, vol. 2 (London: John Murray, 1832–1833), 156.

36. See Charles Darwin, *The Works of Charles Darwin*, vols. 21 and 22, *The Descent of Man, and Selection in Relation to Sex*, ed. P. H. Barrett and R. B. Freeman, vol. 1 (New York: New York University Press, 1989), 188–198.

37. Johann Gottfried Herder, *Outlines of a Philosophy of the History of Man*, trans. T. Churchill (London: Printed for J. Johnson, by L. Hansard, 1800), 314.

38. Quoted from Gruber, "Ethnographic Salvage," 1293.

39. See Sloan, "Gaze of Natural History," 112–151.

40. On *castas* and the doctrine of temperaments, see Carlos López Beltrán, "Hippocratic Bodies: Temperament and Castas in Spanish America (1570–1820)," *Journal of Spanish Cultural Studies* 8, no. 2 (2007): 253–289; on skin color in the Middle Ages, Maaike van der Lugt, "La peau noire dans la science médiévale," *Micrologus* 13 (2005): 439–475.

41. See Mazzolini, "Il colore della pelle."

42. Carl Linnaeus, *Critica Botanica in quo nomina plantarum generica, specifica, et variantia examini subjiciuntur* (Leiden: Wishoff, 1737), 152–155.

43. Johann Friedrich Blumenbach, "On the Natural Variety of Mankind," in *Anthropological Treatises of Johann Friedrich Blumenbach*, ed. and trans. T. Bendyshe (London: Published for the Anthropological Society by Longman, Green, Longman, Roberts, & Green, 1865), 207–222.

44. Buffon, *Natural History*, vol. 3, 178.

45. See Michael Hagner, *Homo cerebralis: Der Wandel vom Seelenorgan zum Gehirn* (Berlin: Berlin Verlag, 1997).

46. On the history of physical anthropology, see Steven J. Gould, *The Mismeasure of Man* (New York: Norton, 1981), George W. Stocking, ed., *Bones, Bodies, and Behaviour: Essays on Biological Anthropology* (Madison: University of Wisconsin Press, 1988), and Uwe Hoßfeld, *Geschichte der biologischen Anthropologie in Deutschland: Von den Anfängen bis in die Nachkriegszeit* (Stuttgart: Franz Steiner Verlag, 2005).

47. See Lorraine Daston and Peter Galison, "The Image of Objectivity," *Representations* 40 (1992): 81.

48. See Theodore M. Porter, *The Rise of Statistical Thinking: 1820–1900* (Princeton, NJ: Princeton University Press, 1986); and Libby Schweber, *Disciplining Statistics: Demography and Vital Statistics in France and England, 1830–1885* (Durham, NC: Duke University Press, 2006).

49. Quoted from Porter, *Rise of Statistical Thinking*, 129.

50. See Schwartz Cowan, "Nature and Nurture."

51. On Galton's experiments, see Porter, *Rise of Statistical Thinking*, 128–146, and Gayon, *Darwinism's Struggle for Survival*, 105–146.

52. The hypothesis of a law governing ancestral influences is present throughout Galton's oeuvre, first discussed in his "Hereditary Talent and Character," but with an almost exclusive reference to male ancestors.

53. Karl Pearson, "Mathematical Contributions to the Theory of Evolution: On the Law of Ancestral Heredity," *Proceedings of the Royal Society of London* 62 (1898): 411–412; see Donald A. MacKenzie, "Statistical Theory and Social Interests," *Social Studies of Science* 8 (1978): 54–58; Kevles, *In the Name of Eugenics*, chap. 2; Porter, *Rise of Statistical Thinking*, 258–259; and Gayon, *Darwinism's Struggle for Survival*, 132–146.

54. See Christine Hanke, *Zwischen Auflösung und Fixierung: Zur Konstitution von "Rasse" und "Geschlecht" in der physischen Anthropologie um 1900* (Bielefeld: Transcript, 2007).

55. Ludger Müller-Wille, ed., *Franz Boas among the Inuit on Baffin Island*

*(1883–1884): Journals and Letters*, trans. W. Barr (Toronto: University of Toronto Press, 1998).

56. Franz Boas, "Mixed Races," *Science* 17 (1891): 179.

57. See Franz Boas, "The Correlation of Anatomical or Physiological Measurements," *American Anthropologist* 7 (1894): 313–324.

58. Franz Boas, *The Mind of Primitive Man* (New York: Macmillan, 1911). On Boas's criticism of the race concept, see Stocking, *Race, Culture, and Evolution*, chap. 2.

59. See Elazar Barkan, *The Retreat of Scientific Racism: Changing Concepts of Race in Britain and the United States between the World Wars* (Cambridge: Cambridge University Press, 1992), chap. 6.

60. Franz Boas, *Kultur und Rasse* (Leipzig: Veit, 1914), 98–113.

61. See the collected works in Franz Boas, *Race, Language and Culture* (Chicago: University of Chicago Press, 1996).

62. See Maria Kronfeldner, "'If There Is Nothing Beyond the Organic . . .': Heredity and Culture at the Boundaries of Anthropology in the Work of Alfred L. Kroeber," *NTM: Zeitschrift für Geschichte der Wissenschaften, Technik und Medizin* 17 (2009): 107–133.

63. Karl Pearson, *Nature and Nurture: The Problem of the Future* (London: Cambridge University Press, 1913).

64. Karl Pearson, "Mathematical Contributions to the Theory of Evolution III: Regression, Heredity and Panmixia," *Proceedings of the Royal Society of London, Series A, Containing Papers of a Mathematical and Physical Character*, 187 (1896): 255.

65. See Kyung-Man Kim, *Explaining Scientific Consensus: The Case of Mendelian Genetics* (New York: Guilford Press, 1994).

66. See MacKenzie, "Statistical Theory and Social Interests."

67. On the concept of heritability and its aporiae, see Sahotra Sarkar, *Genetics and Reductionism* (Cambridge: Cambridge University Press, 1998), chap. 4.

68. Udny Yule, "Mendel's Laws and Their Probable Relation to Intra-Racial Heredity," *New Phytologist* 1 (1902): 227.

69. Ibid., 206.

70. See Jean Gayon, "From Measurement to Organization: A Philosophical Scheme for the History of the Concept of Heredity," in *The Concept of the Gene in Development and Evolution: Historical and Epistemological Perspectives*, ed. P. Beurton, R. Falk, and H.-J. Rheinberger (Cambridge: Cambridge University Press, 2000), 76–77, and Müller-Wille and Orel, "From Linnaean Species to Mendelian Factors," 211–213.

71. See Gayon, *Darwinism's Struggle for Survival*, 289–298.

72. For a detailed description of the entanglement of genealogical concepts and figures of thought in literature, art, philosophy, and science in the nineteenth and twentieth centuries, see Weigel, *Genea-Logik: Generation, Tradition und Evolution zwischen Kultur- und Naturwissenschaften* (Munich: Fink, 2006).

73. Parnes, "On the Shoulders of Generations," 317; see Parnes, Vedder and Willer, *Das Konzept der Generation*, chaps. 8 and 9, on the further development of the concept in the nineteenth and twentieth centuries.

74. Karl Mannheim, "The Problem of Generations," in *Essays on the Sociology of Knowledge by Karl Mannheim*, ed. P. Kecskemeti (New York: Routledge & Kegan Paul, 1952), 276–320.

75. Parnes, "On the Shoulders of Generations," 324.

76. Galton, *Hereditary Genius*, 50–51.

77. Quoted from Elisabeth Tooker, "Lewis H. Morgan and His Contemporaries," *American Anthropologist* 94 (1992): 364.

78. Lewis H. Morgan, *Systems of Consanguinity and Affinity of the Human Family* (Washington, DC: Smithsonian Institution, 1871), 10–11.

79. See Tooker, "Lewis H. Morgan."

80. See George W. Stocking, *After Tylor: British Social Anthropology 1888–1951* (Madison: University of Wisconsin Press, 1995), 208.

81. See Adam Kuper, "On Human Nature: Darwin and the Anthropologists," in *Nature and Society in Historical Perspective*, ed. M. Teich, R. Porter, and B. Gustafsson (Cambridge: Cambridge University Press, 1997), 274–290; Lawrence Krader, *Ethnologie und Anthropologie bei Marx* (Munich: Hanser, 1973), chap. 2.

82. See Cartron, "Degeneration and 'Alienism.'"

83. See Ian Robert Dowbiggin, *Inheriting Madness: Professionalization and Psychiatric Knowledge in Nineteenth-Century France* (Berkeley: University of California Press, 1991), and Dowbiggin, *Keeping America Sane: Psychiatry and Eugenics in the United States, 1880–1940* (Ithaca, NY: Cornell University Press, 1997). On the concept of diathesis, see Olby, "Constitutional and Hereditary Disorders."

84. Bernd Gausemeier, "Pedigree vs. Mendelism: Concepts of Heredity in Psychiatry Before and After 1900," in *Conference: Heredity in the Century of the Gene (A Cultural History of Heredity IV)*, preprint 343 (Berlin: Max Planck Institute for the History of Science, 2008), 150–152.

85. See Garland E. Allen, "The Eugenics Record Office, Cold Spring Harbor, 1910–1940," *Osiris*, 2nd series, 2 (1986): 225–264.

86. See Phillip K. Wilson, "Pedigree Charts as Tools to Visualize Inherited Disease in Progressive Era America," in *Conference: Heredity in the Century of the Gene*, 163–190.

87. We would like to thank Bernd Gausemeier for the suggestion of this comparison.

88. See Tamara Hareven, "The Search for Generational Memory: Tribal Rites in Industrial Society," *Daedalus* 107 (1978): 137–149.

89. "Die Liebe gedeiht am meisten bei Ahnenverlust und Ebenbürtigkeit"; quoted from Bernd Gausemeier, "Auf der 'Brücke zwischen Natur- und Geschichtswissenschaft': Ottokar Lorenz und die Neuerfindung der Genealogie um 1900," in *Wissensobjekt Mensch: Praktiken der Humanwissenschaften im 20. Jahrhundert*, ed. F. Vienne and C. Brandt (Berlin: Kulturverlag Kadmos, 2008), 137–164; Ottokar Lorenz, *Lehrbuch der gesammten wissenschaftlichen Genealogie* (Berlin: Wilhelm Hertz, 1898).

90. See Volker Roelcke, "Programm und Praxis der psychiatrischen Genetik an der Deutschen Forschungsanstalt für Psychiatrie unter Ernst Rüdin: Zum

Verhältnis von Wissenschaft, Politik und Rasse-Begriff vor und nach 1933," in *Rassenforschung an Kaiser-Wilhelm-Instituten vor und nach 1933*, ed. H.-W. Schmuhl (Göttingen: Wallstein, 2003), 38–67.

91. Quoted from Gausemeier, "Pedigree vs. Mendelism," 153.

92. See ibid., 155–158.

93. See Stocking, *Race, Culture, and Evolution*, chap. 2; Doris Kaufmann, "'Rasse und Kultur': Die amerikanische Kulturanthropologie um Franz Boas (1858–1942) in der ersten Hälfte des 20. Jahrhunderts—ein Gegenentwurf zur Rassenforschung in Deutschland," in Schmuhl, *Rassenforschung*, 318–319.

94. See Benoit Massin, "From Virchow to Fischer: Physical Anthropology and 'Modern Race Theories' in Wilhelmine Germany," in *Volksgeist as Method and Ethic: Essays on Boasian Ethnography and the German Anthropological Tradition*, ed. G. W. Stocking (Madison: University of Wisconsin Press, 1996), 123.

95. Boas, *Race, Language and Culture*, 32.

96. See Hans-Jörg Rheinberger and Peter McLaughlin, "Darwin's Experimental Natural History," *Journal of the History of Biology* 17 (1984): 345–368.

97. See Veronika Lipphardt, "Zwischen 'Inzucht' und 'Mischehe': Demographisches Wissen in der Debatte um die 'Biologie der Juden,'" *Tel Aviver Jahrbuch für Deutsche Geschichte* 35 (2007): 45–66.

98. On the aporiae that motivated the debate between polygenists and monogenists well into the twentieth century, see Peter J. Bowler, *Theories of Human Evolution: A Century of Debate 1844–1944* (Oxford: Basil Blackwell, 1986); Marianne Sommer, *Bones and Ochre: The Curious Afterlife of the Red Lady of Paviland* (Cambridge, MA: Harvard University Press, 2007), chap. 7–12.

99. See Stocking, *Victorian Anthropology*, 62–69.

100. See Henrika Kuklick, "The British Tradition," in *A New History of Anthropology*, ed. H. Kuklick (Oxford: Wiley Blackwell, 2007), 58–60.

101. Edward B. Tylor, *Primitive Culture: Researches Into the Development of Mythology, Philosophy, Religion, Language, Art and Custom*, 3rd ed., vol. 2 (New York: Holt, 1889), 453.

102. We would like to thank Norton Wise for this simile.

103. UNESCO, ed., *The Race Concept: Results of an Inquiry* (Paris: UNESCO House, 1952).

104. Theodosius Dobzhansky, *Genetics and the Origins of Species* (New York: Columbia University Press, 1937), 62–63; see Jean Gayon, "Do Biologists Need the Expression 'Human Races'? UNESCO 1950–1951," in *Bioethical and Ethical Issues Surrounding the Trials and Code of Nuremberg: Nuremberg Revisited*, ed. Jacques Rozenberg, 23–48 (Lewiston, NY: Edwin Mellen Press, 2003). Dobzhansky's statement seems all the more ambivalent because the work of racist scientists like Ernst Rüdin, who worked as an adviser for the National Socialist regime, is an integral part of the prehistory of population genetics; on Rüdin, see Roelcke, "Programm und Praxis," 47, and Weingart, Bayertz, and Kroll, *Rasse, Blut und Gene*, 460–464.

105. Weingart, Bayertz und Kroll, *Rasse, Blut und Gene*, 15.

CHAPTER SIX

1. Bernard, *Leçons*. See also Manfred D. Laubichler, "'Allgemeine Biologie' als selbständige Grundwissenschaft und die allgemeinen Grundlagen des Lebens," in *Der Hochsitz des Wissens: Das Allgemeine als wissenschaftlicher Wert*, ed. M. Hagner and M. D. Laubichler (Zürich: Diaphanes, 2006), 185–206.

2. See Robert C. Olby, *Mendel, Mendelism and Genetics, MendelWeb*, http://www.mendelweb.org/MWolby.html, 1997 (accessed October 2, 2011), for a critical overview.

3. See, e.g., Mikulás Teich, "Fermentation Theory and Practice: The Beginnings of Pure Yeast Cultivation and English Brewing, 1883–1913," *Technology and Culture* 8 (1983): 117–133, and Teich, *Bier, Wissenschaft und Wirtschaft in Deutschland 1800–1914: Ein Beitrag zur deutschen Industrialisierungsgeschichte* (Vienna: Böhlau, 2000).

4. See Thurtle, *Emergence of Genetic Rationality*; Christophe Bonneuil, "Producing Identity, Industrializing Purity: Elements for a Cultural History of Genetics," in *Conference: Heredity in the Century of the Gene*, 81–110; Müller-Wille, "Hybrids, Pure Cultures, and Pure Lines."

5. Alexander Powell, Maureen A. O'Malley, Staffan Müller-Wille, Jane Calvert, and John Dupré, "Disciplinary Baptisms: A Comparison of the Naming Stories of Genetics, Molecular Biology, Genomics and Systems Biology," *History and Philosophy of the Life Sciences* 29 (2007): 5–32.

6. See Olby, "Mendel No Mendelian?" and, more recently, Müller-Wille and Orel, "From Linnaean Species to Mendelian Factors."

7. See Olby, *Origins of Mendelism*, 205–207, and Sander Gliboff, "Gregor Mendel and the Laws of Evolution," *History of Science* 37 (1999): 217–235. Mendel's correspondence with Nägeli was published in 1905 by Carl Correns; see Correns, "Gregor Mendels Briefe an Carl Nägeli 1866–1873: Ein Nachtrag zu den veröffentlichten Bastardisierungsversuchen Mendels," in *Gesammelte Abhandlungen zur Vererbungswissenschaft aus periodischen Schriften 1899–1924* (Berlin: Julius Springer, 1924), 1233–1297.

8. Quoted from Ilse Jahn, *Geschichte der Biologie* (Berlin: Spektrum Akademischer Verlag, 2000), 389.

9. See Olby, *Origins of Mendelism*, chap. 2.

10. Charles Naudin, "Nouvelles recherches sur l'hybridité dans les végétaux: Mémoire présenté à l'Académie en décembre 1861," *Nouvelles Archives du Muséum* 1 (1865): 149–150.

11. Gregor Mendel, "Experiments in Plant Hybrids," trans. E. R. Sherwood, in *Gregor Mendel's Experiments on Plant Hybrids: A Guided Study*, ed. A. F. Corcos and F. V. Monaghan (New Brunswick, NJ: Rutgers University Press, 1993), 59.

12. Ibid., 63 and 82.

13. Ibid., 160.

14. Gregor Mendel, "On Hieracium-Hybrids Obtained by Artificial Fertilisation," in William Bateson, *Mendel's Principles of Heredity: A Defense* (Placitas, NM: Geneticas Heritage Press, 1996), 96–103, and Correns, "Gregor Mendels Briefe an Carl Nägeli"; see Müller-Wille and Orel, "From Linnaean Species to Mendelian Factors."

15. Foucault, *Archaeology of Knowledge*, 224.

16. Vítězslav Orel, *Gregor Mendel: The First Geneticist* (Oxford: Oxford University Press, 1996).

17. See Teich, *Bier, Wissenschaft und Wirtschaft* for Germany in particular; also Thomas Wieland, "*Wir beherrschen den pflanzlichen Organismus besser . . .*": *Wissenschaftliche Pflanzenzüchtung in Deutschland 1889–1945* (Munich: Deutsches Museum, 2004), and Wieland, "Scientific Theory and Agricultural Practice: Plant Breeding in Germany from the Late 19th to the Early 20th Century," *Journal of the History of Biology* 39 (2006): 309–343, as well as Jonathan Harwood, *Technology's Dilemma: Agricultural Colleges between Science and Practice in Germany, 1860–1934* (Bern: Peter Lang, 2005).

18. Roger J. Wood and Vítězslav Orel, "Scientific Breeding in Central Europe during the Early Nineteenth Century: Background to Mendel's Later Work," *Journal of the History of Biology* 38 (2005): 239–272.

19. See Charles Massy, *The Australian Merino: The Story of a Nation*, 2nd ed. (Sydney: Random House Australia, 2007), 205.

20. See Wood and Orel, *Genetic Prehistory in Selective Breeding*.

21. See Gayon and Zallen, "The Role of the Vilmorin Company."

22. Quoted from Nils Roll-Hansen, "Sources of Johannsen's Genotype Theory," in *A Cultural History of Heredity III*, 47.

23. See Margaret Derry, *Bred for Perfection: Shorthorn Cattle, Collies, and Arabian Horses since 1800* (Baltimore: Johns Hopkins University Press, 2003).

24. See Staffan Müller-Wille, "Early Mendelism and the Subversion of Taxonomy: Epistemological Obstacles as Institutions," *Studies in History and Philosophy of Biological and Biomedical Sciences* 36 (2005): 465–487.

25. Diane B. Paul and Barbara A. Kimmelman, "Mendel in America: Theory and Practice 1900–1919," in *The American Development of Biology*, ed. K. Benson, J. Maienschein, and R. Rainger (Philadelphia: University of Pennsylvania Press, 1988), 281–310; Paolo Palladino, "Between Craft and Science: Plant Breeding, Mendelian Genetics, and British Universities, 1900–1920," *Technology and Culture* 34 (1993): 300–323; Barbara A. Kimmelman, "Mr. Blakeslee Builds His Dream House: Agricultural Institutions, Genetics, and Careers, 1900–1915," *Journal of the History of Biology* 39 (2006): 241–280; Wieland, "*Wir beherrschen den pflanzlichen Organismus besser*," and Wieland, "Scientific Theory and Agricultural Practice."

26. See Jane Maienschein, *Whose View of Life? Embryos, Cloning, and Stem Cells* (Cambridge, MA: Harvard University Press, 2003); Christina Brandt, "Clones, Pure Lines and Heredity: The Work of Victor Jollos," in *Conference: Heredity in the Century of the Gene*, 139–148.

27. Müller-Wille, "Hybrids, Pure Cultures, and Pure Lines," 801.

28. Eugenius Warming and Wilhelm Johannsen, *Lehrbuch der allgemeinen Botanik*, trans. E. P. Meinecke (Berlin: Bornträger, 1909), 355.

29. See the list of breweries at the end of Emil Christian Hansen, *Practical Studies in Fermentation: Being Contributions to the Life History of Micro-Organisms* (London: Spon and Chamberlain, 1896); concerning Hansen, see Teich, "Fermentation Theory and Practice."

30. See Andrew Mendelsohn, "Message in a Bottle: The Business of Vaccines

and the Nature of Heredity after 1880," in *A Cultural History of Heredity III*, 85–100; Bonneuil, "Producing Identity, Industrializing Purity"; Thurtle, *Emergence of Genetic Rationality*.

31. See Daniel J. Kevles, *A History of Patenting Life in the United States: With Comparative Attention to Europe and Canada* (Brussels: European Community, 2002).

32. See Garland E. Allen, "Mendel and Modern Genetics: The Legacy for Today," *Endeavour* 27 (2003): 63–68.

33. See Peter Ruckenbauer, "E. von Tschermak-Seysenegg and the Austrian Contribution to Plant Breeding," *Vorträge für Pflanzenzüchtung* 48 (2000): 31–46.

34. See Ida H. Stamhuis, "Hugo de Vries's Transitions in Research Interest and Method," in *A Cultural History of Heredity III*, 115–136.

35. See Ilse Jahn, "Zur Geschichte der Wiederentdeckung der Mendelschen Gesetze," *Wissenschaftliche Zeitschrift der Friedrich-Schiller Universität Jena*, Mathematisch-naturwissenschaftliche Reihe 7, no. 2–3 (1958): 215–227; Joan W. Bennett and Herman J. Phaff, "Early Biotechnology: The Delft Connection," *ASM News* 59 (1993): 401–404.

36. See Hugo de Vries, "Sur la loi de disjonction des hybrides," *Comptes rendus de l'Académie des Sciences* 130 (1900): 845–847, and Carl Correns, "Mendels Regeln über das Verhalten der Nachkommenschaft der Rassenbastarde," *Berichte der Deutschen Botanischen Gesellschaft* 18 (1900): 158–168.

37. William Bateson, "Problems of Heredity as a Subject for Horticultural Investigation," *Journal of the Royal Horticultural Society* 25 (1900): 54. See Robert C. Olby, "William Baleson's Introduction of Mendelism to Britain: A Reassessment," *British Journal for the History of Science* 20 (1987): 399–420.

38. William Bateson, "Hybridisation and Cross-Breeding as a Method of Scientific Investigation," *Journal of the Royal Horticultural Society* 24 (1899): 59.

39. William Bateson, *The Methods and Scope of Genetics: An Inaugural Lecture Delivered 23 October 1908* (Cambridge: Cambridge University Press, 1908), 48.

40. Max Hartmann, *Allgemeine Biologie* (Jena: Gustav Fischer, 1927), 5–11.

41. See also Nils Roll-Hansen, "The Genotype Theory of Wilhelm Johannsen and Its Relation to Plant Breeding and the Study of Evolution," *Centaurus* 22 (1978): 201–235, and Roll-Hansen, "The Crucial Experiment of Wilhelm Johannsen," *Biology and Philosophy* 4 (1989): 303–329; Hans-Jörg Rheinberger, *An Epistemology of the Concrete: Twentieth Century Histories of Life* (Durham: Duke University Press, 2010), chap. 4.

42. Wilhelm Johannsen, "The Genotype Conception of Heredity," *American Naturalist* 45 (1911): 131.

43. Wilhelm Johannsen, *Elemente der exakten Erblichkeitslehre: Mit Grundzügen der biologischen Variationsstatistik*, 3rd ed. (Jena: Gustav Fischer, 1926), vi.

44. Johannsen, "Genotype Conception of Heredity," 130.

45. Ibid., 133.

46. Ibid., 133.

47. Ibid., 133 and 150.

48. Ibid., 139.

49. See Frederick B. Churchill, "William Johannsen and the Genotype Concept," *Journal of the History of Biology* 7 (1974): 5–30.

50. Carl Correns, Autobiographical sketch, typescript, Archive of the Max Planck Society, Abt. III, Rep. 17, no. 1, p. 3.

51. For Correns's encounter with Mendel's paper, see Hans-Jörg Rheinberger, "When Did Carl Correns Read Gregor Mendel's Paper? A Research Note," *Isis* 86 (1995): 612–616. See also Hans-Jörg Rheinberger, "*Pisum*: Carl Correns's Experiments on Xenia, 1896–99," chap. 4 in *Epistemology of the Concrete.*

52. Hans-Jörg Rheinberger, "Mendelian Inheritance in Germany between 1900–1910: The Case of Carl Correns (1864–1933)," *Comptes rendus de l'Académie des Sciences, Série III, Sciences de la vie* 323 (2000): 1089–1096.

53. Carl Correns, "Die Ergebnisse der neuesten Bastardforschungen für die Vererbungslehre," in *Gesammelte Abhandlungen zur Vererbungswissenschaft,* 276.

54. Ibid., 277; original emphasis.

55. Ibid., 279; original emphasis.

56. Ibid., 279.

57. Ibid., 281.

58. See Bonneuil, "Producing Identity, Industrializing Purity."

59. See Hans-Jörg Rheinberger, "Carl Correns and the Early History of Genetic Linkage," in *Classical Genetic Research and Its Legacy: The Mapping Cultures of Twentieth Century Genetics,* ed. H.-J. Rheinberger and J.-P. Gaudillière (London: Routledge, 2004), 21–33.

60. Carl Correns, "Zur Kenntnis der Rolle von Kern und Plasma bei der Vererbung," in *Gesammelte Abhandlungen zur Vererbungswissenschaft,* 655.

61. Johannsen, "Genotype Conception of Heredity," 153.

62. Wilhelm Johannsen, "Some Remarks about Units in Heredity," *Hereditas* 4 (1923): 137.

63. Johannsen, "Genotype Conception of Heredity," 159.

64. See Garland E. Allen, *Thomas Hunt Morgan: The Man and His Science* (Princeton, NJ: Princeton University Press, 1978).

65. See Robert E. Kohler, *Lords of the Fly: Drosophila Genetics and the Experimental Life* (Chicago: University of Chicago Press, 1994), which describes Morgan's fly room as a "breeding reactor."

66. For maize, see Lee B. Kass and Christophe Bonneuil, "Mapping and Seeing: Barbara McClintock and the Linking of Genetics and Cytology in Maize Genetics, 1928–1935," in Rheinberger and Gaudillière, *Classical Genetic Research and Its Legacy,* 91–118, and Lee B. Kass, Christophe Bonneuil, and Edward H. Coe, "Cornfests, Cornfabs and Cooperation: The Origins and Beginnings of the Maize Genetics Cooperation News Letter," *Genetics* 169 (2005): 1787–1797; for the mouse, see Jean-Paul Gaudillière, "Mapping as Technology: Genes, Mutant Mice, and Biomedical Research (1910–65)," in Rheinberger and Gaudillière, *Classical Genetic Research and Its Legacy,* and Karen Rader, *Making Mice: Standardizing Animals for American Biomedical Research, 1900–1955* (Princeton, NJ: Princeton University Press, 2004).

67. See Sara Schwartz, "The Differential Concept of the Gene: Past and Present," in Beurton, Falk, and Rheinberger, *Concept of the Gene,* 26–39, and C. Kenneth Waters, "What Was Classical Genetics?" *Studies in History and Philosophy of Science* 35 (2004): 83–109.

68. See Hugo de Vries, *The Mutation Theory: Experiments and Observations on the Origin of Species in the Vegetable Kingdom*, 2 vols., trans. J. B. Framer and A. D. Darbishire (Chicago: Open Court, 1909–1910).

69. See Kohler, *Lords of the Fly*.

70. Herbert S. Jennings, "Pure Lines in the Study of Genetics in Lower Organisms," *American Naturalist* 45 (1911): 80.

71. See Luis Campos, "Genetics without Genes: Blakeslee, Datura and 'Chromosomal Mutations,'" in *Conference: Heredity in the Century of the Gene*, 21–23.

72. See Ernst Peter Fischer and Carol Lipson, *Thinking about Science: Max Delbrück and the Origins of Molecular Biology* (New York: Norton, 1988).

73. See Rader, *Making Mice*, and Ilana Löwy and Jean-Paul Gaudillière, "Disciplining Cancer: Mice and the Practice of Genetic Purity," in *The Invisible Industrialist: Manufactures and the Production of Scientific Knowledge*, ed. Jean-Paul Gaudillière and Ilana Löwy (London: MacMillan, 1998), 209–249.

74. Alexander von Schwerin, "Tiere vermehren—Institutionen vergrößern: Die Versuchstiere der Notgemeinschaft Deutscher Wissenschaft, 1920–1931," *Verhandlungen zur Geschichte und Theorie der Biologie* 11 (2005): 358–359; see also Schwerin, *Experimentalisierung des Menschen: Der Genetiker Hans Nachtsheim und die vergleichende Erbpathologie 1920–1945* (Göttingen: Wallstein, 2004).

75. See Brandt, "Clones, Pure Lines and Heredity."

76. See Alexander von Schwerin, "Seeing, Breeding and the Organisation of Variation: Erwin Baur and the Culture of Mutations in the 1920s," in *Conference: Heredity in the Century of the Gene*, 259–278.

77. On Muller, see Elof A. Carlson, *Genes, Radiation, and Society: The Life and Work of H. J. Muller* (Ithaca, NY: Cornell University Press, 1981); Schwartz, *In Pursuit of the Gene*.

78. See Frederick L. Holmes, *Reconceiving the Gene: Seymour Benzer's Adventures in Phage Genetics* (New Haven, CT: Yale University Press, 2006).

79. See Evelyn Fox Keller, *A Feeling for the Organism: The Life and Work of Barbara McClintock* (San Francisco: Freeman, 1983); Nathaniel C. Comfort, *The Tangled Field: Barbara McClintock's Search for the Patterns of Genetic Control* (Cambridge, MA: Harvard University Press, 2001).

80. For Boveri's experiments, see Laubichler and Davidson, "Boveri's Long Experiment": Helga Satzinger, "Theodor and Marcella Boveri: Chromosomes and Cytoplasm in Heredity and Development," *Nature Reviews Genetics* 9 (2008): 231.

81. See Campos, "Genetics without Genes."

82. Hermann J. Muller, "The Development of the Gene Theory," in *Genetics in the 20th Century: Essays on the Progress of Genetics during Its First 50 Years*, ed. L. C. Dunn (New York: Macmillan, 1951), 95–96.

83. See Martha Richmond, "The Cell as the Basis for Heredity, Development, and Evolution: Richard Goldschmidt's Program of Physiological Genetics," in Laubichler and Maienschein, *From Embryology to Evo-Devo*, 169–211.

84. See Fischer and Lipson, *Thinking about Science*; see also Lily E. Kay, *The Molecular Vision of Life: Caltech, the Rockefeller Foundation, and the Rise of the New Biology* (New York: Oxford University Press, 1993), chapter 4.

85. Alfred Kühn, "Über eine Gen-Wirkkette der Pigmentbildung bei Insekten," *Nachrichten der Akademie der Wissenschaften in Göttingen, mathematisch-physikalische Klasse* (1941), 258.

86. Lenny Moss, *What Genes Can't Do* (Cambridge, MA: MIT Press, 2003).

87. Thomas Hunt Morgan, "The Relation of Genetics to Physiology and Medicine," Nobel Lecture, presented in Stockholm on June 4, 1934, *Les Prix Nobel en 1933*, 3.

88. See Sahotra Sarkar, ed., *The Founders of Evolutionary Genetics* (Dordrecht: Kluwer Academic Publishers, 1992).

89. Gayon, *Darwinism's Struggle for Survival*, 297.

90. See Ernst Mayr and William B. Provine, eds., *The Evolutionary Synthesis: Perspectives on the Unification of Biology* (Cambridge, MA: Harvard University Press, 1980), and Vassiliki Betty Smocovitis, *Unifying Biology: The Evolutionary Synthesis and Evolutionary Biology* (Princeton, NJ: Princeton University Press, 1996).

91. For the American context, see Jane Maienschein, *Transforming Traditions in American Biology, 1880–1915* (Baltimore: Johns Hopkins University Press, 1991).

92. See Harwood, *Styles of Scientific Thought.*

93. See, e.g., Mayr, *Growth of Biological Thought.*

94. See, e.g., Allen, "T. H. Morgan and the Split between Embryology and Genetics"; Mayr and Provine, *Evolutionary Synthesis.*

95. See Thomas Junker and Uwe Hoßfeld, *Die Entdeckung der Evolution: Eine revolutionare Theorie und ihre Geschichte* (Darmstadt: Wissenschaftliche Buchgesellschaft, 2001); Smocovitis, *Unifying Biology.*

96. Bateson, *Methods and Scope of Genetics*, 1.

97. On Goldschmidt, see Helga Satzinger, "Racial Purity, Stable Genes and Sex Difference: Gender in the Making of Genetic Concepts by Richard Goldschmidt and Fritz Lenz, 1916–1936," in *The Kaiser Wilhelm Society under National Socialism*, ed. S. Heim, C. Sachse, and M. Walker (Cambridge: Cambridge University Press, 2009).

98. See Weindling, *Health, Race and German Politics*; Schwerin, *Experimentalisierung des Menschen*; Hans-Walter Schmuhl, *Rassenforschung an Kaiser-Wilhelm-Instituten vor und nach 1933* (Göttingen: Wallstein, 2003), and Schmuhl, *Crossing Boundaries: The Kaiser-Wilhelm-Institute for Anthropology, Human Heredity, and Eugenics, 1927–1945* (Dordrecht: Springer, 2008).

99. See Ilana Löwy and Jean-Paul Gaudillière, "Mendelian Factors and Human Disease: A Conversation," in *Conference: Heredity in the Century of the Gene*, 19–26; Jean-Paul Gaudillière and Ilana Löwy, eds., *Heredity and Infection: The History of Disease Transmission* (London: Routledge, 2001); Diane B. Paul, *The Politics of Heredity: Essays on Eugenics, Biomedicine and the Nature-Nurture Debate* (New York: State University of New York Press, 1998), esp. chap. 9.

100. See Palladino, "Between Craft and Science", and Robert C. Olby, "Scientists and Bureaucrats in the Establishment of the John Innes Institute under William Baleson," *Annals of Science* 46 (1989): 497–510, for England; Christophe Bonneuil, "Mendelism, Plant Breeding and Experimental Cultures: Agriculture

and the Development of Genetics in France," *Journal of the History of Biology* 39 (2006): 281–308, for France.

101. See Bert Theunissen, "Breeding Dutch Dairy Cows (1900–1950): Heredity without Mendelism," in *Conference: Heredity in the Century of the Gene*, 27–50 for the Netherlands; Jennifer Marie, "The Importance of Place: A History of Genetics in 1930s Britain," Ph.D. diss., University College London, 2004 for Great Britain; Schwerin, *Experimentalisierung des Menschen* for Germany.

102. See Sapp, *Beyond the Gene*; Judy Johns Schloegel, "Herbert Spencer Jennings, Heredity, and Protozoa as Model Organisms, 1908–1918," in *Conference: Heredity in the Century of the Gene*, 129–138.

103. Georges Canguilhem, "Qu'est-ce que une idéologie scientifique?" in *Idéologie et rationalité dans l'histoire des sciences de la vie*, ed. G. Canguilhem (Paris: Vrin, 1981), 33–45.

104. James R. Griesemer, "Reproduction and the Reduction of Genetics," in Beurton, Falk, and Rheinberger, *Concept of the Gene in Development and Evolution*, 240–285, and Griesemer, "Tracking Organic Processes: Representations and Research Styles in Classical Embryology and Genetics," in Laubichler and Maienschein, *From Embryology to Evo-Devo*, 375–433.

105. See, most recently, Nils Roll-Hansen, *The Lysenko Effect: The Politics of Science* (Amherst, NY: Humanity Books, 2005).

CHAPTER SEVEN

1. Fox Keller, *Century of the Gene*.

2. Stubbe, *History of Genetics*, 290.

3. See Michel Morange, *A History of Molecular Biology*, trans. N. Cobb (Cambridge, MA: Harvard University Press, 1998).

4. See, classically, Horace Freeland Judson, *The Eighth Day of Creation: Makers of the Revolution in Biology*, expanded ed. (New York: Cold Spring Harbor Laboratory Press, 1996).

5. See Robert C. Olby, *The Path to the Double Helix* (Seattle: University of Washington Press, 1974).

6. See Frieder W. Lichtenthaler, "Hundert Jahre Schlüssel-Schloss-Prinzip: Was führte Emil Fischer zu dieser Analogie?" *Angewandte Chemie* 106 (1994): 2456–2467.

7. See Robert E. Kohler, *From Medical Chemistry to Biochemistry: The Making of a Biomedical Discipline* (Cambridge: Cambridge University Press, 1982).

8. See Gaston Bachelard, *The Formation of the Scientific Mind*, trans. Mary MacAllester Jones (Manchester: Clinamen Press, 2002).

9. For an exemplary case study, see Bruno J. Strasser, *La fabrique d'une nouvelle science: La biologie moléculaire à l'âge atomique (1945–1964)* (Florence: Olschki, 2006).

10. See Jean-Paul Gaudillière and Ilana Löwy, *The Invisible Industrialist: Manufactures and the Production of Scientific Knowledge* (London: Macmillan, 1998); Karl Grandin, Nina Wormbs, and Sven Widmalm, eds., *The Science-Industry Nexus: History, Policy, Implications* (Sagamore Beach, MA: Science History Publications, 2004).

11. See Jean-Paul Gaudillière, "Biochemie und Industrie: Der 'Arbeitskreis Butenandt-Schering' im Nationalsozialismus," in *Adolf Butenandt und die Kaiser-Wilhelm-Gesellschaft: Wissenschaft, Industrie und Politik in 'Dritten Reich,'* ed. W. Schieder and A. Trunk (Göttingen: Wallstein, 2004), 198–246.

12. See Christina Ratmoko, *Damit die Chemie stimmt: Die Anfänge der industriellen Herstellung von weiblichen und männlichen Sexualhormonen 1914–1938* (Zürich: Chronos Verlag, 2010).

13. See Viviane Quirke, *Collaboration in the Pharmaceutical Industry: Changing Relationships in Britain and France 1935–1965* (London: Routledge, 2006); Nelly Oudshoorn, *Beyond the Natural Body: An Archeology of Sex Hormones* (London: Routledge, 1994).

14. Sven Widmalm, "A Machine to Work In: The Ultracentrifuge and the Modernist Laboratory Ideal," in *Taking Place: The Spatial Contexts of Science, Technology, and Business*, ed. E. Baraldi, H. Fors, and A. Houltz (Sagamore Beach, MA: Science History Publications, 2006), 59–80.

15. See Nicolas Rasmussen, *Picture Control: The Electron Microscope and the Transformation of Biology in America* (Stanford, CA: Stanford University Press, 1997).

16. See Olby, *Path to the Double Helix*, chap. 4.

17. Peter Keating and Alberto Cambrosio, *Biomedical Platforms: Realigning the Normal and the Pathological in Late-Twentieth-Century Medicine* (Cambridge, MA: MIT Press, 2003).

18. See Hans-Jörg Rheinberger, "Cytoplasmic Particles in Brussels (Jean Brachet, Hubert Chantrenne, Raymond Jeener) and at Rockefeller (Albert Claude), 1935–1955," *History and Philosophy of the Life Sciences* 19 (1997): 47–67; Olka Amsterdamska, "From Pneumonia to DNA: The Research Career of Oswald T. Avery," *Historical Studies in the Physical and Biological Sciences* 24 (1993): 1–40.

19. See Rheinberger, *Toward a History of Epistemic Things*.

20. See Jean-Paul Gaudillière, *Inventer la biomédecine: La France, l'Amérique et la production des savoirs du vivant, 1945–1965* (Paris: La Découverte, 2002).

21. David Cantor, ed., *Cancer in the Twentieth Century* (Baltimore: Johns Hopkins University Press, 2008).

22. See Bernward Joerges and Terry Shinn, eds., *Instrumentation between Science, State and Industry* (Dordrecht: Kluwer Academic Publishers, 2001), as well as Shinn and Joerges, "The Transverse Science and Technology Culture: Dynamics and Roles of Research-Technology," *Social Science Information* 41 (2002): 207–251.

23. See Rheinberger, *Epistemology of the Concrete*, chap. 9.

24. Richard Burian, "Underappreciated Pathways toward Molecular Genetics as Illustrated by Jean Brachet's Cytochemical Embryology," in *The Philosophy and History of Molecular Biology: New Perspectives*, ed. S. Sarkar (Dordrecht: Kluwer, 1996), 67.

25. Kay, *Molecular Vision of Life*, 46.

26. The concept of "protoplasm" itself goes back to Max Schultze; see his "Über Muskelkörperchen und das, was man eine Zelle zu nennen habe," *Archiv für Anatomie, Physiologie und Wissenschaftlicher Medicin* (1861): 1–27.

27. Boelie Elzen, "Two Ultracentrifuges: A Comparative Study of the Social Construction of Artefacts," *Social Studies of Science* 16 (1986): 621–662.

28. See Lily E. Kay, "The Tiselius Electrophoresis Apparatus and the Life Sciences, 1930–1940," *History and Philosophy of the Life Sciences* 10 (1988): 51–72.

29. See Rheinberger, "Cytoplasmic Particles in Brussels."

30. See Kay, *Molecular Vision of Life.*

31. See Soraya de Chadarevian, *Designs for Life: Molecular Biology after World War II* (Cambridge: Cambridge University Press, 2002).

32. See Olby, *Path to the Double Helix.*

33. See Rheinberger, *Epistemology of the Concrete,* chap. 11.

34. See Pnina Abir-Am, "From Multi-disciplinary Collaboration to Transnational Objectivity: International Space as Constitutive of Molecular Biology, 1930–1970," in *Denationalizing Science: The International Context of Scientific Practice,* ed. E. Crawford, T. Shinn, and S. Sorlin (Dordrecht: Kluwer Academic Publishers, 1993): 153–186.

35. Angela N. Creager, for instance, has shown this in detail for Wendell Stanley's work on TMV; see *The Life of a Virus: Tobacco Mosaic Virus as an Experimental Model, 1930–1965* (Chicago: University of Chicago Press, 2002).

36. Kay, *Molecular Vision of Life,* 6. See also Robert E. Kohler, "The Management of Science: The Experience of Warren Weaver and the Rockefeller Foundation Programme in Molecular Biology," *Minerva* 14, no. 3 (1976): 279–306; Pnina G. Abir Am, "The Discourse of Physical Power and Biological Knowledge in the 1930s: A Reappraisal of the Rockefeller Foundation's Policy in Molecular Biology," *Social Studies of Science* 12 (1982): 341–382, and "The Rockefeller Foundation and the Rise of Molecular Biology," *Nature Reviews Molecular Cell Biology* 3 (2002): 65–70.

37. Olby, *Path to the Double Helix,* 440.

38. Kay, *Molecular Vision of Life,* 8.

39. Kay, *Molecular Vision of Life;* Eric J. Vettel, *Biotech: The Countercultural Origins of an Industry* (Philadelphia: University of Pennsylvania Press, 2006).

40. For case studies, see Rheinberger, "Cytoplasmic Particles in Brussels"; Gaudillière, *Inventer la biomédecine;* Creager, *Life of a Virus.*

41. See Ilana Löwy, "Variances in Meaning in Discovery Accounts," *Historical Studies in the Physical and Biological Sciences* 21 (1990): 87–121.

42. Richard Burian, "Technique, Task Definition, and the Transition from Genetics to Molecular Genetics: Aspects of the Work on Protein Synthesis in the Laboratories of J. Monod and P. Zamecnik," *Journal of the History of Biology* 26 (1993): 387–407; Rheinberger, *Toward a History of Epistemic Things.*

43. Herbert C. Friedmann, "From Friedrich Wöhler's Urine to Eduard Buchner's Alcohol," in *New Beer in an Old Bottle: Eduard Buchner and the Growth of Biochemical Knowledge,* ed. A. Cornish-Bowden (Valencia: Universitat de Valencia, 1997), 108.

44. See Rheinberger, *Epistemology of the Concrete,* chap. 9.

45. See Angela N. Creager, "The Industrialization of Radioisotopes by the U.S. Atomic Energy Commission," in Grandin et al., *Science-Industry Nexus,* 141–168; John Krige, "Atoms for Peace, Scientific Internationalism, and Scien-

tific Intelligence," *Osiris* 21 (2006): 161–181; Alison Kraft, "Between Medicine and Industry: Medical Physics and the Rise of the Radioisotope 1945–65," *Contemporary British History* 20 (2006): 1–35.

46. Engelbert Broda, *Radioactive Isotopes in Biochemistry* (Amsterdam: Elsevier, 1960), 2.

47. See Morange, *History of Molecular Biology*, chap. 1.

48. See Angela N. Creager and María Jesús Santesmases, eds., "Radiobiology in the Atomic Age," special issue, *Journal of the History of Biology* 39, no. 4 (2006).

49. See Rheinberger, *Epistemology of the Concrete*, chap. 9.

50. Pnina G. Abir Am, "The Strategy of Large Versus Small Scale Investments: The Rockefeller Foundation's International Network of Protein Research, 1930–1960," in *American Foundations and Large Scale Research*, ed. G. Gemelli (Bologna: Clueb, 2001), 71–90; Hans-Jörg Rheinberger, "Internationalism and The History of Molecular Biology," *Annals of the History and Philosophy of Biology* 11 (2007): 249–254.

51. See Comfort, *Tangled Field*.

52. See Hans-Jörg Rheinberger, "A History of Protein Biosynthesis and Ribosome Research," in *Protein Synthesis and Ribosome Structure: Translating the Genome*, ed. K. H. Nierhaus and D. N. Wilson (Weinheim: Wiley-VCH, 2004), 1–51.

53. See Muriel Lederman and Richard M. Burian, "The Right Organism for the Job," *Journal of the History of Biology* 26 (1993): 235–367; Gabriel Gachelin, ed., *Les organismes modèles dans la recherche médicale* (Paris: Presses Universitaires de France, 2006); Rachel A. Ankeny, "Wormy Logic: Model Organisms as Case-Based Reasoning," in A. N. H. Creager, E. Lunbeck, and M. N. Wise, eds., *Science without Laws: Model Systems, Cases, Exemplary Narratives* (Chapel Hill, NC: Duke University Press, 2007), 46–58; Ankeny, "Historiographic Considerations on Model Organisms: Or, How the Muresucracy May Be Limiting Our Understanding of Contemporary Genetics and Genomics," *History and Philosophy of the Life Sciences* 32 (2010): 91–104; Rachel A. Ankeny and Sabina Leonelli, "What Is So Special about Model Organisms?," *Studies in History and Philosophy of Science: part A*, 42 (2011): 313–323; Sabina Leonelli and Rachel A. Ankeny, "Re-thinking Organisms: The Impact of Databases on Model Organism Biology," *Studies in History and Philosophy of the Biological and Biomedical Sciences* 43(2012): 29–36.

54. See Lily E. Kay, "W. M. Stanley's Crystallization of the Tobacco Mosaic Virus, 1930–1940," *Isis* 77 (1986): 450–472; Creager, *Life of a Virus*.

55. See Creager, *Life of a Virus*; Rheinberger, *Epistemology of the Concrete*, chap. 7.

56. See Lily E. Kay, "Selling Pure Science in Wartime: The Biochemical Genetics of G. W. Beadle," *Journal for the History of Biology* 22 (1989): 73–101; Robert E. Kohler, "Systems of Production: *Drosophila*, *Neurospora*, and Biochemical Genetics," *Historical Studies in the Physical and Biological Sciences* 22 (1991): 87–130.

57. Paul Berg and Maxine Singer, *George Beadle: An Uncommon Farmer; The Emergence of Genetics in the 20th Century* (Cold Spring Harbor: Cold

Spring Harbor Laboratory Press, 2003). Beadle came from the school of Roy Emerson in Cornell.

58. William C. Summers, *Félix d'Hérelle and the Origins of Molecular Biology* (New Haven, CT: Yale University Press, 1999).

59. See Edward J. Yoxen, "The Role of Schrödinger in the Rise of Molecular Biology," *History of Science* 17 (1979): 17–52, and "Giving Life a New Meaning: The Rise of the Molecular Biology Establishment," in *Scientific Establishments and Hierarchies*, ed. N. Elias, H. Martins, and R. Whitley (Dordrecht: Reidel, 1982), 123–143; Abir Am, "Discourse of Physical Power"; Lily E. Kay, "Conceptual Models and Analytical Tools: The Biology of Physicist Max Delbrück," *Journal of the History of Biology* 18 (1985): 207–246; Fischer and Lipson, *Thinking about Science*; Robert C. Olby, "From Physics to Biophysics," *History and Philosophy of the Life Sciences* 11 (1989): 305–309; Evelyn Fox Keller, "Physics and the Emergence of Molecular Biology: A History of Cognitive and Political Synergy," *Journal of the History of Biology* 23 (1990): 389–409; John Cairns, Gunther S. Stent, and James D. Watson, eds., *Phage and the Origins of Molecular Biology* (Plainview, NY: Cold Spring Harbor Laboratory Press, 1992).

60. Erwin Schrödinger, *What Is Life?* (Cambridge: Cambridge University Press, 1967), 21, 5.

61. Gunther S. Stent, "That Was the Molecular Biology That Was," *Science* 160 (1968): 390–395.

62. See Jan A. Witkowski, *Illuminating Life: Selected Papers from Cold Spring Harbor, 1903–1969* (Cold Spring Harbor: Cold Spring Harbor Laboratory Press, 1999); Witkowski, *75 Years in Science at the Cold Spring Harbor Symposium on Quantitative Biology* (Cold Spring Harbor: Cold Spring Harbor Laboratory Press, 2010).

63. See Kevles, *In the Name of Eugenics*, chap. 3; Jan A. Witkowski and John R. Inglis, eds., *Davenport's Dream: 21st Century Reflections on Heredity and Eugenics* (New York: Cold Spring Harbor Laboratory Press: 2008).

64. Kay, *Molecular Vision of Life*, 104.

65. Amsterdamska, "From Pneumonia to DNA."

66. Michel Morange, "La révolution silencieuse de la biologie moléculaire: D'Avery à Hershey," *Le Débat* 22 (1982): 66.

67. Pnina Abir-Am, "From Biochemistry to Molecular Biology: DNA and the Acculturated Journey of the Critic of Science Erwin Chargaff," *History and Philosophy of the Life Sciences* 2 (1980): 3–60.

68. See Thomas D. Brock, *The Emergence of Bacterial Genetics* (Cold Spring Harbor: Cold Spring Harbor Laboratory Press, 1990).

69. See Angela N. H. Creager, "Mapping Genes in Microorganisms," in *From Molecular Genetics to Genomics*, edited by J.-P. Gaudillière and H.-J. Rheinberger (London: Routledge, 2004): 9–41.

70. Denis Thieffry, "*Escherichia coli* as a model system with which to study cell differentiation," *History and Philosophy of the Life Sciences* 18 (1996): 163–193.

71. James D. Watson, *The Double Helix* (London: Weidenfeld and Nich-

olson, 1968); Olby, *Path to the Double Helix*; Chadarevian, *Designs for Life*, chap. 6.

72. See Rheinberger, *Toward a History of Epistemic Things*; Lily E. Kay, *Who Wrote the Book of Life? A History of the Genetic Code* (Stanford, CA: Stanford University Press, 2000).

73. Robert C. Olby, "The Molecular Revolution in Biology," in *Companion to the History of Modern Science*, ed. R. C. Deby, G. N. Cantor, J. R. R. Christie, and M. J. S. Hodge (London: Routledge, 1990), 503–520.

74. James D. Watson and Francis H. C. Crick, "Molecular Structure of Nucleic Acids: A Structure for Deoxyribose Nucleic Acid," *Nature* 171 (1953): 737.

75. Robert C. Olby, "Francis Crick, DNA, and the Central Dogma," in *The Twentieth-Century Sciences*, ed. G. Holton (New York: Norton, 1972), 227–280.

76. Francis H. C. Crick, "On Protein Synthesis," *Symposium of the Society for Experimental Biology* 12 (1958): 138–163.

77. Paul Rabinow, *Anthropologie der Vernunft: Studien zu Wissenschaft und Lebensführung* (Frankfurt/M.: Suhrkamp, 2004), 63.

78. Paul Rabinow, "Epochs, Presents, Events," in *Living and Working with the New Medical Technologies: Intersections of Inquiry*, ed. Margaret Lock, Allan Young, and Alberto Cambrosio (Cambridge: Cambridge University Press, 2000), 44.

79. See Jacob, *Logic of Life*; Jacques Monod, *Chance and Necessity: An Essay on the Natural Philosophy of Modern Biology*, trans. A. Wainhouse (New York: Knopf, 1971).

80. Norbert Wiener, *Cybernetics; or, Control and Communication in the Animal and the Machine* (Cambridge, MA: MIT Press, 1996), 39.

81. Andrew Pickering, *The Cybernetic Brain: Sketches of Another Future* (Chicago: University of Chicago Press, 2010).

82. See, e.g., Judson, *Eighth Day of Creation*; Sahotra Sarkar, "Biological Information: A Sceptical Look at Some Central Dogmas of Molecular Biology," in Sarkar, *Philosophy and History of Molecular Biology*, 187–231; Kay, *Who Wrote the Book of Life?*; John Maynard Smith, "The Concept of Information in Biology," *Philosophy of Science* 67 (2000): 177–194.

83. See Dorothy Nelkin and M. Susan Lindee, *The DNA Mystique: The Gene as a Cultural Icon* (New York: Freeman, 1995).

84. See Robert N. Proctor, "Three Roots of Human Recency: Molecular Anthropology, the Refigured Acheulean, and the UNESCO Response to Auschwitz," *Current Anthropology* 44 (2003): 213–229.

85. See Gordon Wolstenholme, ed., *Man and His Future: A CIBA Foundation Volume* (Boston: Little, Brown, 1963).

86. Linus Pauling, Harvey A. Itano, S. J. Singer, and Ibert C. Wells, "Sickle-Cell Anemia, a Molecular Disease," *Science* 110 (1949): 543; Bruno J. Strasser, "'Sickle Cell Anemia, a Molecular Disease,'" *Science* 286 (1999): 1488–1490; Bruno J. Strasser, "Linus Pauling's 'Molecular Diseases,' Between History and Memory," *American Journal of Medical Genetics* 115 (2002): 83–93.

CHAPTER EIGHT

1. Vettel, *Biotech*, ix.

2. See Horace F. Judson, "A History of the Science and Technology Behind Gene Mapping and Sequencing," in *The Code of Codes: Scientific and Social Issues in the Human Genome Project*, edited by D. J. Kevles and L. Hood (Cambridge, MA: Harvard University Press, 1992): 37–80.

3. Lindley Darden, *Reasoning in Biological Discoveries: Essays on Mechanisms, Interfield Relations, and Anomaly Resolution* (Cambridge: Cambridge University Press, 2006), chap. 10.

4. Mathias Grote, "Hybridizing Bacteria, Crossing Methods, Cross-checking Arguments: The Transition from Episomes to Plasmids (1961–1969)," *History and Philosophy of the Life Sciences* 30 (2008): 407–430.

5. See Morange, *History of Molecular Biology*, chap. 16.

6. See Hans-Jörg Rheinberger, "Beyond Nature and Culture: Modes of Reasoning in the Age of Molecular Biology and Medicine," in *Living and Working with the New Medical Technologies: Intersections of Inquiry*, ed. M. Lock, A. Young, and A. Cambrosio (Cambridge: Cambridge University Press, 2000), 19–30.

7. Doogab Yi, "Cancer, Viruses, and Mass Migration: Paul Berg's Venture into Eukaryotic Biology and the Advent of Recombinant DNA Research and Technology, 1967–1980," *Journal of the History of Biology* 41 (2008): 589–636.

8. Waclaw Szybalski and Ann Skalka, "Editorial: Nobel Prizes and Restriction Enzymes," *Gene* 4 (1978): 181.

9. Sheila Jasanoff, "Biotechnology and Empire: The Global Power of Seeds and Science," *Osiris* 21 (2006): 284.

10. See Morange, *History of Molecular Biology*, chap. 17.

11. See Paul Rabinow, *Making PCR: A Story of Biotechnology* (Chicago: Chicago University Press, 1996). On Sauger's sequencing procedure see Miguel García-Sancho, "A New Insight into Sauger's Development of sequencing— From Proteins to DNA, 1943–1977," *Journal of the History of Biology* 43 (2010): 265–323.

12. Kaushik Sunder Rajan, *Biocapitalism: The Constitution of Postgenomic Life* (Durham, NC: Duke University Press, 2006), 16.

13. On the early phase of genome sequencing, see Daniel J. Kevles and Leroy Hood, eds., *The Code of Codes: Scientific and Social Issues in the Human Genome Project* (Cambridge, MA: Harvard University Press, 1992).

14. David A. Jackson, "DNA: Template for an Economic Revolution," in *DNA: The Double Helix, Perspective and Prospective at Forty Years*, ed. D. A. Chambers (New York: New York Academy of Sciences, 1995), 364.

15. Documented in James D. Watson and John Tooze, *The DNA Story: A Documentary History of Gene Cloning* (San Francisco: W. H. Freeman, 1981).

16. See Susan Wright, *Molecular Politics: Developing American and British Regulatory Policy for Genetic Engineering* (Chicago: University of Chicago Press, 1994); Herbert Gottweis, *Governing Molecules: The Discursive Politics of Genetic Engineering in Europe and the United States* (Cambridge, MA: MIT Press, 1998).

17. Jean-Paul Gaudillière, "New Wine in Old Bottles? The Biotechnology

Problem in the History of Molecular Biology," in *Workshop, History and Epistemology of Molecular Biology and Beyond: Problems and Perspectives*, preprint 310 (Berlin: Max Planck Institute for the History of Science, 2006), 81–93.

18. See Vettel, *Biotech.*

19. On the history of patenting living beings in the United States, see Kevles, *History of Patenting Life.*

20. Sally Smith Hughes, "Making Dollars out of DNA: The First Major Patent in Biotechnology and the Commercialization of Molecular Biology 1974–1980," *Isis* 92 (2001), 541–575.

21. See, e.g., Daniel J. Kevles, "Out of Eugenics: The Historical Politics of the Human Genome," in Kevles and Hood, *Code of Codes*, 3–36.

22. Gottweis, *Governing Molecules*, 4; for more recent developments, see Sheila Jasanoff, *Designs on Nature: Science and Democracy in Europe and the United States* (Princeton, NJ: Princeton University Press, 2005); Barnes and Dupré, *Genomes.*

23. Rajan, *Biocapitalism*, 5.

24. See Sheldon Krimsky, *Biotechnics and Society: The Rise of Industrial Genetics* (New York: Praeger, 1991).

25. See Robert Cook-Deegan, *The Gene Wars: Science, Politics, and the Human Genome* (New York: W. W. Norton, 1994).

26. Ibid., 98.

27. See Rachel A. Ankeny, "Model Organisms as Models: Understanding the 'Lingua Franca' of the Human Genome Project," *Philosophy of Science* 68 (2001): 251–261.

28. See Kevles and Hood, *The Code of Codes*; Stephen Hilgartner, "The Human Genome Project," in *Handbook of Science and Technology Studies*, ed. Sheila Jasanoff et al. (Thousand Oaks, CA: Sage Publications, 1995).

29. See Paul Rabinow, *French DNA: Trouble in Purgatory* (Chicago: University of Chicago Press, 1999); Alain Kaufmann, "Mapping the Human Genome at Généthon Laboratory: The French Muscular Dystrophy Association and the Politics of the Gene," in *From Molecular Genetics to Genomics: The Mapping Cultures of Twentieth-Century Genetics*, ed. J.-P. Gaudillière and H.-J. Rheinberger (London: Routledge, 2004), 129–157.

30. Cook-Deegan, *Gene Wars.*

31. See Chadarevian, *Designs for Life.*

32. See Adam Bostanci, "Sequencing Human Genomes," in Gaudillière and Rheinberger, *From Molecular Genetics to Genomics*, 158–179.

33. Bostanci, "Two Drafts, One Genome? Human Diversity and Human Genome Research," *Science as Culture* 15 (2006): 183–198.

34. Figures as of October 2011. In the summer of 2008, the corresponding figures were 416 phages, 1,450 viruses, 51 archaebacteria, 569 bacteria, and 73 eukaryotes.

35. See Rabinow, *French DNA;* Kaufmann, "Mapping the Human Genome."

36. Charles Cantor, "The Challenges to Technology and Informatics," in Kevles and Hood, *Code of Codes*, 107.

37. Stephen Hilgartner, "Biomolecular Databases: New Communication Regimes for Biology?" *Science Communication* 17 (1995): 240–263; "Data Access

Policy in Genome Research," in *Private Science*, ed. A. Thackray (Philadelphia: University of Pennsylvania Press, 1998); and "Making Maps and Making Social Order: Governing American Genome Centers, 1988–93," in Gaudillière and Rheinberger, *From Molecular Genetics to Genomics*.

38. See Bruno Strasser, "Collecting and Experimenting: The Moral Economies of Biological Research," 1960s–1980s," in *Workshop, History and Epistemology*, 105–123; Bruno Strasser, "Collecting, Comparing, and Computing Sequences: The Making of Margaret O. Dayhoff's Atlas of Protein Sequence and Structure, 1954–1966," *Journal of the History of Biology* 43 (2010): 623–660.

39. See, e.g., Sabina Leonelli, "Documenting the Emergence of Bio-Ontologies: Or, Why Studying Bioinformatics Requires HPSSB," *History and Philosophy of the Life Sciences* 32 (2010): 105–125; Sabina Leonelli, "Making Sense of Data-Driven Research in the Biological and Biomedical Sciences," *Studies in History and Philosophy of the Biological and Biomedical Sciences* 43 (2012): 1–3.

40. See Peter Galison and Mario Biagioli, eds., *Scientific Authorship: Credit and Intellectual Property in Science* (London: Routledge, 2002); Philip Mirowski and Mirjam Sent, eds., *Science Bought and Sold: Essays in the Economics of Science* (Chicago: University of Chicago Press, 2002); Mario Biagioli, "The Instability of Authorship: Credit and Responsibility in Contemporary Biomedicine," in Mirowski and Sent, *Science Bought and Sold*, 486–514.

41. François Gros, *Les secrets du gène*, 2nd ed. (Paris: Ed. Odile Jacob, 1991), chap. 7; Rheinberger, *Epistemology of the Concrete*, chap. 8.

42. Lenny Moss, "Redundancy, Plasticity, and Detachment: The Implications of Comparative Genomics for Evolutionary Thinking," *Philosophy of Science* 73 (2006): 930–946.

43. See Fox Keller, *Century of the Gene*.

44. See Gísli Pálsson and Paul Rabinow, "Iceland: The Case of a National Human Genome Project," *Anthropology Today* 15 (1999): 14–18.

45. See Skúli Sigurdsson, "Yin-Yang Genetics, or the HSD deCODE Controversy," *New Genetics and Society* 20 (2001): 103–117; Mike Fortun, *Promising Genomics: Iceland and deCODE Genetics in a World of Speculation* (Berkeley: University of California Press, 2008). An extensive bibliography can be found at http://www.raunvis.hi.is/~sksi/hsd_dec.html.

46. Gísli Pálsson, "Decoding Relations and Disease: The Icelandic Biogenetic Project," in Gaudillière and Rheinberger, *From Molecular Genetics to Genomics*, 180–198.

47. See Reardon, *Race to the Finish*.

48. See Trudy R. Turner, ed., *Biological Anthropology and Ethics: From Repatriation to Genetic Identity* (Albany: SUNY Press, 2005); Marianne Sommer, "DNA and Cultures of Remembrance: Anthropological Genetics, Biohistories, and Biosocialities," *Biosocieties* 5 (2010): 366–390.

49. See Sheila Jasanoff, *Science at the Bar: Science and Technology in American Law* (Cambridge, MA: Harvard University Press, 1996); Michael Lynch and Simon A. Cole, "Science and Technology Studies on Trial: Dilemmas of Expertise," *Social Studies of Science* 35 (2005): 269–311; David Lazer, ed., *DNA*

*and the Criminal Justice System: The Technology of Justice* (Cambridge, MA: MIT Press, 2004); Stephen Hilgartner, *Science on Stage: Expert Advice as Public Drama* (Stanford, CA: Stanford University Press, 2000).

50. See Bernd Müller-Röber et al., eds., *Grüne Gentechnologie: Aktuelle Entwicklungen in Wissenschaft und Wirtschaft* (Munich: Elsevier Spektrum Akademischer Verlag, 2007).

51. See Christophe Bonneuil, "Les recherches sur les impacts agro-écologiques des OGM dans les années 1990," in *Maîtrise des risques, prévention et principe de precaution*, 81–94 (Paris: INRS Edition, 2002); Jasanoff, *Designs on Nature*; Gaudillière, "New Wine in Old Bottles?"

52. See Daniel J. Kevles, "Of Mice and Money: The Story of the World's First Animal Patent," *Daedalus* (Spring 2002): 78–88.

53. See Daniel J. Kevles, "The Advent of Animal Patents: Innovation and Controversy in the Engineering and Ownership of Life," in *Intellectual Property Rights and Patenting in Animal Breeding and Genetics*, ed. S. Newman and M. Rothschild (New York: CABI Publishing, 2002), 17–30.

54. See Daniel J. Kevles, "Breeding, Biotechnology, and Agriculture: The Establishment and Protection of Intellectual Property in Animals since the Late Eighteenth Century," in *Workshop, History and Epistemology*, 69–79.

55. See Michel Morange, *La part des gènes* (Paris: Odile Jacob, 1998), chap. 7.

56. François Jacob, *The Possible and the Actual*, trans. B. E. Spillmann (Seattle: University of Washington Press, 1982), chap. 2.

57. Thomas C. Caskey, "DNA-based Medicine: Prevention and Therapy," in Kevles and Hood, *Code of Codes*, 123.

58. See Keith Wailoo and Stephen Pemberton, *The Troubled Dream of Genetic Medicine: Ethnicity and Innovation in Tay-Sachs, Cystic Fibrosis, and Sickle Cell Disease* (Baltimore: Johns Hopkins University Press, 2006); see also Melbourne Tapper, *In the Blood: Sickle Cell Anemia and the Politics of Race*, Critical Histories (Philadelphia: University of Pennsylvania Press, 1998).

59. See Wailoo and Pemberton, *Troubled Dream of Genetic Medicine*, chap. 2.

60. Ibid., 170.

61. Eric Lander, "Interview," *Newsweek*, October 14, 2007, 46.

62. See, e.g., Jörg Schmidtke, Bernd Müller-Röber, Wolfgang van den Daele, Ferdinand Hucho, Kristian Köchy, Karl Sperling, Jens Reich, Hans-Jörg Rheinberger, Anna M. Wobus, Mathias Boysen, and Silke Domasch, eds., *Gendiagnostik in Deutschland: Status quo und Problemerkundung* (Limburg: Forum W—Wissenschaftlicher Verlag, 2007).

63. See Jean-Paul Gaudillière and Ilana Löwy, "Medicine, Markets and Public Health: Contemporary Testing for Breast Cancer Predispositions," in *Medicine, Markets and Mass Media: Producing Health in the Twentieth Century*, ed. V. Berridge and K. Loughlin (London: Routledge, 2005).

64. Norbert Paul, *The Making of Molecular Medicine: Historical, Theoretical, and Ethical Dimensions* (Habilitation, Heinrich-Heine-Universität Düsseldorf, 2003); Virginia Berridge and Kelly Laughlin, eds., *Medicine, the Market and the Mass Media: Producing Health in the Twentieth Century* (London: Routledge, 2005).

65. W. Bodmer, "Where Will Genome Analysis Lead Us?" Forty Years On?" in Chambers, *DNA*, 414.

66. See Ernst Ludwig Winnacker, "Das Gen und das Ganze," *Die Zeit*, May 2, 1997, 34.

67. See Richards, *The Romantic Conception of Life*.

68. See Anne Harrington, *Reenchanted Science: Holism in German Culture from Wilhelm II to Hitler* (Princeton: Princeton University Press, 1996).

69. On the early history of nucleic acid hybridization, see Edna Suárez Díaz and Victor H. Anaya-Muñoz, "History, Objectivity, and the Construction of Molecular Phylogenies," *Studies in History and Philosophy of Biological and Biomedical Sciences* 39 (2008): 451–468.

70. Eva Jablonka and Marion J. Lamb, *Evolution in Four Dimensions: Genetic, Epigenetic, Behavioral, and Symbolic Variation in the History of Life* (Cambridge, MA: MIT Press, 2005); Eva Neumann-Held and Christoph Rehmann-Sutter, eds., *Genesis in Development: Re-Reading the Molecular Paradigm* (Durham, NC: Duke University Press, 2006).

71. See Leonelli, "Documenting the Emergence of Bio-Ontologies."

72. See Marcel Weber, "Walking on the Chromosome: *Drosophila* and the Molecularization of Development," in Gaudillière and Rheinberger, *From Molecular Genetics to Genomics*, 63–78; Scott F. Gilbert, ed., *A Conceptual History of Modern Developmental Biology* (Baltimore: Johns Hopkins University Press, 1994).

73. See Morange, *La part des gènes*; for the longer history, see Laubichler and Maienschein, *From Embryology to Evo-Devo*.

74. See Soraya de Chadarevian, "Of Worms and Programmes: *Caenorhabditis Elegans* and the Study of Development," *Studies in History and Philosophy of Biological and Biomedical Sciences* 29 (1998): 81–105, and Rachel A. Ankeny, "The Natural History of C. *elegans* Research," *Nature Reviews Genetics* 2 (2001): 474–478.

75. See Morange, *History of Molecular Biology*, and Michel Morange, "The Developmental Gene Concept: History and Limits," in Beurton, Falk, and Rheinberger, *Concept of the Gene*, 193–215.

76. See Moss, "Redundancy, Plasticity, and Detachment."

77. Evelyn Fox Keller, *Making Sense of Life: Explaining Biological Development with Models, Metaphors, and Machines* (Cambridge, MA: Harvard University Press, 2002); Scott F. Gilbert, *Developmental Biology*, 9th ed. (Sunderland, MA: Sinanex Associates, 2010).

78. See Maienschein, *Whose View of Life?*

79. See Sally Morgan, *From Sea Urchins to Dolly the Sheep: Discovering Cloning* (Chicago: Heinemann Library, 2006).

80. Herbert Gottweis and Robert Triendl, "South Korean Policy Failure and the Hwang Debacle," *Nature Biotechnology* 24 (2006), 141–143.

81. Herbert Gottweis, "Human Embryonic Stem Cells, Cloning, and the Transformation of Biopolitics," in *Biotechnology: Between Commerce and Civil Society*, ed. N. Stehr (New Brunswick, NJ: Transaction, 2004), 239–265; Herbert Gottweis, Brian Salter, and Catherine Waldby, *The Global Politics of Human Embryonic Stem Cell Science: Regenerative Medicine in Transition* (Basingstoke: Palgrave Macmillan, 2009).

82. See Waters, "What Was Classical Genetics?"

83. Morange, *La part des gènes*, 43.

84. Kay, *Who Wrote the Book of Life?*; Nelkin and Lindee, *The DNA Mystique.*

85. See Rheinberger and Gaudillière, *Classical Genetic Research and Its Legacy.*

86. Gaudillière and Rheinberger, *From Molecular Genetics to Genomics.*

## Bibliography

Abir-Am, Pnina G. "From Biochemistry to Molecular Biology: DNA and the Acculturated Journey of the Critic of Science Erwin Chargaff." *History and Philosophy of the Life Sciences* 2 (1980): 3–60.

———. "The Discourse of Physical Power and Biological Knowledge in the 1930s: A Reappraisal of the Rockefeller Foundation's Policy in Molecular Biology." *Social Studies of Science* 12 (1982): 341–382.

———. "From Multi-disciplinary Collaboration to Transnational Objectivity: International Space as Constitutive of Molecular Biology, 1930–1970." In *Denationalizing Science: The International Context of Scientific Practice*, edited by Elisabeth T. Crawford, Terry Shinn, and Sverker Sörlin, 153–186. Dordrecht: Kluwer Academic Publishers, 1993.

———. "The Strategy of Large Versus Small Scale Investments: The Rockefeller Foundation's International Network of Protein Research, 1930–1960." In *American Foundations and Large Scale Research*, edited by Giuliana Gemelli, 71–90. Bologna: Clueb, 2001.

———. "The Rockefeller Foundation and the Rise of Molecular Biology." *Nature Reviews Molecular Cell Biology* 3 (2002): 65–70.

Adams, Mark B., ed. *The Wellborn Science: Eugenics in Germany, France, Brazil, and Russia*. New York: Oxford University Press, 1990.

Adelmann, Howard B. *Marcello Malpighi and the Evolution of Embryology*. 5 vols. Ithaca, NY: Cornell University Press, 1966.

Allen, Garland E. *Thomas Hunt Morgan: The Man and His Science*. Princeton, NJ: Princeton University Press, 1978.

———. "The Eugenics Record Office, Cold Spring Harbor, 1910–1940." *Osiris*, 2nd series, 2 (1986): 225–264.

———. "T. H. Morgan and the Split between Embryology and Genetics, 1910–1926." In *A History of Embryology*, edited by Timothy Horder, Jan A. Witkowski, and Christopher C. Wylie, 113–144. Cambridge: Cambridge University Press, 1986.

———. "Mendel and Modern Genetics: The Legacy for Today." *Endeavour* 27 (2003): 63–68.

Aly, Götz, and Susanne Heim. *Architects of Annihilation: Auschwitz and the Logic of Destruction*. Translated by A. G. Blunden. London: Weidenfeld & Nicolson, 2002. Originally published as *Vordenker der Vernichtung: Auschwitz und die deutschen Pläne für eine neue europäische Ordnung* (Frankfurt/M.: Fischer, 1991).

Amsterdamska, Olga. "From Pneumonia to DNA: The Research Career of Oswald T. Avery." *Historical Studies in the Physical and Biological Sciences* 24 (1993): 1–40.

Amundson, Ron. *The Changing Role of the Embryo in Evolutionary Thought*. Cambridge: Cambridge University Press, 2005.

Andersen, Otto. "Denmark." In *European Demography and Economic Growth*, edited by W. R. Lee, 79–122. London: Croom Helm, 1979.

Ankeny, Rachel A. "Model Organisms as Models: Understanding the 'Lingua Franca' of the Human Genome Project." *Philosophy of Science* 68 (2001): 251–261.

———. "The Natural History of *C. elegans* Research." *Nature Reviews Genetics* 2 (2001): 474–478.

———. "Wormy Logic: Model Organisms as Case-Based Reasoning." In *Science without Laws: Model Systems, Cases, Exemplary Narratives*, edited by Angela N. H. Creager, Elizabeth Lunbeck, and M. Norton Wise, 46–58. Chapel Hill, NC: Duke University Press, 2007.

———. "Historiographic Considerations on Model Organisms: Or, How the Muresucracy May Be Limiting Our Understanding of Contemporary Genetics and Genomics." *History and Philosophy of the Life Sciences* 32 (2010): 91–104.

Ankeny, Rachel A., and Sabina Leonelli. "What Is So Special about Model Organisms?" *Studies in History and Philosophy of Science, part A*, 42, no. 2 (2011): 313–323.

Arendt, Hannah. *The Origins of Totalitarianism*. New York: Schocken Books, 2004. Original 1955.

Aristotle. *Lectures on the Science of Nature*. Books 1–4. Translated by Charles Glenn Wallis. Annapolis: The St. John's Bookstore, 1940.

———. *Generation of Animals*. Translated by Arthur Leslie Peck. Cambridge, MA: Harvard University Press, 1990.

———. *History of Animals*. Translated by Arthur Leslie Peck and David M. Balme. Cambridge, MA: Harvard University Press, 1991–1993.

———. *On Coming-to-Be and Passing-Away*. Translated by Edward Seymour Forster. Cambridge, MA: Harvard University Press, 1992.

Bachelard, Gaston. *The New Scientific Spirit*. Translated by Arthur Goldhammer. Boston: Beacon Press, 1984. Originally published as *Le nouvel esprit scientifique* (Paris: Alcan, 1934).

———. *The Formation of the Scientific Mind*. Translated by Mary MacAllester Jones. Manchester: Clinamen Press, 2002. Originally published as *La formation de l'esprit scientifique* (Paris: Vrin, 1938).

Bäckstedt, Eva. "Rasbiologens barnbarn gör upp med arvet." *Svenska Dagbladet Kultur*, March 2, 2002, 4–5.

Bajema, Carl J. *Artificial Selection and the Development of Evolutionary Theory*. Stroudsburg, PA: Hutchinson Ross, 1982.

Barahona, Ana, Edna Suárez Díaz, and Hans-Jörg Rheinberger, eds. *The Hereditary Hourglass: Genetics and Epigenetics, 1868–2000*. Preprint 392. Berlin: Max Planck Institute for the History of Science, 2010.

Barkan, Elazar. *The Retreat of Scientific Racism: Changing Concepts of Race in Britain and the United States between the World Wars*. Cambridge: Cambridge University Press, 1992.

Barnes, Barry, and John Dupré. *Genomes and What to Make of Them*. Chicago: University of Chicago Press, 2008.

Bateson, William. "Hybridisation and Cross-Breeding as a Method of Scientific Investigation." *Journal of the Royal Horticultural Society* 24 (1899): 59–66.

———. "Problems of Heredity as a Subject for Horticultural Investigation." *Journal of the Royal Horticultural Society* 25 (1900): 54–61.

———. *The Methods and Scope of Genetics: An Inaugural Lecture Delivered 23 October 1908*. Cambridge: Cambridge University Press, 1908.

Bayon, Henry Peter. "William Harvey (1578–1657): His Application of Biological Experiment, Clinical Observation, and Comparative Anatomy to the Problems of Generation." *Journal of the History of Medicine and Allied Sciences* 2 (1947): 51–96.

Bennett, Joan Wennstrom, and Herman Jan Phaff. "Early Biotechnology: The Delft Connection." *ASM News* 59 (1993): 401–404.

Berg, Paul, and Maxine Singer. *George Beadle: An Uncommon Farmer; The Emergence of Genetics in the 20th Century*. Cold Spring Harbor: Cold Spring Harbor Laboratory Press, 2003.

Bernard, Claude. *De la physiologie générale*. Bruxelles: Culture et Civilisation, 1965. Originally published 1872.

———. *Leçons sur les phénomènes de la vie communs aux animaux et aux végétaux*. Paris: Vrin, 1966. Originally published 1878.

Berridge, Virginia, and Kelly Laughlin, eds., *Medicine, the Market and the Mass Media: Producing Health in the Twentieth Century*. London: Routledge, 2005.

Biagioli, Mario. "The Instability of Authorship: Credit and Responsibility in Contemporary Biomedicine." In *Science Bought and Sold: Essays in the Economics of Science*, edited by Philip Mirowski and Mirjam Sent, 486–514. Chicago: University of Chicago Press, 2002.

Blumenbach, Johann Friedrich. "On the Natural Variety of Mankind." In

*Anthropological Treatises of Johann Friedrich Blumenbach*, edited and translated by Thomas Bendyshe, 145–276. London: Published for the Anthropological Society by Longman, Green, Longman, Roberts, & Green, 1865. Originally published as *Über die natürlichen Verschiedenheiten im Menschengeschlechte* (Leipzig: Breitkopf und Härtel, 1795).

Boas, Franz. "Mixed Races." *Science* 17 (1891): 179.

———. "The Correlation of Anatomical or Physiological Measurements." *American Anthropologist* 7 (1894): 313–324.

———. *The Mind of Primitive Man*. New York: Macmillan, 1911.

———. *Kultur und Rasse*. Leipzig: Veit, 1914.

———. *Race, Language and Culture*. Chicago: University of Chicago Press, 1996.

Bodmer, Walter. "Where Will Genome Analysis Lead Us Forty Years On?" In *DNA: The Double Helix; Perspective and Prospective at Forty Years*, edited by Donald A. Chambers, 414–426. New York: New York Academy of Sciences, 1995.

Bonnet, Charles. *Considérations sur les corps organisés, où l'on traite de leur origine, de leur développement, de leur réproduction*. 2 vols. Amsterdam: Rey, 1762.

Bonneuil, Christophe. "Les recherches sur les impacts agro-écologiques des OGM dans les années 1990." In *Maîtrise des risques, prévention et principe de précaution*, 81–94. Paris: INRS Edition, 2002.

———. "Mendelism, Plant Breeding and Experimental Cultures: Agriculture and the Development of Genetics in France." *Journal of the History of Biology* 39 (2006): 281–308.

———. "Producing Identity, Industrializing Purity: Elements for a Cultural History of Genetics." In *Conference: Heredity in the Century of the Gene (A Cultural History of Heredity IV)*, preprint 343, 81–110. Berlin: Max Planck Institute for the History of Science, 2008.

Bostanci, Adam. "Sequencing Human Genomes." In *From Molecular Genetics to Genomics: The Mapping Cultures of Twentieth-Century Genetics*, edited by Jean-Paul Gaudillière and Hans-Jörg Rheinberger, 158–179. London: Routledge, 2004.

———. "Two Drafts, One Genome? Human Diversity and Human Genome Research." *Science as Culture* 15 (2006): 183–198.

Boveri, Theodor. *Zellen-Studien*. Vol. 2. Jena: Gustav Fischer, 1888.

———. "An Organism Produced Sexually Without Characteristics of the Mother." *American Naturalist* 27 (1893): 222–232.

Bowler, Peter J. "Preformation and Pre-Existence in the Seventeenth Century: A Brief Analysis." *Journal of the History of Biology* 4 (1971): 221–244.

———. *Theories of Human Evolution: A Century of Debate 1844–1944*. Oxford: Basil Blackwell, 1986.

———. *The Mendelian Revolution: The Emergence of Hereditarian Concepts in Modern Science and Society*. Baltimore: Johns Hopkins University Press, 1989.

———. *The Eclipse of Darwinism*. Baltimore: Johns Hopkins University Press, 1992.

————. "Variation from Darwin to the Modern Synthesis." In *Variation: A Central Concept in Biology*, edited by Benedikt Hallgrimsson and Brian K. Hall, 414–426. New York: Elsevier, 2005.

Bracegirdle, Brian. *A History of Microtechnique: The Evolution of the Microtome and the Development of Tissue Preparation*. 2nd ed. Lincolnwood: Science Heritage Ltd., 1986.

Bracton, Henry de. *On the Laws and Customs of England*. 4 vols. Edited by George E. Woodbridge and translated by Samuel L. Thorne. Cambridge, MA: Belknap Press, 1968–1977. Originally published as *De Legibus et Consuetudinibus Angliae*, 1569.

Brandt, Christina. "Clones, Pure Lines and Heredity: The Work of Victor Jollos." In *Conference: Heredity in the Century of the Gene (A Cultural History of Heredity IV)*, preprint 343, 139–148. Berlin: Max Planck Institute for the History of Science, 2008.

Brantlinger, Patrick. *Dark Vanishings: Discourse on the Extinction of Primitive Races, 1800–1930*. Ithaca, NY: Cornell University Press, 2003.

Braude, Benjamin. "Sons of Noah." *William and Mary Quarterly*, 3rd Ser., 54 (1997): 103–142.

Broberg, Gunnar, and Nils Roll-Hansen, eds. *Eugenics and the Welfare State*. East Lansing: Michigan State University Press, 1996.

Brock, Thomas D. *The Emergence of Bacterial Genetics*. Cold Spring Harbor: Cold Spring Harbor Laboratory Press, 1990.

Broda, Engelbert. *Radioactive Isotopes in Biochemistry*. Amsterdam: Elsevier, 1960.

Browne, Janet. *Charles Darwin: The Power of Place; Volume II of a Biography*. New York: Knopf, 2002.

Buffon, Georges Louis Leclerc, Comte de. *Natural History, General and Particular, by the Count de Buffon*. Translated by William Smellie. 8 vols. Edinburgh: Printed for William Creech, 1780.

————. *Natural History, General and Particular, by the Count de Buffon, translated into English*. 2nd ed. Translated by William Smellie. 9 vols. London: W. Strahan and T. Cadell, 1785.

————. *De l'homme*. Edited by Michele Duchet. Paris: L'Harmattan, 1971.

Bulmer, Michael. *Francis Galton: Pioneer of Heredity and Biometry*. Baltimore: Johns Hopkins University Press, 2003.

Burian, Richard. "Technique, Task Definition, and the Transition from Genetics to Molecular Genetics: Aspects of the Work on Protein Synthesis in the Laboratories of J. Monod and P. Zamecnik." *Journal of the History of Biology* 26 (1993): 387–407.

————. "Underappreciated Pathways toward Molecular Genetics as Illustrated by Jean Brachet's Cytochemical Embryology." In *The Philosophy and History of Molecular Biology: New Perspectives*, edited by Sahotra Sarkar, 67–85. Dordrecht: Kluwer, 1996.

Burton, Robert. *The Anatomy of Melancholy*, 1621. Projekt Gutenberg Edition, http://www.gutenberg.org/files/10800/10800-h/10800-h.htm (accessed March 4, 2010).

Buss, Allen R. "Galton and the Birth of Differential Psychology and Eugenics:

Social, Political, and Economic Forces." *Journal of the History of the Behavioral Sciences* 12 (1976): 47–58.

Cairns, John, Gunther S. Stent, and James D. Watson, eds. *Phage and the Origins of Molecular Biology*. Plainview, NY: Cold Spring Harbor Laboratory Press, 1992.

Campbell, Chloe. *Race and Empire: Eugenics in Colonial Kenya*. Manchester: Manchester University Press, 2007.

Campos, Luis. "Genetics without Genes: Blakeslee, *Datura* and 'Chromosomal Mutations.'" In *Conference: Heredity in the Century of the Gene (A Cultural History of Heredity IV)*, preprint 343, 21–23. Berlin: Max Planck Institute for the History of Science, 2006.

Canguilhem, Georges. "Qu'est-ce que une idéologie scientifique?" In *Idéologie et rationalité dans l'histoire des sciences de la vie*, edited by Georges Canguilhem, 33–45, 2nd ed. Paris: Vrin, 1981. 1st ed. 1977.

Cantor, Charles. "The Challenges to Technology and Informatics." In *The Code of Codes: Scientific and Social Issues in the Human Genome Project*, edited by Daniel J. Kevles and Leroy Hood, 98–111. Cambridge, MA: Harvard University Press, 1992.

Cantor, David, ed., *Cancer in the Twentieth Century*. Baltimore: Johns Hopkins University Press, 2008.

Carlson, Elof A. *The Gene: A Critical History*. Philadelphia: Saunders, 1966.

———. *Genes, Radiation, and Society: The Life and Work of H. J. Muller*. Ithaca, NY: Cornell University Press, 1981.

———. *Mendel's Legacy: The Origin of Classical Genetics*. Cold Spring Harbor: Cold Spring Harbor Laboratory Press, 2004.

Carol, Anne. *Histoire de l'eugénisme en France: Les médecins et la procréation, XIXe–XXe siècle*. Paris: Seuil, 1995.

Cartron, Laure. "Degeneration and 'Alienism' in Early Nineteenth-Century France." In *Heredity Produced: At the Crossroads of Biology, Politics and Culture, 1500–1870*, edited by Staffan Müller-Wille and Hans-Jörg Rheinberger, 155–174. Cambridge, MA: MIT Press, 2007.

Caskey, Thomas C. "DNA-Based Medicine: Prevention and Therapy." In *The Code of Codes: Scientific and Social Issues in the Human Genome Project*, edited by Daniel J. Kevles and Leroy Hood, 112–135. Cambridge, MA: Harvard University Press, 1992.

Céard, Jean. *La nature et les prodiges: L'insolite au XVIe siècle*. 2nd ed. Geneva: Droz, 1996. 1st ed. 1977.

Chadarevian, Soraya de. "Of Worms and Programmes: *Caenorhabditis elegans* and the Study of Development." *Studies in History and Philosophy of Biological and Biomedical Sciences* 29 (1998): 81–105.

———. *Designs for Life: Molecular Biology after World War II*. Cambridge: Cambridge University Press, 2002.

Churchill, Frederick B. "Hertwig, Weismann and the Meaning of Reduction Division circa 1890." *Isis* 61 (1970): 429–457.

———. "William [*sic*] Johannsen and the Genotype Concept." *Journal of the History of Biology* 7 (1974): 5–30.

————. "From Heredity Theory to 'Vererbung': The Transmission Problem, 1850–1915." *Isis* 78 (1987): 337–364.

————. "Living with the Biogenetic Law: A Reappraisal." In *From Embryology to Evo-Devo: A History of Developmental Evolution*, edited by Manfred D. Laubichler and Jane Maienschein, 37–81. Cambridge, MA: MIT Press, 2007.

*Civil Code.* http://www.napoleon-series.org/research/government/c_code.html (accessed February 19, 2010).

Cohen, I. Bernhard. "Harrington and Harvey: A Theory of the State Based on the New Philosophy." *Journal of the History of Ideas* 55 (1994): 187–210.

Cole, Francis Joseph. *Early Theories of Sexual Generation.* Oxford: Clarendon Press, 1930.

Coleman, William. *Biology in the Nineteenth Century: Problems of Form, Function, and Transformation.* New York: John Wiley, 1971.

Comfort, Nathaniel C. *The Tangled Field: Barbara McClintock's Search for the Patterns of Genetic Control.* Cambridge, MA: Harvard University Press, 2001.

Cook-Deegan, Robert. *The Gene Wars: Science, Politics, and the Human Genome.* New York: W. W. Norton, 1994.

Correns, Carl. Autobiographical sketch. Typescript. Archive of the Max Planck Society, Abt. III, Rep. 17, no. 1.

————. "Mendels Regel über das Verhalten der Nachkommenschaft der Rassenbastarde." *Berichte der Deutschen Botanischen Gesellschaft* 18 (1900): 158–168.

————. "Die Ergebnisse der neuesten Bastardforschungen für die Vererbungslehre." In *Gesammelte Abhandlungen zur Vererbungswissenschaft aus periodischen Schriften 1899–1924*, 264–286. Berlin: Julius Springer, 1924. Original 1901.

————. "Gregor Mendels Briefe an Carl Nägeli 1866–1873: Ein Nachtrag zu den veröffentlichten Bastardisierungsversuchen Mendels." In *Gesammelte Abhandlungen zur Vererbungswissenschaft aus periodischen Schriften 1899–1924*, 1233–1297. Berlin: Julius Springer, 1924. Original 1905.

————. "Zur Kenntnis der Rolle von Kern und Plasma bei der Vererbung." In *Gesammelte Abhandlungen zur Vererbungswissenschaft aus periodischen Schriften 1899–1924*, 648–656. Berlin: Julius Springer, 1924. Original 1909.

Cowan, Ruth Schwartz. "Nature and Nurture: The Interplay of Biology and Politics in the Work of Francis Galton." In *Studies in the History of Biology*, vol. 1, edited by William Coleman and Camille Limoges, 133–208. Baltimore: Johns Hopkins University Press, 1977.

Creager, Angela N. H. *The Life of a Virus: Tobacco Mosaic Virus as an Experimental Model, 1930–1965.* Chicago: University of Chicago Press, 2002.

————. "The Industrialization of Radioisotopes by the U.S. Atomic Energy Commission." In *The Science-Industry Nexus: History, Policy, Implications*, edited by Karl Grandin, Nina Wormbs, and Sven Widmalm, 141–168. Sagamore Beach, MA: Science History Publications, 2004.

————. "Mapping Genes in Microorganisms." In *From Molecular Genetics to Genomics: The Mapping Cultures of Twentieth-Century Genetics*, edited by

Jean-Paul Gaudillière and Hans-Jörg Rheinberger, 9–41. London: Routledge, 2004.

Creager, Angela N. H. and María Jesús Santesmases, eds. "Radiobiology in the Atomic Age." Special issue, *Journal of the History of Biology* 39, no. 4 (2006).

Cremer, Thomas. *Von der Zellenlehre zur Chromosomentheorie: Naturwissenschaftliche Erkenntnis und Theoriewechsel in der frühen Zell- und Vererbungsforschung.* Berlin: Springer, 1985.

Crick, Francis H. C. "On Protein Synthesis." *Symposium of the Society for Experimental Biology* 12 (1958): 138–163.

Crosby, Alfred W. *Ecological Imperialism: The Biological Expansion of Europe, 900–1900.* Cambridge University Press, Cambridge, 1986.

Darden, Lindley. *Reasoning in Biological Discoveries: Essays on Mechanisms, Interfield Relations, and Anomaly Resolution.* Cambridge: Cambridge University Press, 2006.

Darlington, Cyril Dean. Introduction to *Hereditary Genius: An Inquiry into Its Laws and Consequences*, by Francis Galton. London: Collins, 1962.

Darwin, Charles. *On the Origin of Species: A Facsimile of the First Edition.* Introduction by Ernst W. Mayr. Cambridge, MA: Harvard University Press, 1966. Original 1859.

———. *The Works of Charles Darwin.* Vols. 19 and 20. *The Variation of Animals and Plants under Domestication.* 2 vols. Edited by Paul H. Barrett and Richard B. Freeman. New York: New York University Press, 1988.

———. *The Works of Charles Darwin.* Vols. 21 and 22. *The Descent of Man, and Selection in Relation to Sex.* 2 vols. Edited by Paul H. Barrett and Richard B. Freeman. New York: New York University Press, 1989.

Daston, Lorraine, and Peter Galison. "The Image of Objectivity." *Representations* 40 (1992): 81.

Daston, Lorraine, and Katharine Park. *Wonder and the Order of Nature, 1150–1750.* New York: Zone Books, 1998.

De Renzi, Silvia. "Resemblance, Paternity, and Imagination in Early Modern Courts." In *Heredity Produced: At the Crossroads of Biology, Politics and Culture, 1500–1870*, edited by Staffan Müller-Wille and Hans-Jörg Rheinberger, 61–83. Cambridge, MA: MIT Press, 2007.

Derry, Margaret. *Bred for Perfection: Shorthorn Cattle, Collies, and Arabian Horses since 1800.* Baltimore: Johns Hopkins University Press, 2003.

Descartes, René. *Oeuvres de Descartes.* 11 vols. Edited by Charles Adam and Paul Tannery. Paris: Vrin, 1986.

———. *The Philosophical Writings of Descartes.* Translated by John Cottingham, Robert Stoothoff, and Dugald Murdoch. 3 vols. Cambridge: Cambridge University Press, 1998.

Detlefsen, Karen. "Explanation and Demonstration in the Haller-Wolff Debate." In *The Problem of Animal Generation in Early Modern Philosophy*, edited by Justin E. H. Smith, 235–261. Cambridge: Cambridge University Press, 2006.

de Vries, Hugo. "Sur la loi de disjonction des hybrides." *Comptes rendus de l'Académie des Sciences* 130 (1900): 845–847.

———. *Intracellular Pangenesis*. Translated by C. Stuart Gager. Chicago: Open Court Publishing, 1910. Originally published as *Intracellulare Pangenesis* (Jena: G. Fischer, 1889).

———. *The Mutation Theory: Experiments and Observations on the Origin of Species in the Vegetable Kingdom*. 2 vols. Translated by John Bretland Farmer and Arthur Dukinfield Darbishire. Chicago: Open Court, 1909–1910. Reprint, New York: Kraus Reprint, 1969. Originally published as *Die Mutations theorie* (Leipzig: Veit, 1901–1903).

Di Meo, Antonio. "Il concetto di 'circolazione': Storia di una rivoluzione transdisciplinare." In *Le rivoluzioni nelle scienze della vita*, edited by Guido Cimino and Bernardino Fantini, 31–84. Firenze: Leo S. Olschki, 1995.

Dobzhansky, Theodosius. *Genetics and the Origin of Species*. New York: Columbia University Press, 1937.

Dorr, Gregory Michael. *Segregation's Science: Eugenics and Society in Virginia*. Charlottesville: University of Virginia Press, 2008.

Dowbiggin, Ian Robert. *Inheriting Madness: Professionalization and Psychiatric Knowledge in Nineteenth-Century France*. Berkeley: University of California Press, 1991.

———. *Keeping America Sane: Psychiatry and Eugenics in the United States, 1880–1940*. Ithaca, NY: Cornell University Press, 1997.

Drayton, Richard. *Nature's Government: Science, Imperial Britain, and the "Improvement" of the World*. New Haven, CT: Yale University Press, 2000.

Dröscher, Ariane. "Edmund B. Wilson's *The Cell* and Cell Theory between 1896 and 1925." *History and Philosophy of the Life Sciences* 24 (2002): 357–389.

Duchesneau, François. *Les modèles du vivant de Descartes à Leibniz*. Paris: Vrin, 1998.

———. "Charles Bonnet's Neo-Leibnizian Theory of Organic Bodies." In *The Problem of Animal Generation in Early Modern Philosophy*, edited by Justin E. H. Smith, 285–314. Cambridge: Cambridge University Press, 2006.

———. "The Delayed Linkage of Heredity with the Cell Theory." In *Heredity Produced: At the Crossroads of Biology, Politics and Culture, 1500–1870*, edited by Staffan Müller-Wille and Hans-Jörg Rheinberger, 293–314. Cambridge, MA: MIT Press, 2007.

Dunn, Leslie C. *A Short History of Genetics: The Development of Some of the Main Lines of Thought; 1864–1939*. New York: McGraw-Hill, 1965.

Duret, Claude. *Histoire admirable des plantes et herbes esmerveillables et miraculeuses en nature*. Paris: Nicolas Buon, 1605.

Efron, John M. *Defenders of the Race: Jewish Doctors and Race Science in Fin-de-Siècle Europe*. New Haven, CT: Yale University Press, 1994.

Eigen, Sara Paulson. "A Mother's Love, a Father's Line: Law, Medicine and the 18th-Century Fictions of Patrilineal Genealogy." In *Genealogie als Denkform in Mittelalter und Früher Neuzeit*, edited by Kilian Heck and Bernhard Jahn, 87–107. Tübingen: Niemeyer, 2000.

Elzen, Boelie. "Two Ultracentrifuges: A Comparative Study of the Social Construction of Artefacts." *Social Studies of Science* 16 (1986): 621–662.

Engelhardt, Dietrich v. "Luca Ghini (um 1490–1556) und die Botanik des 16.

Jahrhunderts: Leben, Initiativen, Kontakte, Resonanz." *Medizinhistorisches Journal* 30 (1995): 3–49.

Falcon, Andrea. *Aristotle and the Science of Nature: Unity Without Uniformity.* Cambridge: Cambridge University Press, 2005.

Falk, Raphael. "Mendel's Impact." *Science in Context* 19 (2006): 215–236.

Farley, John. *The Spontaneous Generation Controversy from Descartes to Oparin.* Baltimore: Johns Hopkins University Press, 1974.

Feinberg, Harvey M., and Joseph B. Solodow. "Out of Africa." *Journal of African History* 43 (2002): 255–261.

Fernel, Jean François. *Medicina ad Henricum II Galliarum regem christianissimum.* 3 vols. Paris: Andreas Wechel, 1554.

Fischer, Ernst Peter, and Carol Lipson. *Thinking about Science: Max Delbrück and the Origins of Molecular Biology.* New York: Norton, 1988. Originally published as Fischer, Ernst Peter, *Das Atom des Biologen: Max Delbrück und der Ursprung der Molekulargenetik* (Munich: Piper, 1988).

Foote, Edward T. "Harvey: Spontaneous Generation and the Egg." *Annals of Science* 25 (1969): 139–163.

Fortun, Mike. *Promising Genomics: Iceland and deCODE Genetics in a World of Speculation.* Berkeley: University of California Press, 2008.

Foucault, Michel. *The Archaeology of Knowledge and the Discourse on Language.* Translated by A. M. Sheridan Smith. New York: Pantheon Books, 1972. Originally published as *L'archéologie du savoir* (Paris: Gallimard, 1969) and *L'ordre de discours: Leçon inaugurale au collège de France prononcée le 2 décembre 1970* (Paris: Gallimard, 1971).

———. *The History of Sexuality: Volume 1: An Introduction.* Translated by Robert Hurtley. New York: Vintage Books, 1990. Originally published as *Histoire de la Sexualité, I: La volonté de savoir* (Paris: Gallimard, 1976).

———. "Faire vivre et laisser mourir: La naissance du racisme." *Les Temps Modernes* 535 (1991): 37–61.

———. *The Order of Things: An Archaeology of the Human Sciences.* London: Routledge, 1997. Originally published as *Les mots et les choses* (Paris: Gallimard, 1966).

———. *Society Must Be Defended: Lectures at the Collège de France, 1975–1976.* Translated by David Macey. New York: Picador, 2003. Originally published as *Il faut défendre la société: Cours au Collège de France (1975–76)* (Paris: Gallimard, 1996).

Fox Keller, Evelyn. *A Feeling for the Organism: The Life and Work of Barbara McClintock.* San Francisco: Freeman, 1983.

———. "Physics and the Emergence of Molecular Biology: A History of Cognitive and Political Synergy." *Journal of the History of Biology* 23 (1990): 389–409.

———. *The Century of the Gene.* Cambridge, MA: Harvard University Press, 2000.

———. *Making Sense of Life: Explaining Biological Development with Models, Metaphors, and Machines.* Cambridge, MA: Harvard University Press, 2002.

Franklin, Rosalind, and Raymond Gosling. "Molecular Configuration in Sodium Thymonucleate." *Nature* 171 (1953), p. 740.

Friedmann, Herbert C. "From Friedrich Wöhler's Urine to Eduard Buchner's Alcohol." In *New Beer in an Old Bottle: Eduard Buchner and the Growth of Biochemical Knowledge*, edited by Athel Cornish-Bowden, 67–122. Valencia: Universitat de Valencia, 1997.

Fritsche, Johannes. "The Biological Precedents for Medieval Impetus Theory and Its Aristotelian Character." *British Journal for the History of Science* 44 (2010): 1–27.

Froggatt, Peter, and Norman C. Nevin. "The Law of 'Ancestral Heredity' and the Mendelian-Ancestrian Controversy in England, 1889–1906." *Journal of Medical Genetics* 8 (1971): 1–36.

Fuchs, Thomas. *Die Mechanisierung des Herzens: Harvey und Descartes; Der vitale und der mechanische Aspekt des Kreislaufs.* Frankfurt/M.: Suhrkamp, 1992.

Gachelin, Gabriel, ed. *Les organismes modèles dans la recherche médicale.* Paris: Presses Universitaires de France, 2006.

Galison, Peter, and Mario Biagioli, eds. *Scientific Authorship: Credit and Intellectual Property in Science.* London: Routledge, 2002.

Galton, Francis. "Hereditary Talent and Character." *Macmillan's Magazine* (1865): 157–166, 318–327.

———. *Hereditary Genius: An Inquiry into Its Laws and Consequences.* London: Macmillan, 1869.

———. "On Blood Relationship." *Proceedings of the Royal Society* 20 (1872): 394–402.

———. "Hereditary Improvement." *Fraser's Magazine* 7 (1873): 116–130.

———. "A Theory of Heredity." *Journal of the Anthropological Institute* 5 (1876): 329–348.

———. *Inquiries into Human Faculty and Its Development.* London: Macmillan, 1883.

———. *Natural Inheritance.* London: Macmillan, 1889.

———. "A Diagram of Heredity." *Nature* 57 (1898): 293.

Gans, Eduard. *Das Erbrecht in weltgeschichtlicher Entwickelung.* 4 vols. Berlin: Maurer, 1824–1835.

García-Sancho, Miguel. "A New Insight into Sanger's Development of Sequencing—From Proteins to DNA, 1943–1977." *Journal of the History of Biology* 43 (2010): 265–323.

Gardener, Bryan A. *Black's Law Dictionary.* 7th ed. St. Paul, MN: West Publishing, 1999.

Gärtner, Carl Friedrich. *Versuche und Beobachtungen über die Bastarderzeugung im Pflanzenreich.* Stuttgart: At the expense of the author, 1849.

Gasking, Elizabeth. *Investigations into Generation 1651–1828.* London: Hutchinson, 1967.

Gaudillière, Jean-Paul. *Inventer la biomédecine: La France, l'Amérique et la production des savoirs du vivant, 1945–1965.* Paris: La Découverte, 2002.

———. "Biochemie und Industrie: Der 'Arbeitskreis Butenandt-Schering' im Nationalsozialismus." In *Adolf Butenandt und die Kaiser-Wilhelm-Gesellschaft: Wissenschaft, Industrie und Politik in 'Dritten Reich,'* edited by Wolfgang Schieder and Achim Trunk, 198–246. Göttingen: Wallstein, 2004.

———. "Mapping as Technology: Genes, Mutant Mice, and Biomedical Research (1910–65)." In *Classical Genetic Research and Its Legacy: The Mapping Cultures of Twentieth Century Genetics*, edited by Hans-Jörg Rheinberger and Jean-Paul Gaudillière, 173–204. London: Routledge, 2004.

———. "New Wine in Old Bottles? The Biotechnology Problem in the History of Molecular Biology." In *Workshop, History and Epistemology of Molecular Biology and Beyond: Problems and Perspectives*, preprint 310, 81–93. Berlin: Max Planck Institute for the History of Science, 2006.

Gaudillière, Jean-Paul, and Ilana Löwy, eds. *The Invisible Industrialist: Manufactures and the Production of Scientific Knowledge*. London: Macmillan, 1998.

———, eds. *Heredity and Infection: The History of Disease Transmission*. London: Routledge, 2001.

———. "Medicine, Markets and Public Health: Contemporary Testing for Breast Cancer Predispositions." In *Medicine, Markets and Mass Media: Producing Health in the Twentieth Century*, edited by Virginia Berridge and Kelly Loughlin, 266–287. London: Routledge, 2005.

Gaudillière, Jean-Paul, and Hans-Jörg Rheinberger, eds. *From Molecular Genetics to Genomics: The Mapping Cultures of Twentieth Century Genetics*. London: Routledge, 2004.

Gausemeier, Bernd. "Auf der 'Brücke zwischen Natur- und Geschichtswissenschaft': Ottokar Lorenz und die Neuerfindung der Genealogie um 1900." In *Wissensobjekt Mensch: Praktiken der Humanwissenschaften im 20. Jahrhundert*, edited by Florence Vienne and Christina Brandt, 137–164. Berlin: Kulturverlag Kadmos, 2008.

———. "Pedigree vs. Mendelism: Concepts of Heredity in Psychiatry Before and After 1900." In *Conference: Heredity in the Century of the Gene (A Cultural History of Heredity IV)*, preprint 343, 149–162. Berlin: Max Planck Institute for the History of Science, 2008.

Gayon, Jean. "Entre force et structure: Genèse du concept naturaliste de l'hérédité." In *Le paradigme de la filiation*, edited by Jean Gayon and Jean-Jacques Wunenburger, 61–75. Paris: L'Harmattan, 1995.

———. *Darwinism's Struggle for Survival: Heredity and the Hypothesis of Natural Selection*. Cambridge, MA: Cambridge University Press, 1998.

———. "From Measurement to Organization: A Philosophical Scheme for the History of the Concept of Heredity." In *The Concept of the Gene in Development and Evolution: Historical and Epistemological Perspectives*, edited by Peter J. Beurton, Raphael Falk, and Hans-Jörg Rheinberger, 69–90. Cambridge: Cambridge University Press, 2000.

———. "Do Biologists Need the Expression 'Human Races'? UNESCO 1950–1951." In *Bioethical and Ethical Issues Surrounding the Trials and Code of Nuremberg: Nuremberg Revisited*, edited by Jacques Rozenberg, 23–48. Lewiston, NY: Edwin Mellen Press, 2003.

Gayon, Jean, and Jean-Jacques Wunenburger. "Présentation." In *Le paradigme de la filiation*, edited by Jean Gayon and Jean-Jacques Wunenburger, 1–15. Paris: L'Harmattan, 1995.

Gayon, Jean, and Doris T. Zallen. "The Role of the Vilmorin Company in the

Promotion and Diffusion of the Experimental Science of Heredity in France, 1840–1920." *Journal of the History of Biology* 31 (1998): 241–262.

Gilbert, Scott F., ed. *A Conceptual History of Modern Developmental Biology.* Baltimore: Johns Hopkins University Press, 1994.

———. *Developmental Biology.* 9th ed. Sunderland, MA: Sinauer Associates, 2010. 1st ed. 1985.

Glass, Bentley. "The Germination of the Idea of Biological Species." In *Forerunners of Darwin, 1745–1859,* edited by Bentley Glass, Owsei Temkin, and William L. Strauss. Baltimore: Johns Hopkins University Press, 1968. Original 1959.

Glass, Bentley, Owsei Temkin, and William L. Strauss, eds. *Forerunners of Darwin, 1745–1859.* Baltimore: Johns Hopkins University Press, 1959.

Gliboff, Sander. "Gregor Mendel and the Laws of Evolution." *History of Science* 37 (1999): 217–235.

Goltz, Dietlinde. "Der leere Uterus: Zum Einfluß von Harveys 'De generatione animalium' auf die Lehren von der Konzeption." *Medizinhistorisches Journal* 21 (1986): 242–268.

Goody, Jack. *The Development of the Family and Marriage in Europe.* Cambridge: Cambridge University Press, 1983.

Gottweis, Herbert. *Governing Molecules: The Discursive Politics of Genetic Engineering in Europe and the United States.* Cambridge, MA: MIT Press, 1998.

———. "Human Embryonic Stem Cells, Cloning, and the Transformation of Biopolitics." In *Biotechnology: Between Commerce and Civil Society,* edited by Nico Stehr, 239–265. New Brunswick, NJ: Transaction, 2004.

Gottweis, Herbert, Brian Salter, and Catherine Waldby. *The Global Politics of Human Embryonic Stem Cell Science: Regenerative Medicine in Transition.* Basingstoke: Palgrave Macmillan, 2009.

Gottweis, Herbert, and Robert Triendl. "South Korean Policy Failure and the Hwang Debacle." *Nature Biotechnology* 24 (2006): 141–143.

Gould, Steven J. *The Mismeasure of Man.* New York: Norton, 1981.

Grandin, Karl, Nina Wormbs, and Sven Widmalm, eds. *The Science-Industry Nexus: History, Policy, Implications.* Sagamore Beach, MA: Science History Publications, 2004.

Gregory, Andrew. "Harvey, Aristotle and the Weather Cycle." *Studies in the History and Philosophy of the Biological and Biomedical Sciences* 32 (2001): 153–168.

Griesemer, James R. "Reproduction and the Reduction of Genetics." In *The Concept of the Gene in Development and Evolution: Historical and Epistemological Perspectives,* edited by Peter J. Beurton, Raphael Falk, and Hans-Jörg Rheinberger, 240–285. Cambridge Studies in Philosophy and Biology. Cambridge: Cambridge University Press, 2000.

———. "Tracking Organic Processes: Representations and Research Styles in Classical Embryology and Genetics." In *From Embryology to Evo-Devo: A History of Developmental Evolution,* edited by Manfred D. Laubichler and Jane Maienschein, 375–433. Cambridge, MA: MIT Press, 2007.

Gros, François. *Les secrets du gène*. 2nd ed. Paris: Ed. Odile Jacob, 1991. 1st ed. 1986.

Grote, Mathias. "Hybridizing Bacteria, Crossing Methods, Cross-checking Arguments: The Transition from Episomes to Plasmids (1961–1969)." *History and Philosophy of the Life Sciences* 30 (2008): 407–430.

Gruber, Jacob W. "Ethnographic Salvage and the Shaping of Anthropology." *American Anthropologist* 72 (1970): 1289–1299.

Hagner, Michael. *Homo cerebralis: Der Wandel vom Seelenorgan zum Gehirn*. Berlin: Berlin Verlag, 1997.

Hanke, Christine. *Zwischen Auflösung und Fixierung: Zur Konstitution von "Rasse" und "Geschlecht" in der physischen Anthropologie um 1900*. Bielefeld: Transcript, 2007.

Hansen, Emil Christian. *Practical Studies in Fermentation: Being Contributions to the Life History of Micro-Organisms*. London: Spon and Chamberlain, 1896.

Hareven, Tamara. "The Search for Generational Memory: Tribal Rites in Industrial Society." *Daedalus* 107 (1978): 137–149.

Harrington, Anne. *Reenchanted Science: Holism in German culture From Wilhelm II to Hitler*. Princeton: Princeton University Press, 1996.

Hartmann, Max. *Allgemeine Biologie*. Jena: Gustav Fischer, 1927.

Harvey, William. *Exercitationes de generatione animalium*. Amsterdam: Elzevir, 1651.

———. *The Works of William Harvey*. Translated by Robert Willis. London: Sydenham Society, 1847.

Harwood, Jonathan. *Styles of Scientific Thought: The German Genetics Community, 1900–1933*. Chicago: University of Chicago Press, 1993.

———. *Technology's Dilemma: Agricultural Colleges between Science and Practice in Germany, 1860–1934*. Bern: Peter Lang, 2005.

Heck, Kilian, and Bernhard Jahn, eds. *Genealogie als Denkform in Mittelalter und Früher Neuzeit*. Tübingen: Niemeyer, 2000.

Heinrich, Klaus. *Tertium datur: Eine religionsphilosophische Einführung in die Logik*. Basel: Stroemfeld/Roter Stern, 1983.

Herder, Johann Gottfried. *Outlines of a Philosophy of the History of Man*. Translated by Thomas Churchill. London: Printed for Joseph Johnson, by Luke Hansard, 1800. Originally published as *Ideen zur Philosophie der Geschichte der Menschheit* (Riga: Hartknoch, 1784–1791)

Hilgartner, Stephen. "Biomolecular Databases: New Communication Regimes for Biology?" *Science Communication* 17 (1995): 240–263.

———. "The Human Genome Project." In *Handbook of Science and Technology Studies*, edited by Sheila Jasanoff, Gerald E. Markle, James C. Petersen, and Trevor Pinch. Thousand Oaks, CA: Sage Publications, 1995.

———. "Data Access Policy in Genome Research." In *Private Science: Biotechnology and the Rise of the Molecular Sciences*, edited by Arnold Thackray. Philadelphia: University of Pennsylvania Press, 1998.

———. *Science on Stage: Expert Advice as Public Drama*. Stanford: Stanford University Press, 2000.

———. "Making Maps and Making Social Order: Governing American Genome

Centers, 1988–93." In *From Molecular Genetics to Genomics: The Mapping Cultures of Twentieth-Century Genetics*, edited by Jean-Paul Gaudillière and Hans-Jörg Rheinberger. London: Routledge, 2004.

Hill, Christopher. "William Harvey and the Idea of Monarchy." *Past and Present* 27 (1964): 54–72.

Hippocrates. "De genitura." In *Hippocratis Coi Opera qvae Graece et latine extant*, edited and translated by Girolamo Mercuriale, 10–16. Venice: Industria ac sumptibus Iuntarum, 1588.

———. *Hippocratic Writings*. Edited by Geoffrey Ernest Richard Lloyd. Harmondsworth, UK: Penguin Books, 1950.

Hodge, Jonathan. "Darwin as a Lifelong Generation Theorist." In *The Darwinian Heritage*, edited by David Kohn, 204–244. Princeton, NJ: Princeton University Press, 1985.

Holmes, Frederick L. *Reconceiving the Gene: Seymour Benzer's Adventures in Phage Genetics*. New Haven, CT: Yale University Press, 2006.

Hoßfeld, Uwe. *Geschichte der biologischen Anthropologie in Deutschland: Von den Anfängen bis in die Nachkriegszeit*. Stuttgart: Franz Steiner Verlag, 2005.

Ibrahim, Annie. "La notion de moule intérieur dans les théories de la génération au XVIIIe siècle." *Archives de Philosophie* 50 (1987): 555–580.

Jablonka, Eva, and Marion J. Lamb. *Evolution in Four Dimensions: Genetic, Epigenetic, Behavioral, and Symbolic Variation in the History of Life*. Cambridge, MA: MIT Press, 2005.

Jackson, David A. "DNA. Template for an Economic Revolution." In *DNA: The Double Helix, Perspective and Prospective at Forty Years*, edited by D. A. Chambers. New York: New York Academy of Sciences, 1995.

Jacob, François. *The Possible and the Actual*. Translated by Betty E. Spillman. Seattle: University of Washington Press, 1982. Originally published as *Le jeu des possibles: Essai sur la diversité du vivant* (Paris: Fayard, 1981).

———. *The Logic of Life: A History of Heredity*. Translated by Betty E. Spillman. Princeton, NJ: Princeton University Press, 1993. Originally published as *La logique du vivant* (Paris: Gallimard, 1970).

Jahn, Ilse. "Zur Geschichte der Wiederentdeckung der Mendelschen Gesetze." *Wissenschaftliche Zeitschrift der Friedrich-Schiller Universität Jena*, Mathematisch-naturwissenschaftliche Reihe 7, no. 2–3 (1958): 215–227.

———. *Geschichte der Biologie*. Berlin: Spektrum Akademischer Verlag, 2000.

James, William. "Two Reviews of *The Variation of Animals and Plants under Domestication* by Charles Darwin." In *The Works of William James: Essays, Comments, and Reviews*, edited by Frederick H. Burkhardt, Fredson Bowers, and Ignas K. Skrupskelis, 229–239. Cambridge, MA: Harvard University Press, 1987.

Jasanoff, Sheila. *Science at the Bar: Science and Technology in American Law*. Cambridge, MA: Harvard University Press, 1996.

———. *Designs on Nature: Science and Democracy in Europe and the United States*. Princeton, NJ: Princeton University Press, 2005.

———. "Biotechnology and Empire: The Global Power of Seeds and Science." *Osiris* 21 (2006): 273–292.

Jennings, Herbert S. "Pure Lines in the Study of Genetics in Lower Organisms." *American Naturalist* 45 (1911): 79–89.

Joerges, Bernward, and Terry Shinn, eds. *Instrumentation between Science, State and Industry.* Dordrecht: Kluwer Academic Publishers, 2001.

Johannsen, Wilhelm. "The Genotype Conception of Heredity." *American Naturalist* 45, no. 531 (1911): 129–159.

———. "Some Remarks about Units in Heredity." *Hereditas* 4 (1923): 133–141.

———. *Elemente der exakten Erblichkeitslehre: Mit Grundzügen der biologischen Variationsstatistik.* 3rd ed. Jena: Gustav Fischer, 1926. 1st ed. 1909.

Jordanova, Ludmilla. "Interrogating the Concept of Reproduction in the Eighteenth Century." In *Conceiving the New World Order*, edited by Faye D. Ginsburg and Rayna Rapp, 369–386. Berkeley: University of California Press, 1995.

Jucovy, Peter M. "Circle and Circulation: The Language and Imagery of William Harvey's Discovery." *Perspectives in Biology and Medicine* 20 (1976): 92–107.

Judson, Horace Freeland. "A History of the Science and Technology Behind Gene Mapping and Sequencing." In *The Code of Codes: Scientific and Social Issues in the Human Genome Project*, edited by Daniel J. Kevles and Leroy Hood, 37–80. Cambridge, MA: Harvard University Press, 1992.

———. *The Eighth Day of Creation: Makers of the Revolution in Biology.* Expanded ed. New York: Cold Spring Harbor Laboratory Press, 1996. 1st ed. 1979.

Junker, Thomas, and Uwe Hoßfeld. *Die Entdeckung der Evolution: Eine revolutionare Theorie und ihre Geschichte.* Darmstadt: Wissenschaftliche Buchgesellschaft, 2001.

Kant, Immanuel. *Kant's Critique of Judgement.* 2nd ed. Translated by John Henry Bernard. London: Macmillan, 1914. Originally published as *Kritik der Urteilskraft* 1790.

———. "Idea for a Universal History with a Cosmopolitan Purpose." In *Kant: Political Writings*, 2nd ed., 41–53. Cambridge Texts in the History of Political Thought. Cambridge: Cambridge University Press, 1991. Originally published as "Idee zu einer allgemeinen Geschichte in weltbürgerlicher Absicht," 1784.

———. "Determination of the Concept of a Human Race." Translated by H. Wilson and Günter Zöller. In *The Cambridge Edition of the Works of Immanuel Kant*, vol. 7, *Anthropology, History, and Education*, edited by Robert B. Loudon and Günter Zöller, 143–153. Cambridge: Cambridge University Press, 2007. Originally published as "Bestimmung des Begriffs einer Menschenrace," 1785.

———. "On the Use of Teleological Principles in Philosophy." Translated by Günter Zöller. In *The Cambridge Edition of the Works of Immanuel Kant*, vol. 7, *Anthropology, History, and Education*, edited by Robert B. Loudon and Günter Zöller, 192–218. Cambridge: Cambridge University Press, 2007. Originally published as "Über den Gebrauch teleologischer Principien in der Philosophie," 1788.

Kass, Lee B., and Christophe Bonneuil. "Mapping and Seeing: Barbara McClintock and the Linking of Genetics and Cytology in Maize Genetics, 1928–1935." In *Classical Genetic Research and Its Legacy: The Mapping Cultures of Twentieth Century Genetics*, edited by Hans-Jörg Rheinberger and Jean-Paul Gaudillière, 91–118. London: Routledge, 2004.

Kass, Lee B., Christophe Bonneuil, and Edward H. Coe. "Cornfests, Cornfabs and Cooperation: The Origins and Beginnings of the Maize Genetics Cooperation News Letter." *Genetics* 169 (2005): 1787–1797.

Kaufmann, Alain. "Mapping the Human Genome at Généthon Laboratory: The French Muscular Dystrophy Association and the Politics of the Gene." In *From Molecular Genetics to Genomics: The Mapping Cultures of Twentieth-Century Genetics*, edited by Jean-Paul Gaudillière and Hans-Jörg Rheinberger, 129–157. London: Routledge, 2004.

Kaufmann, Doris. "'Rasse und Kultur': Die amerikanische Kulturanthropologie um Franz Boas (1858–1942) in der ersten Hälfte des 20. Jahrhunderts—ein Gegenentwurf zur Rassenforschung in Deutschland." In *Rassenforschung an Kaiser-Wilhelm-Instituten vor und nach 1933*, edited by Hans-Walter Schmuhl, 309–327. Göttingen: Wallstein, 2003.

Kay, Lily E. "Conceptual Models and Analytical Tools: The Biology of Physicist Max Delbrück." *Journal of the History of Biology* 18 (1985): 207–246.

———. "W. M. Stanley's Crystallization of the Tobacco Mosaic Virus, 1930–1940." *Isis* 77 (1986): 450–472.

———. "The Tiselius Electrophoresis Apparatus and the Life Sciences, 1930–1940." *History and Philosophy of the Life Sciences* 10 (1988): 51–72.

———. "Selling Pure Science in Wartime: The Biochemical Genetics of G. W. Beadle." *Journal for the History of Biology* 22 (1989): 73–101.

———. *The Molecular Vision of Life: Caltech, the Rockefeller Foundation, and the Rise of the New Biology*. New York: Oxford University Press, 1993.

———. *Who Wrote the Book of Life? A History of the Genetic Code*. Stanford, CA: Stanford University Press, 2000.

Keating, Peter, and Alberto Cambrosio. *Biomedical Platforms: Realigning the Normal and the Pathological in Late-Twentieth-Century Medicine*. Cambridge, MA: MIT Press, 2003.

Keller, Eve. "Making Up for Losses: The Workings of Gender in William Harvey's 'De generatione animalium.'" In *Inventing Maternity: Politics, Science, and Literature, 1650–1865*, edited by Susan C. Greenfield and Carol Barash, 34–56. Lexington: University of Kentucky Press, 1999.

Kevles, Daniel J. *In the Name of Eugenics: Genetics and the Use of Human Heredity*. Cambridge, MA: Harvard University Press, 1985.

———. "Out of Eugenics: The Historical Politics of the Genome." In *The Code of Codes: Scientific and Social Issues in the Human Genome Project*, edited by Daniel J. Kevles and Leroy Hood, 3–36. Cambridge, MA: Harvard University Press, 1992.

———. "The Advent of Animal Patents: Innovation and Controversy in the Engineering and Ownership of Life." In *Intellectual Property Rights and Patenting in Animal Breeding and Genetics*, edited by S. Newman and M. Rothschild, 17–30. New York: CABI Publishing, 2002.

————. *A History of Patenting Life in the United States: With Comparative Attention to Europe and Canada.* Brussels: European Community, 2002.

————. "Of Mice and Money: The Story of the World's First Animal Patent." *Daedalus* (Spring 2002): 78–88.

————. "Breeding, Biotechnology, and Agriculture: The Establishment and Protection of Intellectual Property in Animals since the Late Eighteenth Century." In *Workshop, History and Epistemology of Molecular Biology and Beyond: Problems and Perspectives,* preprint 310, 69–79. Berlin: Max Planck Institute for the History of Science, 2006.

Kevles, Daniel J., and Leroy Hood, eds. *The Code of Codes: Scientific and Social Issues in the Human Genome Project.* Cambridge, MA: Harvard University Press, 1992.

Kielmeyer, Carl Friedrich. *Über die Verhältnisse der organischen Kräfte unter einander in der Reihe der verschiedenen Organisationen.* With an introduction by Kai Torsten Kanz. Marburg: Basilisken-Presse, 1993. Original 1793.

Kim, Kyung-Man. *Explaining Scientific Consensus: The Case of Mendelian Genetics.* New York: Guilford Press, 1994.

Kimmelman, Barbara A. "Mr. Blakeslee Builds His Dream House: Agricultural Institutions, Genetics, and Careers, 1900–1915." *Journal of the History of Biology* 39 (2006): 241–280.

Koch, Lene. "The Meaning of Eugenics: Reflections on the Government of Genetic Knowledge in the Past and the Present." *Science in Context* 17 (2004): 315–331.

Koerner, Lisbet. *Linnaeus: Nature and Nation.* Cambridge, MA: Harvard University Press, 1999.

Kohler, Robert E. "The Management of Science: The Experience of Warren Weaver and the Rockefeller Foundation Programme in Molecular Biology." *Minerva* 14 (1976): 279–306.

————. *From Medical Chemistry to Biochemistry: The Making of a Biomedical Discipline.* Cambridge: Cambridge University Press, 1982.

————. "Systems of Production: *Drosophila, Neurospora,* and Biochemical Genetics." *Historical Studies in the Physical and Biological Sciences* 22 (1991): 87–130.

————. *Lords of the Fly: Drosophila Genetics and the Experimental Life.* Chicago: University of Chicago Press, 1994.

Kölreuter, Joseph Gottlieb. *Vorläufige Nachricht von einigen, das Geschlecht der Pflanzen betreffenden Versuchen und Beobachtungen nebst Fortsetzungen 1, 2 und 3 (1761–1766).* Leipzig: Wilhelm Engelmann, 1893.

Koyré, Alexandre. "Galileo and the Scientific Revolution of the Seventeenth Century." *Philosophical Review* 52 (1943): 333–348.

Krader, Lawrence. *Ethnologie und Anthropologie bei Marx.* Munich: Hanser, 1973.

Kraft, Alison. "Between Medicine and Industry: Medical Physics and the Rise of the Radioisotope 1945–65." *Contemporary British History* 20 (2006): 1–35.

Krige, John. "Atoms for Peace, Scientific Internationalism, and Scientific Intelligence." *Osiris* 21 (2006): 161–181.

Krimsky, Sheldon. *Biotechnics and Society: The Rise of Industrial Genetics.* New York: Praeger, 1991.

Kronfeldner, Maria. "'If There Is Nothing Beyond the Organic . . .': Heredity and Culture at the Boundaries of Anthropology in the Work of Alfred L. Kroeber." *NTM: Zeitschrift für Geschichte der Wissenschaften, Technik und Medizin* 17 (2009): 107–133.

Kühl, Stefan. *Die Internationale der Rassisten: Aufstieg und Niedergang der internationalen Bewegung für Eugenik und Rassenhygiene im 20. Jahrhundert.* Frankfurt/M.: Campus, 1997.

Kühn, Alfred. "Über eine Gen-Wirkkette der Pigmentbildung bei Insekten." *Nachrichten der Akademie der Wissenschaften in Göttingen, mathematisch-physikalische Klasse* (1941): 231–261.

Kuklick, Henrika. "The British Tradition." In *A New History of Anthropology,* edited by Henrika Kuklick, 52–78. Oxford: Wiley Blackwell, 2007.

Kuper, Adam. "On Human Nature: Darwin and the Anthropologists." In *Nature and Society in Historical Perspective,* edited by Mikulás Teich, Roy Porter, and Bo Gustafsson, 274–290. Cambridge: Cambridge University Press, 1997.

La Berge, Ann F. *Mission and Method: The Early Nineteenth-Century French Public Health Movement.* Cambridge: Cambridge University Press, 1992.

Lagier, Raphaël. *Les races humaines selon Kant.* Paris: Presses Universitaires de France, 2004.

Lander, Eric. "Interview." *Newsweek,* October 14, 2007, 46–47.

Larson, James L. *Interpreting Nature: The Science of Living Form from Linnaeus to Kant.* Baltimore: Johns Hopkins University Press, 1994.

Laubichler, Manfred D. "'Allgemeine Biologie' als selbständige Grundwissenschaft und die allgemeinen Grundlagen des Lebens." In *Der Hochsitz des Wissens: Das Allgemeine als wissenschaftlicher Wert,* edited by Michael Hagner and Manfred D. Laubichler, 185–206. Zürich: Diaphanes, 2006.

Laubichler, Manfred D., and Eric H. Davidson. "Boveri's Long Experiment: Sea Urchin Merogones and the Establishment of the Role of Nuclear Chromosomes in Development." *Developmental Biology* 314, no. 1 (2007): 1–11.

Laubichler, Manfred D. and Jane Maienschein, eds. *From Embryology to Evo-Devo: A History of Developmental Evolution.* Cambridge, MA: MIT Press, 2007.

Lauremberg, Peter. *Horticultura, libris II. comprehensa; huic nostro coelo et solo accomodata.* Frankfurt/M.: Merian, 1631.

Lazer, David, ed. *DNA and the Criminal Justice System: The Technology of Justice.* Cambridge, MA: MIT Press, 2004.

Lederman, Muriel, and Richard M. Burian. "The Right Organism for the Job." *Journal of the History of Biology* 26 (1993): 235–367.

Lefèvre, Wolfgang. "Inheritance of Acquired Characters: Heredity and Evolution in Late Nineteenth-Century Germany." In *A Cultural History of Heredity III: 19th and Early 20th Centuries,* preprint 294, 53–66. Berlin: Max Planck Institute for the History of Science, 2005.

Lennox, James G. "The Comparative Study of Animal Development: William Harvey's Aristotelianism." In *The Problem of Animal Generation in Early*

*Modern Philosophy*, edited by Justin E. H. Smith, 21–46. Cambridge: Cambridge University Press, 2006.

Lenoir, Timothy. "Kant, Blumenbach, and Vital Materialism in German Biology." *Isis* 71 (1980): 77–108.

———. *The Strategy of Life: Teleology and Mechanics in Nineteenth-Century Germany*. Chicago: University of Chicago Press, 1982.

Leonelli, Sabina. "Documenting the Emergence of Bio-Ontologies: Or, Why Studying Bioinformatics Requires HPSSB." *History and Philosophy of the Life Sciences* 32 (2010): 105–125.

———. "Making Sense of Data-Driven Research in the Biological and Biomedical Sciences." *Studies in History and Philosophy of the Biological and Biomedical Sciences* 43 (2012): 1–3.

Leonelli, Sabina, and Rachel Ankeny. "Rethinking Organisms: The Epistemic Impact of Databases on Model Organism Biology." *Studies in History and Phiolosophy of the Biological and Biomedical Sciences* 43 (2012): 29–36.

Lepenies, Wolf. *Das Ende der Naturgeschichte: Wandel kultureller Selbstverständlichkeiten in den Wissenschaften des 18. und 19. Jh.* Munich: Hanser, 1976.

Lesky, Erna. *Die Zeugungs- und Vererbungslehren der Antike und ihr Nachwirken.* Abhandlungen der Geistes- und Sozialwissenschaftlichen Klasse der Akademie der Wissenschaften und der Literatur in Mainz, no. 19 (1950). Wiesbaden: Franz Steiner, 1951.

———. "Harvey und Aristoteles." *Sudhoffs Archiv* 41 (1957): 289–316 and 349–378.

Lévi-Strauss, Claude. *The Savage Mind*. London: Weidenfeld and Nicolson, 1974. Originally published as *La pensée sauvage* (Paris: Plon, 1962).

———. *Totemism*. Translated by Rodney Needham. London: Merlin Press, 1991. Originally published as *Le Totemisme aujourd'hui* (Paris: P. U. F., 1962).

Lichtenthaler, Frieder W. "Hundert Jahre Schlüssel-Schloss-Prinzip: Was führte Emil Fischer zu dieser Analogie?" *Angewandte Chemie* 106 (1994): 2456–2467.

Limoges, Camille. "Introduction." In *C. Linné: L' équilibre de la nature: L'histoire des sciences; textes et études*, translated by Bernard Jasmin, 7–24. Paris: Vrin, 1972.

Linnaeus, Carl. *Systema Naturae, sive Regna Tria Naturae systematice proposita per classes, ordines, genera, et species*. Leiden: De Groot, 1735.

———. *Critica botanica in quo nomina plantarum generica, specifica, et variantia examini subjiciuntur*. Leiden: Wishoff, 1737.

———. *Amoenitates academicae, seu Dissertationes variae Physicae, Medicae, Botanicae antehac seorsim editae*. 7 vols. Stockholm: Godofredus Kiesewetter, 1749–1769.

———. *Auserlesene Abhandlungen der Naturgeschichte, Physik und Arzney-wissenschaft*. Edited and translated by E. J. T. Hoepfner. 3 vols. Leipzig: Adam Friedrich Böhme, 1776–1778.

———. *Miscellaneous Tracts Relating to Natural History, Husbandry, and Physick*. Translated by Benjamin Stillingfleet. London: J. Dodsley; Leigh and Sotheby; T. Payne, 1791.

————. *Select Dissertations from the Amoenitates Academicae.* Translated by Fitz John Brand. London: 1781. Reprint, New York: Arno Press Inc, 1977.

————. *Linnaeus' Philosophia Botanica.* Translated by Stephen Freer. Oxford: Oxford University Press, 2003. Original 1751.

Lipphardt, Veronika. "Zwischen 'Inzucht' und 'Mischehe': Demographisches Wissen in der Debatte um die 'Biologie der Juden.'" *Tel Aviver Jahrbuch für Deutsche Geschichte* 35 (2007): 45–66.

Lippmann, Edmund O. *Urzeugung und Lebenskraft: Zur Geschichte dieser Probleme von den ältesten Zeiten an bis zu den Anfängen des 20. Jahrhunderts.* Berlin: Julius Springer, 1933.

Locke, John. *An Essay Concerning Human Understanding.* Oxford: Clarendon Press, 1979. Original 1690.

López Beltrán, Carlos. "Natural Things and Non-natural Things: The Boundaries of the Hereditary in the 18th Century." In *Conference: A Cultural History of Heredity I: 17th and 18th Centuries,* preprint 222, 69–84. Berlin: Max Planck Institute for the History of Science, 2003.

————. "In the Cradle of Heredity: French Physicians and *l'hérédité naturelle* in the Early Nineteenth Century." *Journal of the History of Biology* 37 (2004): 39–72.

————. *El sesgo hereditario: Ámbitos históricos del concepto de herencia biológica.* Mexico City: Universidad Nacional Autónoma de México, 2004.

————. "Hippocratic Bodies: Temperament and Castas in Spanish America (1570–1820)." *Journal of Spanish Cultural Studies* 8, no. 2 (2007): 253–289.

————. "The Medical Origins of Heredity." In *Heredity Produced: At the Crossroads of Biology, Politics and Culture, 1500–1870,* edited by Staffan Müller-Wille and Hans-Jörg Rheinberger, 105–132. Cambridge, MA: MIT Press, 2007.

Lorenz, Ottokar. *Lehrbuch der gesammten wissenschaftlichen Genealogie.* Berlin: Wilhelm Hertz, 1898.

Löwy, Ilana. "Variances in Meaning in Discovery Accounts." *Historical Studies in the Physical and Biological Sciences* 21 (1990): 87–121.

Löwy, Ilana, and Jean-Paul Gaudillière. "Disciplining Cancer: Mice and the Practice of Genetic Purity." In *The Invisible Industrialist: Manufactures and the Production of Scientific Knowledge,* edited by Jean-Paul Gaudillière and Ilana Löwy, 209–249. London: MacMillan, 1998.

————. "Mendelian Factors and Human Disease: A Conversation." In *Conference: Heredity in the Century of the Gene (A Cultural History of Heredity IV),* preprint 343, 19–26. Berlin: Max Planck Institute for the History of Science, 2008.

Lucas, Prosper. *Traité philosophique et physiologique de l'hérédité naturelle.* 2 vols. Paris: J. B. Baillière, 1847–1850.

Ludes, Francis, J., Arnold O. Ginnow, Lawrence J. Culligan, Robert J. Owens, and Matthew J. Canavan, eds. *Corpus Juris Secundum: A Contemporary Statement of American Law as Derived from Reported Cases and Legislation.* Vol. 39A. Saint Paul, MN: Thomson-West, 2003.

Lugt, Maaike van der. "La peau noire dans la science médiévale." *Micrologus* 13 (2005): 439–475.

―――. "Les maladies héréditaires dans la pensée scolastique (XIIe–XVIe siècle)." In *L'hérédité entre Moyen Age et époque moderne*, edited by Maaike van der Lugt and Charles de Miramon, 273–322. Florence: SISMEL— Edizioni del Galluzzo, 2008.

Lugt, Maaike van der, and Charles de Miramon. "Introduction." In *L'hérédité entre Moyen Age et époque moderne*, edited by Maaike van der Lugt and Charles de Miramon, 3–40. Florence: SISMEL—Edizioni del Galluzzo, 2008.

Lyell, Charles. *Principles of Geology, Being an Attempt to Explain the Former Changes of the Earth's Surface, by Reference to Causes Now in Operation.* 3 vols. London: John Murray, 1832–1833.

Lynch, Michael, and Simon A. Cole. "Science and Technology Studies on Trial: Dilemmas of Expertise." *Social Studies of Science* 35 (2005): 269–311.

Lyon, John, and Phillip Reid Sloan, eds. *From Natural History to the History of Nature: Readings from Buffon and His Critics.* Notre Dame, IN: University of Notre Dame Press, 1981.

MacKenzie, Donald A. "Eugenics in Britain." *Social Studies of Science* 6 (1976): 499–532.

―――. "Statistical Theory and Social Interests." *Social Studies of Science* 8 (1978): 35–83.

―――. *Statistics in Britain, 1865–1930: The Social Construction of Scientific Knowledge.* Edinburgh: Edinburgh University Press, 1981.

Maienschein, Jane. "Heredity/Development in the United States, circa 1900." *History and Philosophy of the Life Sciences* 9 (1987): 79–93.

―――. *Transforming Traditions in American Biology, 1880–1915.* Baltimore: Johns Hopkins University Press, 1991.

―――. *Whose View of Life? Embryos, Cloning, and Stem Cells.* Cambridge, MA: Harvard University Press, 2003.

Malebranche, Nicholas. *Dialogues on Metaphysics and Religion.* Translated by Jonathan Bennett. http://www.earlymoderntexts.com/pdfbits/ml5.pdf, 2007 (accessed September 30, 2011). Original 1688.

Mannheim, Karl. "The Problem of Generations." In *Essays on the Sociology of Knowledge by Karl Mannheim*, edited by Paul Kecskemeti, 276–320. New York: Routledge & Kegan Paul, 1952.

Marie, Jennifer. "The Importance of Place: A History of Genetics in 1930s Britain." Ph.D. diss., University College London, 2004.

Marx, Karl, and Friedrich Engels. *Collected Works.* 50 vols. London: Lawrence & Wishart, 1975–2005.

Massin, Benoit. "From Virchow to Fischer: Physical Anthropology and 'Modern Race Theories' in Wilhelmine Germany." In *Volksgeist as Method and Ethic: Essays on Boasian Ethnography and the German Anthropological Tradition*, edited by George W. Stocking, 79–154. Madison: University of Wisconsin Press, 1996.

Massy, Charles. *The Australian Merino: The Story of a Nation.* 2nd ed. Sydney: Random House Australia, 2007. 1st ed. 1990.

Maupertuis, Pierre Louis Moreau de. *Oeuvres.* 4 vols. Lyon: Jean Marie Bruyset, 1768. Reprint, Hildesheim: Olms, 1965.

Maynard Smith, John. "The Concept of Information in Biology," *Philosophy of Science* 67 (2000): 177–194.

Mayr, Ernst. "The Recent Historiography of Genetics." *Journal for the History of Biology* 6 (1973): 125–154.

———. *The Growth of Biological Thought: Diversity, Evolution and Inheritance.* Cambridge, MA: Belknap Press, 1982.

Mayr, Ernst, and William B. Provine, eds. *The Evolutionary Synthesis: Perspectives on the Unification of Biology.* Cambridge, MA: Harvard University Press, 1980.

Mazzolini, Renato G. *Politisch-Biologische Analogien im Frühwerk Rudolf Virchows.* Marburg: Basilisken-Presse, 1988.

———. "Il colore della pelle e l'origine dell'antropologia fisica (1492–1848)." In *L'epopea delle scoperte*, edited by Renzo Zorzi, 227–239. Firenze: Leo S. Olschki, 1994.

———. "Las Castas: Inter-Racial Crossing and Social Structure (1770–1835)." In *Heredity Produced: At the Crossroads of Biology, Politics and Culture, 1500–1870*, edited by Staffan Müller-Wille and Hans-Jörg Rheinberger, 349–374. Cambridge, MA: MIT Press, 2007.

McCarthy, Thomas A. "On the Way to a World Republic? Kant on Race and Development." In *Politik, Moral und Religion: Gegensätze und Ergänzungen*, edited by Lothar R. Waas, 223–242. Berlin: Duncker & Humblot, 2004.

McLaughlin, Peter. "Blumenbach und der Bildungstrieb: Zum Verhältnis von epigenetischer Embryologie und typologischem Artbegriff." *Medizinhistorisches Journal* 17 (1982): 357–372.

———. *Kant's Critique of Teleology in Biological Explanation.* Lewiston, NY: Edwin Mellen Press, 1990.

———. *What Functions Explain: Functional Explanation and Self-Reproducing Systems.* Cambridge: Cambridge University Press, 2001.

———. "Spontaneous versus Equivocal Generation in Early Modern Science." *Annals of the History and Philosophy of Biology* 10 (2005): 79–88.

———. "Kant on Heredity and Adaptation." In *Heredity Produced: At the Crossroads of Biology, Politics and Culture, 1500–1870*, edited by Staffan Müller-Wille and Hans-Jörg Rheinberger, 277–291. Cambridge, MA: MIT Press, 2007.

Menand, Louis. *The Metaphysical Club: A Story of American Ideas.* New York: Farrar, Straus and Giroux, 2002.

Mendel, Gregor. "Experiments in Plant Hybrids." Translated by Eva R. Sherwood. In *Gregor Mendel's Experiments on Plant Hybrids: A Guided Study*, edited by Alain F. Corcos and Floyd V. Monaghan, 57–174. New Brunswick, NJ: Rutgers University Press, 1993. Originally published as "Versuche über Pflanzen-Hybriden" (1866).

———. "On Hieracium-Hybrids Obtained by Artificial Fertilisation." In William Bateson, *Mendel's Principles of Heredity: A Defense*, 96–103. Placitas, NM: Geneticas Heritage Press, 1996. Originally published as "Uber einige aus küntlicher Befruchtung gewannene Hieracium-Bastarde" (1870).

Mendelsohn, Andrew. "Message in a Bottle: The Business of Vaccines and the Nature of Heredity after 1880." In *A Cultural History of Heredity III: 19th and Early 20th Centuries*, preprint 294, 85–100. Berlin: Max Planck Institute for the History of Science, 2005.

Minelli, Alessandro, ed. *L'Orto botanico di Padova 1545–1995*. Venice: Marsilio, 1995.

Miramon, Charles de. "Aux origines de la noblesse et des princes du sang: France et Angleterre au XIVe siècle." In *L'hérédité entre Moyen Age et époque moderne*, edited by Maaike van der Lugt and Charles de Miramon, 157–210. Florence: SISMEL—Edizioni del Galluzzo, 2008.

Mirowski, Philip, and Mirjam Sent, eds. *Science Bought and Sold: Essays in the Economics of Science*. Chicago: University of Chicago Press, 2002.

Mittwoch, Ursula. "Sex Determination in Mythology and History." *Arquivos Brasileiros de Endocrinologia e Metabologia* 49 (2005): 7–13.

Mocek, Reinhard. "Biology of Liberation: Some Historical Aspects of 'Proletarian Race Hygienics.'" In *From Physico-Theology to Bio-Technology*, edited by Mikuláš Teich and Kurt Bayertz, 224–231. Amsterdam: Rodopi, 1998.

———. *Biologie und soziale Befreiung: Zur Geschichte des Biologismus und der Rassenhygiene in der Arbeiterbewegung*. Frankfurt/M.: Peter Lang, 2002.

Monod, Jacques. *Chance and Necessity: An Essay on the Natural Philosophy of Modern Biology*. Translated by Austryu Wainhouse. New York: Knopf, 1971. Originally published as *Le hasard et la nécessité: essai sur la philosophie naturelle de la biologie moderne*. Paris: Seuil, 1970.

Montaigne, Michel de. *The Complete Essays of Montaigne*. Translated by Donald M. Frame. Stanford, CA: Stanford University Press, 1976. Original 1580.

Morange, Michel. "La révolution silencieuse de la biologie moléculaire: D'Avery à Hershey." *Le Débat* 22 (1982): 62–75.

———. *A History of Molecular Biology*. Translated by Matthew Cobb. Cambridge, MA: Harvard University Press, 1998. Originally published as *Histoire de la biologie moléculaire* (Paris: La Découverte, 1994).

———. *La part des gènes*. Paris: Odile Jacob, 1998.

———. "The Developmental Gene Concept: History and Limits." In *The Concept of the Gene in Development and Evolution: Historical and Epistemological Perspectives*, edited by Peter Beurton, Raphael Falk, and Hans-Jörg Rheinberger, 193–215. Cambridge: Cambridge University Press, 2000.

Morgan, Lewis H. *Systems of Consanguinity and Affinity of the Human Family*. Washington, DC: Smithsonian Institution, 1871.

Morgan, Sally. *From Sea Urchins to Dolly the Sheep: Discovering Cloning*. Chicago: Heinemann Library, 2006.

Morgan, Thomas Hunt. "The Relation of Genetics to Physiology and Medicine: Nobel Lecture, presented in Stockholm on June 4, 1934." In *Les Prix Nobel en 1933*, 1–16. Stockholm: Norstedt & Söner, 1935.

Morgan, Thomas Hunt, Alfred Sturtevant, Herman J. Muller, and Calvin Bridges. *The Mechanism of Mendelian Heredity*. New York: Henry Holt, 1915.

Moss, Lenny. *What Genes Can't Do*. Cambridge, MA: MIT Press, 2003.

———. "Redundancy, Plasticity, and Detachment: The Implications of

Comparative Genomics for Evolutionary Thinking." *Philosophy of Science* 73 (2006): 930–946.

Muller, Hermann J. "The Development of the Gene Theory." In *Genetics in the 20th Century: Essays on the Progress of Genetics during Its First 50 Years*, edited by Leslie Clarence Dunn, 77–100. New York: Macmillan, 1951.

Müller-Röber, Bernd, Ferdinand Hucho, Wolfgang van den Daele, Kristian Köchy, Jens Reich, Hans-Jörg Rheinberger, Karl Sperling, Anna M. Wobus, Mathias Boysen, and Meike Kölsch. *Grüne Gentechnologie: Aktuelle Entwicklungen in Wissenschaft und Wirtschaft*. Munich: Elsevier Spektrum Akademischer Verlag, 2007.

Müller-Sievers, Helmut. *Self-Generation: Biology, Philosophy, and Literature Around 1800*. Stanford, CA: Stanford University Press, 1997.

———. "The Heredity of Poetics." In *Heredity Produced: At the Crossroads of Biology, Politics and Culture, 1500–1870*, edited by Staffan Müller-Wille and Hans-Jörg Rheinberger, 443–465. Cambridge, MA: MIT Press, 2007.

Müller-Wille, Ludger, ed. *Franz Boas Among the Inuit on Baffin Island (1883–1884): Journals and Letters*. Translated by William Barr. Toronto: University of Toronto Press, 1998.

Müller-Wille, Staffan. "'Varietäten auf ihre Arten zurückführen'—Zu Carl von Linnés Stellung in der Vorgeschichte der Genetik." *Theory in Biosciences* 117 (1998): 346–376.

———. *Botanik und weltweiter Handel: Zur Begründung eines Natürlichen Systems der Pflanzen durch Carl von Linné (1707–1778)*. Berlin: Verlag für Wissenschaft und Bildung, 1999.

———. "Genealogie, Naturgeschichte und Naturgesetz bei Linné und Buffon." In *Genealogie als Denkform in Mittelalter und Früher Neuzeit*, edited by Kilian Heck and Bernhard Jahn, 109–119. Tübingen: Niemeyer, 2000.

———. "Nature as a Marketplace: The Political Economy of Linnaean Botany." In *Oeconomies in the Age of Newton*, edited by Neil Di Marchi and Margaret Schabas, 155–173. History of Political Economy, Supplement. Vol. 35. Durham, NC: Duke University Press, 2003.

———. "Ein Anfang ohne Ende: Das Archiv der Naturgeschichte und die Geburt der Biologie." In *Macht des Wissens: Die Entstehung der modernen Wissengesellschaft*, edited by Richard v. Dülmen and Sina Rauschenbach, 587–605. Cologne: Böhlau, 2004.

———. "Early Mendelism and the Subversion of Taxonomy: Epistemological Obstacles as Institutions." *Studies in History and Philosophy of Biological and Biomedical Sciences* 36 (2005): 465–487.

———. "Schwarz, Weiß, Gelb, Rot: Zur Darstellung menschlicher Vielfalt." In *Dingwelten: Das Museum als Erkenntnisort*, edited by Anke te Heesen and Petra Lutz, 161–170. Cologne: Böhlau, 2005.

———. "Hybrids, Pure Cultures, and Pure Lines: From Nineteenth-Century Biology to Twentieth-Century Genetics." *Studies in History and Philosophy of the Biological and Biomedical Sciences* 38 (2007): 796–806.

Müller-Wille, Staffan, and Vítězslav Orel. "From Linnaean Species to Mendelian Factors: Elements of Hybridism, 1751–1870." *Annals of Science* 64 (2007): 171–215.

Müller-Wille, Staffan, and Hans-Jörg Rheinberger, eds. *Heredity Produced: At the Crossroads of Biology, Politics and Culture, 1500–1870.* Cambridge, MA: MIT Press, 2007.

———. *Das Gen im Zeitalter der Postgenomik: Eine wissenschaftshistorische Bestandsaufnahme.* Frankfurt/M.: Suhrkamp, 2009.

Musto, David F. "The Theory of Hereditary Disease of Luis Mercado, Chief Physician to the Spanish Habsburgs." *Bulletin of the History of Medicine* 35 (1961): 346–373.

Nägeli, Carl Wilhelm von. "Die Individualität in der Natur mit vorzüglicher Berücksichtigung des Pflanzenreiches." *Veröffentlichungen des Wissenschaftlichen Vereins in Zürich* 1 (1856): 171–212.

———. *Mechanisch-Physiologische Theorie der Abstammungslehre.* Munich: R. Oldenbourg, 1884.

Naudin, Charles. "Nouvelles recherches sur l'hybridité dans les végétaux: Mémoire présenté à l'Académie en décembre 1861." *Nouvelles Archives du Muséum* 1 (1865): 25–176.

Naudin, Charles, and Ferdinand von Müller. *Manuel de l'acclimateur ou choix des plantes recommandées pour l'agriculture, l'industrie et la médecine et adaptées aux divers climates de l'Europe et des pays tropicaux.* Paris: Société d'Acclimatation, 1887.

Nelkin, Dorothy, and M. Susan Lindee. *The DNA Mystique: The Gene as a Cultural Icon.* New York: Freeman, 1995.

Neumann-Held, Eva, and Christoph Rehmann-Sutter, eds. *Genesis in Development: Re-Reading the Molecular Paradigm.* Durham, NC: Duke University Press, 2006.

Olby, Robert C. "Francis Crick, DNA, and the Central Dogma." In *The Twentieth-Century Sciences: Studies in the Biography of Ideas*, edited by Gerald Holton, 227–280. New York: Norton, 1972.

———. *The Path to the Double Helix.* Seattle: University of Washington Press, 1974.

———. "Mendel No Mendelian?" *History of Science* 17 (1979): 53–72.

———. *Origins of Mendelism.* 2nd ed. Chicago: University of Chicago Press, 1985. 1st ed. 1966.

———. "William Bateson's Introduction of Mendelism to Britain: A Reassessment." *British Journal for the History of Science* 20 (1987): 399–420.

———. "From Physics to Biophysics." *History and Philosophy of the Life Sciences* 11 (1989): 305–309.

———. "Scientists and Bureaucrats in the Establishment of the John Innes Institute under William Bateson." *Annals of Science* Ltb (1989): 497–510.

———. "The Molecular Revolution in Biology." In *Companion to the History of Modern Science*, edited by R. C. Deby, G. N. Cantor, J. R. R. Christie, and M. J. S. Hodge, 503–520. London: Routledge, 1990.

———. "Constitutional and Hereditary Disorders." In *Companion Encyclopedia of the History of Medicine*, vol. 1, edited by William F. Bynum and Roy Porter, 412–437. London: Routledge, 1993.

———. "Mendel, Mendelism and Genetics." *MendelWeb*, http://www.mendelweb.org/MWolby.html, 1997 (accessed October 2, 2011).

Orel, Vítězslav. *Gregor Mendel: The First Geneticist.* Oxford: Oxford University Press, 1996.

Orel, Vítězslav, and Roger J. Wood. "Essence and Origin of Mendel's Discovery." *Comptes rendus de l'Académie des Sciences Paris, Sciences de la vie* 323 (2000): 1037–1041.

Osborne, Michael A. *Nature, the Exotic, and the Science of French Colonialism.* Bloomington: Indiana University Press, 1994.

Oschema, Klaus. "Maison, noblesse et légitimité: Aspects de la notion d'hérédité dans le milieu de la cour bourguignonne (XVe siècle)." In *L'hérédité entre Moyen Age et époque moderne,* edited by Maaike van der Lugt and Charles de Miramon, 211–244. Florence: SISMEL—Edizioni del Galluzzo, 2008.

Osterhammel, Jürgen. *Die Verwandlung der Welt: Eine Geschichte des 19. Jahrhunderts.* Munich: Beck, 2009.

Oudshoorn, Nelly. *Beyond the Natural Body: An Archeology of Sex Hormones.* London: Routledge, 1994.

Pagel, Walter. "The Philosophy of Circles—Cesalpino—Harvey." *Journal of the History of Medicine and Allied Sciences* 12 (1957): 140–157.

Palladino, Paolo. "Between Craft and Science: Plant Breeding, Mendelian Genetics, and British Universities, 1900–1920." *Technology and Culture* 34 (1993): 300–323.

Pálsson, Gísli. "Decoding Relations and Disease: The Icelandic Biogenetic Project." In *From Molecular Genetics to Genomics: The Mapping Cultures of Twentieth-Century Genetics,* edited by Jean-Paul Gaudillière and Hans-Jörg Rheinberger, 180–198. London: Routledge, 2004.

Pálsson, Gísli, and Paul Rabinow. "Iceland: The Case of a National Human Genome Project." *Anthropology Today* 15 (1999): 14–18.

Parnes, Ohad. "On the Shoulders of Generations: The New Epistemology of Heredity in the Nineteenth Century." In *Heredity Produced: At the Crossroads of Biology, Politics and Culture, 1500–1870,* edited by Staffan Müller-Wille and Hans-Jörg Rheinberger, 315–346. Cambridge, MA: MIT Press, 2007.

Parnes, Ohad, Ulrike Vedder, and Stefan Willer. *Das Konzept der Generation: Eine Wissenschafts- und Kulturgeschichte.* Frankfurt/M.: Suhrkamp, 2008.

Paul, Diane B. "Eugenics and the Left." *Journal of the History of Ideas* 45 (1984): 567–590.

———. *Controlling Human Heredity: 1865 to the Present.* Amherst, NY: Prometheus Books, 1995.

———. *The Politics of Heredity: Essays on Eugenics, Biomedicine and the Nature-Nurture Debate.* New York: State University of New York Press, 1998.

Paul, Diane B., and Barbara A. Kimmelman. "Mendel in America: Theory and Practice 1900–1919." In *The American Development of Biology,* edited by Keith R. Benson, Jane Maienschein, and Ronald Rainger, 281–310. Philadelphia: University of Pennsylvania Press, 1988.

Paul, Norbert. *The Making of Molecular Medicine: Historical, Theoretical, and Ethical Dimensions.* Habilitation, Heinrich-Heine-Universität Düsseldorf, 2003.

Pauling, Linus, Harvey A. Itano, Seymour J. Singer, and Ibert C. Wells. "Sickle-Cell Anemia, a Molecular Disease." *Science* 110 (1949): 543–548.

Pearson, Karl. "Contributions to the Mathematical Theory of Evolution." *Philosophical Transactions of the Royal Society of London, Series A*, 185 (1894): 70–110.

———. "Mathematical Contributions to the Theory of Evolution III: Regression, Heredity and Panmixia." *Proceedings of the Royal Society of London, Series A, Containing Papers of a Mathematical and Physical Character* 187 (1896): 253–318.

———. "Mathematical Contributions to the Theory of Evolution: On the Law of Ancestral Heredity." *Proceedings of the Royal Society of London* 62 (1898): 386–412.

———. *Nature and Nurture: The Problem of the Future*. London: Cambridge University Press, 1913.

Pestre, Dominique. *Science, argent et politique: Un essai d'interprétation*. Paris: INRA, 2003.

———. "The Technosciences between Markets, Social Worries and the Political: How to Imagine a Better Future?" In *The Public Nature of Science under Assault: Politics, Markets, Science and the Law*, edited by Helga Nowotny, Hans-Heinrich Trute, Eberhard Schmidt-Aßmann, Dominique Pestre, and Helmut Schulze-Fielitz, 29–52. Berlin: Springer, 2005.

Pethes, Nicolas. "'Victor, l'enfant de la forêt': Experiments on Heredity in Savage Children." In *Heredity Produced: At the Crossroads of Biology, Politics and Culture, 1500–1870*, edited by Staffan Müller-Wille and Hans-Jörg Rheinberger, 399–418. Cambridge, MA: MIT Press, 2007.

Pickering, Andrew. *The Cybernetic Brain: Sketches of Another Future*. Chicago: University of Chicago Press, 2010.

Pinto-Correia, Clara. *The Ovary of Eve: Egg and Sperm and Preformation*. Chicago: University of Chicago Press, 1997.

Pomata, Gianna. "Comments on Session III: Heredity and Medicine." In *Conference: A Cultural History of Heredity II: 18th and 19th Centuries*, preprint 247, 145–151. Berlin: Max Planck Institute for the History of Science, 2003.

Porphyry. *Introduction*. Translated by Jonathan Barnes. Oxford: Clarendon Press, 2003.

Porter, Roy, and George S. Rousseau. *Gout: The Patrician Malady*. New Haven, CT: Yale University Press, 1998.

Porter, Theodore M. *The Rise of Statistical Thinking: 1820–1900*. Princeton, NJ: Princeton University Press, 1986.

———. "The Biometric Sense of Heredity: Statistics, Pangenesis and Positivism." In *A Cultural History of Heredity III: 19th and Early 20th Centuries*, preprint 294, 31–42. Berlin: Max Planck Institute for the History of Science, 2005.

Powell, Alexander, Maureen A. O'Malley, Staffan Müller-Wille, Jane Calvert, and John Dupré. "Disciplinary Baptisms: A Comparison of the Naming Stories of Genetics, Molecular Biology, Genomics and Systems Biology." *History and Philosophy of the Life Sciences* 29 (2007): 5–32.

Proctor, Robert N. *Racial Hygiene: Medicine Under the Nazis.* Cambridge, MA: Harvard University Press, 1988.

——. "Three Roots of Human Recency: Molecular Anthropology, the Refigured Acheulean, and the UNESCO Response to Auschwitz." *Current Anthropology* 44 (2003): 213–229.

"Protokoll über die Verhandlungen bei der Schafzüchter-Versammlung in Brünn am 1. und 2. Mai 1837." *Mittheilungen der k. k. Mährisch-Schlesischen Gesellschaft zur Beförderung des Ackerbaues, der Natur- und Landeskunde in Brünn*, 1837, 201–205, 225–231, and 233–238.

Pyle, Andrew. "Malebranche on Animal Generation: Preexistence and the Microscope." In *The Problem of Animal Generation in Early Modern Philosophy*, edited by Justin E. H. Smith, 194–214. Cambridge: Cambridge University Press, 2006.

Quatrefages de Bréau, Armand de. *The Human Species.* London: Kegan Paul, Trench & Co, 1883. Originally published as *L'Espèce humaine* 1877.

Quirke, Viviane. *Collaboration in the Pharmaceutical Industry: Changing Relationships in Britain and France 1935–1965.* London: Routledge, 2006.

Rabinow, Paul. *Making PCR: A Story of Biotechnology.* Chicago: University of Chicago Press, 1996.

——. *French DNA: Trouble in Purgatory.* Chicago: University of Chicago Press, 1999.

——. "Epochs, Presents, Events." In *Living and Working with the New Medical Technologies: Intersections of Inquiry*, edited by Margaret Lock, Allan Young, and Alberto Cambrosio, 31–46. Cambridge: Cambridge University Press, 2000.

——. *Anthropologie der Vernunft: Studien zu Wissenschaft und Lebensführung.* Translated by Carlo Caduff and Tobias Rees. Frankfurt/M.: Suhrkamp, 2004.

Rader, Karen. *Making Mice: Standardizing Animals for American Biomedical Research, 1900–1955.* Princeton, NJ: Princeton University Press, 2004.

Rajan, Kaushik Sunder. *Biocapital: The Constitution of Postgenomic Life.* Durham, NC: Duke University Press, 2006.

Rasmussen, Nicolas. *Picture Control: The Electron Microscope and the Transformation of Biology in America.* Stanford, CA: Stanford University Press, 1997.

Ratcliff, Marc J. "Duchesne's Strawberries: Between Growers' Practices and Academic Knowledge." In *Heredity Produced: At the Crossroads of Biology, Politics and Culture, 1500–1870*, edited by Staffan Müller-Wille and Hans-Jörg Rheinberger, 205–228. Cambridge, MA: MIT Press, 2007.

Ratmoko, Christina. *Damit die Chemie stimmt: Die Anfänge der industriellen Herstellung von weiblichen und männlichen Sexualhormonen 1914–1938.* Zürich: Chronos Verlag, 2010.

Ray, John. *Historia plantarum species hactenus editas aliasque insuper multas noviter inventas et descriptas complectens.* 3 vols. London: Clark and Faithorne, 1686.

Reardon, Jenny. *Race to the Finish: Identity and Governance in an Age of Genomics.* Princeton, NJ: Princeton University Press, 2005.

Rey, Roselyne. "Génération et hérédité au 18e siècle." In *L'ordre des caractères: Aspects de l'hérédité dans l'histoire des sciences de l'homme*, edited by Jean-Louis Fischer, 7–41. Paris: Sciences en situation, 1989.

Rheinberger, Hans-Jörg. *Experiment—Differenz—Schrift*. Marburg: Basilisken-Presse, 1992.

———. "Morphologie bei Claude Bernard." *Aufsätze und Reden der Senckenbergischen Naturforschenden Gesellschaft* 41 (1994): 137–150.

———. "When Did Carl Correns Read Gregor Mendel's Paper? A Research Note." *Isis* 86 (1995): 612–616.

———. "Cytoplasmic Particles in Brussels (Jean Brachet, Hubert Chantrenne, Raymond Jeener) and at Rockefeller (Albert Claude), 1935–1955." *History and Philosophy of the Life Sciences* 19 (1997): 47–67.

———. *Toward a History of Epistemic Things: Synthesizing Proteins in the Test Tube*. Stanford, CA: Stanford University Press, 1997.

———. "Beyond Nature and Culture: Modes of Reasoning in the Age of Molecular Biology and Medicine." In *Living and Working with the New Medical Technologies: Intersections of Inquiry*, edited by Margaret Lock, Allan Young, and Alberto Cambrosio, 19–30. Cambridge: Cambridge University Press, 2000.

———. "Mendelian Inheritance in Germany between 1900–1910: The Case of Carl Correns (1864–1933)." *Comptes rendus de l'Académie des Sciences, Série III, Sciences de la vie* 323 (2000): 1089–1096.

———. "Carl Correns and the Early History of Genetic Linkage." In *Classical Genetic Research and Its Legacy: The Mapping Cultures of Twentieth Century Genetics*, edited by Hans-Jörg Rheinberger and Jean-Paul Gaudillière, 21–33. London: Routledge, 2004.

———. "A History of Protein Biosynthesis and Ribosome Research." In *Protein Synthesis and Ribosome Structure: Translating the Genome*, edited by Knud H. Nierhaus and Daniel N. Wilson, 1–51. Weinheim: Wiley-VCH, 2004.

———. "Internationalism and the History of Molecular Biology." *Annals of the History and Philosophy of Biology* 11 (2007): 249–254.

———. "Heredity and Its Entities around 1900." *Studies in History and Philosophy of Science* 39 (2008): 370–374.

———. *An Epistemology of the Concrete: Twentieth Century Histories of Life*. Durham, NC: Duke University Press, 2010.

Rheinberger, Hans-Jörg, and Peter McLaughlin. "Darwin's Experimental Natural History." *Journal of the History of Biology* 17 (1984): 345–368.

Richards, Robert J. "Kant and Blumenbach on the Bildungstrieb: A Historical Misunderstanding." *Studies in the History and Philosophy of Biology and the Biomedical Sciences* 31 (2000): 11–32.

———. *The Romantic Conception of Life: Science and Philosophy in the Age of Goethe*. Chicago: University of Chicago Press, 2002.

Richardson, Angelique. *Love and Eugenics in the Late Nineteenth Century: Rational Reproduction and the New Woman*. Oxford: Oxford University Press, 2003.

Richmond, Martha. "The Cell as the Basis for Heredity, Development, and Evolution: Richard Goldschmidt's Program of Physiological Genetics." In

*From Embryology to Evo-Devo: A History of Developmental Evolution*, edited by Manfred Laubichler and Jane Maienschein, 169–211. Cambridge, MA: MIT Press, 2007.

Rignol, Loïc. "Augustin Thierry et la politique de l'histoire: Genèse et principes d'un système de pensée." *Revue d'histoire du XIXe siècle* 25 (2002): 87–100.

Ritterbush, Philip C. *Overtures to Biology: The Speculations of Eighteenth Century Naturalists.* New Haven, CT: Yale University Press, 1964.

Ritvo, Harriet. *The Animal Estate.* Cambridge, MA: Harvard University Press, 1987.

Roelcke, Volker. "Programm und Praxis der psychiatrischen Genetik an der Deutschen Forschungsanstalt für Psychiatrie unter Ernst Rüdin: Zum Verhältnis von Wissenschaft, Politik und Rasse-Begriff vor und nach 1933." In *Rassenforschung an Kaiser-Wilhelm-Instituten vor und nach 1933*, edited by Hans-Walter Schmuhl, 38–67. Göttingen: Wallstein, 2003.

Roger, Jacques. *The Life Sciences in Eighteenth-Century French Thought.* Edited by Keith R. Benson and translated by Robert Ellrich. Stanford, CA: Stanford University Press, 1997. Originally published as *Les sciences de la vie dans la pensée française du XVIIIe siècle*, 1963.

Roll-Hansen, Nils. "The Genotype Theory of Wilhelm Johannsen and Its Relation to Plant Breeding and the Study of Evolution." *Centaurus* 22 (1978): 201–235.

———. "The Crucial Experiment of Wilhelm Johannsen." *Biology and Philosophy* 4 (1989): 303–329.

———. *The Lysenko Effect: The Politics of Science.* Amherst, NY: Humanity Books, 2005.

———. "Sources of Johannsen's Genotype Theory." In *A Cultural History of Heredity III: 19th and Early 20th Centuries*, preprint 294, 43–52. Berlin: Max Planck Institute for the History of Science, 2005.

Roumy, Franck. "La naissance de la notion canonique de consanguinitas et sa réception dans le droit civil." In *L'hérédité entre Moyen Age et époque moderne*, edited by Maaike van der Lugt and Charles de Miramon, 41–66. Florence: SISMEL—Edizioni del Galluzzo, 2008.

Ruckenbauer, Peter. "E. von Tschermak-Seysenegg and the Austrian Contribution to Plant Breeding." *Vorträge für Pflanzenzüchtung* 48 (2000): 31–46.

Russell, Nicholas. *Like Engend'ring Like: Heredity and Animal Breeding in Early Modern England.* Cambridge: Cambridge University Press, 1986.

Sabean, David W. *Power in the Blood: Popular Culture and Village Discourse in Early Modern Germany.* Cambridge: Cambridge University Press, 1984.

———. *Kinship in Neckarhausen, 1700–1870.* Cambridge: Cambridge University Press, 1998.

———. "From Clan to Kindred: Kinship and the Circulation of Property in Premodern and Modern Europe." In *Heredity Produced: At the Crossroads of Biology, Politics and Culture, 1500–1870*, edited by Staffan Müller-Wille and Hans-Jörg Rheinberger, 37–59. Cambridge, MA: MIT Press, 2007.

Sapp, Jan. *Beyond the Gene: Cytoplasmatic Inheritance and the Struggle for Authority in Genetics.* Oxford: Oxford University Press, 1987.

Sarasin, Philipp. *Reizbare Maschinen: Eine Geschichte des Körpers 1765–1914.* Frankfurt/M.: Suhrkamp, 2001.

Sarkar, Sahotra, ed. *The Founders of Evolutionary Genetics.* Dordrecht: Kluwer Academic Publishers, 1992.

———. "Biological Information: A Sceptical Look at Some Central Dogmas of Molecular Biology." In *The Philosophy and History of Molecular Biology: New Perspectives,* edited by Sahotra Sarkar, 187–231. Dordrecht: Kluwer, 1996.

———. *Genetics and Reductionism.* Cambridge: Cambridge University Press, 1998.

Satzinger, Helga. "The Chromosomal Theory of Heredity and the Problem of Gender Equality in the Work of Theodor and Marcella Boveri." In *A Cultural History of Heredity III: 19th and Early 20th Centuries,* preprint 294, 101–114. Berlin: Max Planck Institute for the History of Science, 2005.

———. "Theodor and Marcella Boveri: Chromosomes and Cytoplasm in Heredity and Development." *Nature Reviews Genetics* 9 (2008): 231.

———. "Racial Purity, Stable Genes and Sex Difference: Gender in the Making of Genetic Concepts by Richard Goldschmidt and Fritz Lenz, 1916–1936." In *The Kaiser Wilhelm Society under National Socialism,* edited by Susanne Heim, Carola Sachse, and Mark Walker. Cambridge: Cambridge University Press, 2009.

Schloegel, Judy Johns. "Herbert Spencer Jennings, Heredity, and Protozoa as Model Organisms, 1908–1918." In *Conference: Heredity in the Century of the Gene (A Cultural History of Heredity IV),* preprint 343, 129–138. Berlin: Max Planck Institute for the History of Science, 2008.

Schmidgen, Henning. "Fehlformen des Wissens." In Georges Canguilhem, *Die Herausbildung des Reflexbegriffs im 17. und 18. Jahrhundert,* translated by Henning Schmidgen. Munich: Fink, 2008.

Schmidtke, Jörg, Bernd Müller-Röber, Wolfgang van den Daele, Ferdinand Hucho, Kristian Köchy, Karl Sperling, Jens Reich, Hans-Jörg Rheinberger, Anna M. Wobus, Mathias Boysen, and Silke Domasch, eds. *Gendiagnostik in Deutschland: Status quo und Problemerkundung.* Limburg: Forum W — Wissenschaftlicher Verlag, 2007.

Schmuhl, Hans-Walter, ed. *Rassenforschung an Kaiser-Wilhelm-Instituten vor und nach 1933.* Göttingen: Wallstein, 2003.

———. *Crossing Boundaries: The Kaiser-Wilhelm-Institute for Anthropology, Human Heredity, and Eugenics, 1927–1945.* Dordrecht: Springer, 2008.

Schneider, William H. *Quality and Quantity: The Quest for Biological Regeneration in Twentieth-Century France.* Cambridge: Cambridge University Press, 1990.

Schöner, Erich. *Das Viererschema in der antiken Humoralpathologie.* Wiesbaden: Steiner, 1964.

Schrödinger, Erwin. *What Is Life?* Cambridge: Cambridge University Press, 1967. Originally published as *Was ist Leben?* (1944).

Schultze, Max. "Über Muskelkörperchen und das, was man eine Zelle zu nennen habe." *Archiv für Anatomie, Physiologie und Wissenschaftlicher Medicin* (1861): 1–27.

Schwartz, Sara. "The Differential Concept of the Gene: Past and Present." In *The Concept of the Gene in Development and Evolution: Historical and Epistemological Perspectives*, edited by Peter Beurton, Raphael Falk, and Hans-Jörg Rheinberger, 26–39. Cambridge: Cambridge University Press, 2000.

Schwartz, James. *In Pursuit of the Gene: From Darwin to DNA*. Cambridge, MA: Harvard University Press, 2010.

Schwartz Cowan, Ruth. "Nature and Nurture: The Interplay of Biology and Politics in the Work of Francis Galton." In *Studies in the History of Biology*, vol. 1, edited by William Coleman and Camille Limoges, 133–208. Baltimore: Johns Hopkins University Press, 1977.

Schweber, Libby. *Disciplining Statistics: Demography and Vital Statistics in France and England, 1830–1885*. Durham, NC: Duke University Press, 2006.

Schwerin, Alexander von. *Experimentalisierung des Menschen: Der Genetiker Hans Nachtsheim und die vergleichende Erbpathologie 1920–1945*. Göttingen: Wallstein, 2004.

———. "Tiere vermehren—Institutionen vergrößern: Die Versuchstiere der Notgemeinschaft Deutscher Wissenschaft, 1920–1931." *Verhandlungen zur Geschichte und Theorie der Biologie* 11 (2005): 353–365.

———. "Seeing, Breeding and the Organisation of Variation: Erwin Baur and the Culture of Mutations in the 1920s." In *Conference: Heredity in the Century of the Gene (A Cultural History of Heredity IV)*, preprint 343, 259–278. Berlin: Max Planck Institute for the History of Science, 2008.

Sebright, John Saunders. "The Art of Improving the Breeds of Domestic Animals. In a Letter, addressed to the Right Hon. Sir Joseph Banks." In *Artificial Selection and the Development of Evolutionary Theory*, edited by Carl Jay Bajema, 93–122. Stroudsburg, PA: Hutchinson Ross, 1982. Originally published 1809.

Shinn, Terry, and Bernward Joerges. "The Transverse Science and Technology Culture: Dynamics and Roles of Research-Technology." *Social Science Information* 41 (2002): 207–251.

Sigurdsson, Skúli. "Yin-Yang Genetics, or the HSD deCODE Controversy." *New Genetics and Society* 20 (2001): 103–117.

Sloan, Philip R. "The Gaze of Natural History." In *Inventing Human Science: Eighteenth-Century Domains*, edited by Christopher Fox, Roy Porter, and Robert Wokler, 112–151. Berkeley: University of California Press, 1995.

Smith, Justin E. H. "Imagination and the Problem of Heredity in Mechanist Embryology." In *The Problem of Animal Generation in Early Modern Philosophy*, edited by Justin E. H. Smith, 80–99. Cambridge: Cambridge University Press, 2006.

———. "Introduction." In *The Problem of Animal Generation in Early Modern Philosophy*, edited by Justin E. H. Smith, 1–18. Cambridge: Cambridge University Press, 2006.

———, ed. *The Problem of Animal Generation in Early Modern Philosophy*. Cambridge: Cambridge University Press, 2006.

Smith Hughes, Sally. "Making Dollars out of DNA: The First Major Patent in Biotechnology and the Commercialization of Molecular Biology 1974–1980." *Isis* 92 (2001), 541–575.

Smocovitis, Vassiliki Betty. *Unifying Biology: The Evolutionary Synthesis and Evolutionary Biology*. Princeton, NJ: Princeton University Press, 1996.

Snelders, Stephen, Frans J. Meijman, and Toine Pieters. "Bismarck the Tomcat and Other Tales: Heredity and Alcoholism in the Medical Sphere, the Netherlands 1850–1900." In *A Cultural History of Heredity III: 19th and Early 20th Centuries*, preprint 294, 193–211. Berlin: Max Planck Institute for the History of Science, 2005.

Soloway, Richard A. *Demography and Degeneration: Eugenics and the Declining Birthrate in Twentieth-Century Britain*. Chapel Hill: University of North Carolina Press, 1995.

Sommer, Marianne. *Bones and Ochre: The Curious Afterlife of the Red Lady of Paviland*. Cambridge, MA: Harvard University Press, 2007.

———. "DNA and Cultures of Remembrance: Anthropological Genetics, Biohistories, and Biosocialities." *Biosocieties* 5 (2010): 366–390.

Spary, Emma C. *Utopia's Garden: French Natural History from Old Regime to Revolution*. Chicago: University of Chicago Press, 2000.

Stamhuis, Ida H. "Hugo de Vries's Transitions in Research Interest and Method." In *A Cultural History of Heredity III: 19th and Early 20th Centuries*, preprint 294, 115–136. Berlin: Max Planck Institute for the History of Science, 2005.

Stearn, William T. "Botanical Gardens and Botanical Literature in the Eighteenth Century." In *Catalogue of Botanical Books in the Collection of Rachel McMasters Miller Hunt*, vol. 2, edited by Jane Quinby and Allan Stevenson, 41–140. Pittsburgh, PA: Hunt Botanical Library, 1961.

Stent, Gunther S. "That Was the Molecular Biology That Was." *Science* 160 (1968): 390–395.

Stepan, Nancy. *The Hour of Eugenics: Race, Gender, and Nation in Latin America*. Ithaca, NY: Cornell University Press, 1991.

Sterne, Laurence. *The Life and Opinions of Tristram Shandy, Gentleman—A Sentimental Journey Through France and Italy*. Munich: Günter Jürgensmeier, 2005. Original 1760.

Stevens, Peter F., and Sean P. Cullen. "Linnaeus, the Cortex-Medulla Theory, and the Key to His Understanding of Plant Form and Natural Relationships." *Journal of the Arnold Arboretum* 71 (1990): 179–220.

Stocking, George W. *Race, Culture, and Evolution: Essays in the History of Anthropology*. Chicago: University of Chicago Press, 1982.

———. *Victorian Anthropology*. London: Macmillan, 1987.

———, ed. *Bones, Bodies, and Behavior: Essays on Biological Anthropology*. Madison: University of Wisconsin Press, 1988.

———. "The Turn-of-the-Century Concept of Race." *Modernism/Modernity* 1 (1994): 4–16.

———. *After Tylor: British Social Anthropology 1888–1951*. Madison: University of Wisconsin Press, 1995.

Strasser, Bruno J. "Sickle Cell Anemia, a Molecular Disease." *Science* 286 (1999): 1488–1490.

———. "Linus Pauling's 'Molecular Diseases,' Between History and Memory." *American Journal of Medical Genetics* 115 (2002): 83–93.

———. "Collecting and Experimenting: The Moral Economies of Biological Research, 1960s–1980s." In *Workshop, History and Epistemology of Molecular Biology and Beyond: Problems and Perspectives*, preprint 310, 105–123. Berlin: Max Planck Institute for the History of Science, 2006.

———. *La fabrique d'une nouvelle science: La biologie moléculaire à l'âge atomique (1945–1964)*. Florence: Olschki, 2006.

———. "Collecting, Comparing, and Computing Sequences: The Making of Margaret O. Dayhoff's Atlas of Protein Sequence and Structure, 1954–1966." *Journal of the History of Biology* 43 (2010): 623–660.

Stubbe, Hans. *History of Genetics: From Prehistoric Times to the Rediscovery of Mendel's Laws.* Translated by T. R. W. Waters. Cambridge, MA: MIT Press, 1972. Originally published as *Kurze Geschichte der Genetik bis zur Wiederentdeckung der Vererbungsregeln Gregor Mendels* (Jena: Fischer, 1965).

Suárez Díaz, Edna, and Victor H. Anaya-Muñoz. "History, Objectivity, and the Construction of Molecular Phylogenies." *Studies in History and Philosophy of Biological and Biomedical Sciences* 39 (2008): 451–468.

Summers, William C. *Félix d'Hérelle and the Origins of Molecular Biology.* New Haven, CT: Yale University Press, 1999.

Szybalski, Waclaw, and Ann Skalka. "Editorial: Nobel Prizes and Restriction Enzymes." *Gene* 4 (1978): 181–182.

Tapper, Melbourne. *In the Blood: Sickle Cell Anemia and the Politics of Race.* Critical Histories. Philadelphia: University of Pennsylvania Press, 1998.

Teich, Mikuláš. "Fermentation Theory and Practice: The Beginnings of Pure Yeast Cultivation and English Brewing, 1883–1913." *Technology and Culture* 8 (1983): 117–133.

———. *Bier, Wissenschaft und Wirtschaft in Deutschland 1800–1914: Ein Beitrag zur deutschen Industrialisierungsgeschichte.* Vienna: Böhlau, 2000.

Terrall, Mary. *The Man Who Flattened the Earth: Maupertuis and the Sciences in the Enlightenment.* Chicago: University of Chicago Press, 2002.

———. "Speculation and Experiment in Enlightenment Life Sciences." In *Heredity Produced: At the Crossroads of Biology, Politics and Culture, 1500–1870*, edited by Staffan Müller-Wille and Hans-Jörg Rheinberger, 253–276. Cambridge, MA: MIT Press, 2007.

Theunissen, Bert. "Breeding Dutch Dairy Cows (1900–1950): Heredity without Mendelism." In *Conference: Heredity in the Century of the Gene (A Cultural History of Heredity IV)*, preprint 343, 27–50. Berlin: Max Planck Institute for the History of Science, 2008.

Thieffry, Denis. "*Escherichia coli* as a model system with which to study cell differentiation." *History and Philosophy of the Life Sciences* 18 (1996): 163–193.

Thompson, Warren. "Population." *American Journal of Sociology* 34 (1929): 959–975.

Thurtle, Phillip. *The Emergence of Genetic Rationality: Space, Time, and Information in American Biological Science, 1870–1920.* Seattle: University of Washington Press, 2008.

Tocqueville, Alexis de. *Democracy in America*. 2 vols. New York: Alfred A. Knopf, 1948. Originally published as *De la démocratie en Amérique* (1835/ 1840).

Tooker, Elisabeth. "Lewis H. Morgan and His Contemporaries." *American Anthropologist* 94 (1992): 357–375.

Trembley, Abraham. *Mémoires pour servir à l'histoire d'um genre de polypes d'eau douce, à bras en forme de cornes*. Leiden: Verbeek, 1744.

Turner, Trudy R., ed. *Biological Anthropology and Ethics: From Repatriation to Genetic Identity*. Albany: SUNY Press, 2005.

Tylor, Edward B. *Primitive Culture: Researches Into the Development of Mythology, Philosophy, Religion, Language, Art and Custom*. 3rd ed. 2 vols. New York: Holt, 1889. Original 1873.

UNESCO, ed. *The Race Concept: Results of an Inquiry*. Paris: UNESCO House, 1952.

Vedder, Ulrike. "Continuity and Death: Literature and the Law of Succession in the Nineteenth Century." In *Heredity Produced: At the Crossroads of Biology, Politics and Culture, 1500–1870*, edited by Staffan Müller-Wille and Hans-Jörg Rheinberger, 85–101. Cambridge, MA: MIT Press, 2007.

Vettel, Eric J. *Biotech: The Countercultural Origins of an Industry*. Philadelphia: University of Pennsylvania Press, 2006.

Virchow, Rudolf Ludwig. "Über die Reform der pathologischen und therapeutischen Anschauungen durch die mikroskopischen Untersuchungen." In *Archiv für pathologische Anatomie und Physiologie und für klinische Medicin*, edited by Rudolf L. Virchow and Benno E. H. Reinhardt, 207–255. Berlin: G. Reimer, 1847.

———. *Cellular Pathology as Based upon Physiological and Pathological Histology*. Translated by Frank Chance. London: John Churchill, 1860. Originally published as *Die Cellularpathologie in ihrer Begründung auf physiologische und pathologische Gewebelehre* (Berlin: Hirschwald, 1858).

———. "Der Staat und die Ärzte." In *Rudolf Virchow: Sämtliche Werke*, edited by Christian Andree, vol. 28.1., 50–71. Hildesheim: Georg Olms Verlag, 2006. Original 1849.

Voss, Julia. *Darwin's Pictures: Views of Evolutionary Theory, 1837–1874*. New Haven, CT: Yale University Press, 2010.

Wailoo, Keith, and Stephen Pemberton. *The Troubled Dream of Genetic Medicine: Ethnicity and Innovation in Tay-Sachs, Cystic Fibrosis, and Sickle Cell Disease*. Baltimore: Johns Hopkins University Press, 2006.

Waller, John C. "'The Illusion of an Explanation': The Concept of Hereditary Disease, 1770–1870." *Journal of the History of Medicine* 57 (2002): 410–448.

Warming, Eugenius, and Wilhelm Johannsen. *Lehrbuch der allgemeinen Botanik*. Translated by Emilio Pepe Meinecke. Berlin: Bornträger, 1909. Originally published as *Den almindelige botanik* (Copenhagen: Det Nordiske Førlag, 1900).

Waters, C. Kenneth. "What Was Classical Genetics?" *Studies in History and Philosophy of Science* 35 (2004): 83–109.

Watson, James D. *The Double Helix*. London: Weidenfeld and Nicholson, 1968.

Watson, James D., and Francis H. C. Crick. "Molecular Structure of Nucleic Acids: A Structure for Deoxyribose Nucleic Acid." *Nature* 171 (1953): 737–738.

Watson, James D., and John Tooze. *The DNA Story: A Documentary History of Gene Cloning.* San Francisco: W. H. Freeman, 1981.

Weber, Marcel. "Walking on the Chromosome: *Drosophila* and the Molecularization of Development." In *From Molecular Genetics to Genomics: The Mapping Cultures of Twentieth-Century Genetics*, edited by Jean-Paul Gaudillière and Hans-Jörg Rheinberger, 63–78. London: Routledge, 2004.

Weigel, Sigrid. *Genea-Logik: Generation, Tradition und Evolution zwischen Kultur- und Naturwissenschaften.* Munich: Fink, 2006.

Weindling, Paul. *Health, Race and German Politics between National Unification and Nazism, 1870–1945.* Cambridge: Cambridge University Press, 1989.

Weingart, Peter, Kurt Bayertz, and Jürgen Kroll. *Rasse, Blut und Gene: Geschichte der Eugenik und Rassenhygiene in Deutschland.* Frankfurt/M.: Suhrkamp, 1992. Original 1988.

Weismann, August. "The Continuity of the Germ-Plasm as the Foundation of a Theory of Heredity." In *August Weismann: Essays upon Heredity and Kindred Biological Problems*, edited by Edward B. Poulton, Selmar Schönland, and Arthur E. Shipley, 2nd ed., vol. 1, 161–250. Oxford: Clarendon Press, 1891. Originally published as *Die Continuität des Keimplasmas als Grundlage einer Theorie der Vererbung* (Jena: Fischer, 1885).

———. *The Germ-Plasm: A Theory of Heredity.* Translated by W. Newton Parker. New York: Charles Scribner's Sons, 1893. Originally published as *Das Keimplasma: Eine Theorie der Vererbung* (Jena: Fischer, 1892).

White, John S. "William Harvey and the Primacy of Blood." *Annals of Science* 43 (1986): 239–255.

White, Paul. "Acquired Character: The Heredity Material of the 'Self-Made Man.'" In *Heredity Produced: At the Crossroads of Biology, Politics and Culture, 1500–1870*, edited by Staffan Müller-Wille and Hans-Jörg Rheinberger, 375–398. Cambridge, MA: MIT Press, 2007.

Widmalm, Sven. "A Machine to Work In: The Ultracentrifuge and the Modernist Laboratory Ideal." In *Taking Place: The Spatial Contexts of Science, Technology, and Business*, edited by Enrico Baraldi, Hjalmar Fors, and Anders Houltz, 59–80. Sagamore Beach, MA: Science History Publications, 2006.

Wieland, Thomas. *"Wir beherrschen den pflanzlichen Organismus besser . . .": Wissenschaftliche Pflanzenzüchtung in Deutschland 1889–1945.* Munich: Deutsches Museum, 2004.

———. "Scientific Theory and Agricultural Practice: Plant Breeding in Germany from the Late 19th to the Early 20th Century." *Journal of the History of Biology* 39 (2006): 309–343.

Wiener, Norbert. *Cybernetics; or, Control and Communication in the Animal and the Machine.* Cambridge, MA: MIT Press, 1996. Original 1948.

Wijnands, D. Onno. "Hortus auriaci: The Gardens of Orange and their Place in Late 17th Century Botany and Horticulture." *Journal of Garden History* 8 (1988): 61–86, 271–304.

Willer, Stefan. "Heritage—Appropriation—Interpretation: The Debate on the

Schiller Legacy in 1905." In *A Cultural History of Heredity III: 19th and Early 20th Centuries*, preprint 294, 167–178. Berlin: Max Planck Institute for the History of Science, 2005.

———. "Sui generis: Heredity and Heritage of Genius at the Turn of the Eighteenth Century." In *Heredity Produced: At the Crossroads of Biology, Politics and Culture, 1500–1870*, edited by Staffan Müller-Wille and Hans-Jörg Rheinberger, 419–440. Cambridge, MA: MIT Press, 2007.

Williams, Elizabeth A. *The Physical and the Moral: Anthropology, Physiology, and Philosophical Medicine in France, 1750–1850*. Cambridge: Cambridge University Press, 1994.

Wilson, Edmund Beecher. *The Cell in Development and Heredity*. New York: MacMillan, 1896.

Wilson, Phillip K. "Erasmus Darwin and the 'Noble' Disease (Gout): Conceptualizing Heredity and Disease in Enlightenment England." In *Heredity Produced: At the Crossroads of Biology, Politics and Culture, 1500–1870*, edited by Staffan Müller-Wille and Hans-Jörg Rheinberger, 133–154. Cambridge, MA: MIT Press, 2007.

———. "Pedigree Charts as Tools to Visualize Inherited Disease in Progressive Era America." In *Conference: Heredity in the Century of the Gene (A Cultural History of Heredity IV)*, preprint 343, 163–190. Berlin: Max Planck Institute for the History of Science, 2008.

Winnacker, Ernst Ludwig. "Das Gen und das Ganze." *Die Zeit*, May 2, 1997, 34.

Winther, Rasmus G. "Darwin on Variation and Heredity." *Journal of the History of Heredity* 33 (2000): 425–455.

———. "August Weismann on Germ-Plasm Variation." *Journal of the History of Biology* 34 (2001): 517–555.

Witkowski, Jan A. *Illuminating Life: Selected Papers from Cold Spring Harbor, 1903–1969*. Cold Spring Harbor: Cold Spring Harbor Laboratory Press, 1999.

———. *75 Years in Science at the Cold Spring Harbor Symposium on Quantitative Biology*. Cold Spring Harbor: Cold Spring Harbor Laboratory Press, 2010.

Witkowski, Jan A., and John R. Inglis, eds. *Davenport's Dream: 21st Century Reflections on Heredity and Eugenics*. New York: Cold Spring Harbor Laboratory Press, 2008.

Wolff, Caspar Friedrich. *Theorie von der Generation: In 2 Abh. erkl. u. bewiesen*. Berlin: Birnstiel, 1764.

Wolff, Michael. *Geschichte der Impetustheorie: Untersuchungen zum Ursprung der klassischen Mechanik*. Frankfurt/M.: Suhrkamp, 1978.

Wolstenholme, Gordon, ed. *Man and His Future: A CIBA Foundation Volume*. Boston: Little, Brown, 1963.

Wood, Roger J. "The Sheep Breeders' View of Heredity Before and After 1800." In *Heredity Produced: At the Crossroads of Biology, Politics and Culture, 1500–1870*, edited by Staffan Müller-Wille and Hans-Jörg Rheinberger, 229–250. Cambridge, MA: MIT Press, 2007.

Wood, Roger J., and Vítězslav Orel. *Genetic Prehistory in Selective Breeding: A Prelude to Mendel*. Oxford: Oxford University Press, 2001.

———. "Scientific Breeding in Central Europe during the Early Nineteenth Century: Background to Mendel's Later Work." *Journal of the History of Biology* 38 (2005): 239–272.

Wright, Susan. *Molecular Politics: Developing American and British Regulatory Policy for Genetic Engineering.* Chicago: University of Chicago Press, 1994.

Yi, Doogab. "Cancer, Viruses, and Mass Migration: Paul Berg's Venture into Eukaryotic Biology and the Advent of Recombinant DNA Research and Technology, 1967–1980." *Journal of the History of Biology* 41 (2008): 589–636.

Yoxen, Edward J. "The Role of Schrödinger in the Rise of Molecular Biology." *History of Science* 17 (1979): 17–52.

———. "Giving Life a New Meaning: The Rise of the Molecular Biology Establishment." In *Scientific Establishments and Hierarchies,* edited by Norbert Elias, Herminio Martins, and Richard Whitley, 123–143. Dordrecht: Reidel, 1982.

Yule, Udny. "Mendel's Laws and Their Probable Relation to Intra-Racial Heredity." *New Phytologist* 1 (1902): 193–207 and 222–238.

Zedler, Johann Heinrich. *Großes vollständiges Universallexikon aller Wissenschaften und Künste.* 68 vols. Graz: Akademische Druck- und Verlags-Anstalt, 1993. Original 1732–1750.

Zirkle, Conway. *The Beginnings of Plant Hybridization.* Morris Arboretum Monographs. Vol. 1. Philadelphia: University of Pennsylvania Press, 1935.

———. "The Early History of the Idea of the Inheritance of Acquired Characters and of Pangenesis." *Transactions of the American Philosophical Society,* new series, 35 (1946): 91–151.

———. "Species before Darwin." *Proceedings of the American Philosophical Society* 103 (1959): 636–644.

# Index

*A page number in italics refers to a figure or its caption.*

acclimatization: of exotic species to Europe, 4, 62–63, 73, 130; in Russian plant breeding, 159

acquired characters, inheritance of, 61–63; central dogma of molecular biology and, 183; culture and language and, 113; Darwin's views on, 38, 79; eugenics and, 100; Haeckel on, 92; Lysenkoism and, 183, 186; nineteenth-century physicians' views on, 90; as unanswered question at end of nineteenth century, 93; Weismann's views on, 89, 90

acquired diseases, 56

action at a distance, 23, 26, 28, 35

adaptation to environment, 74–75

agnatic lines of descent, 46, 48, 51

agricultural science, 122, 175

agricultural stations, 133, 135, 158

agriculture: *Erbgut* associated with, 194; maize genetics and, 148; Mendel's connections with, 133; Naudin's hybrids for, 130–31; risks of gene technology in, 197; transgenic animals in, 205; transgenic plants in, 204–5. *See also* breeders

*Agrobacterium tumefaciens*, 204

Ahlströmer, Jonas, 65

alleles, 148

analogical reasoning, premodern, 19

ancestor tables, 32, 46, 121, 122–23

ancestral inheritance, law of, 78, 108–9, *109*, 114–15

Anderson, French, 208, 209

Anderson, Thomas, 164, 167

animalculist theories, 29, 31, 93

*Anlagen*, 58–59, 69. *See also* dispositions

anthropology: Boas's work in, 110, 112–13, 124, 125; cultural versus physical, 113; late eighteenth-century, 42, 44, 58–59, 66–69; polygenism/monogenism debate in, 125; social, 119–20. *See also* race

anthropometry, 106–7. *See also* cephalic index; statistical methods

*Antirrhinum majus* (snapdragon), 150, 151

anti-Semitism, 125

*Arabidopsis thaliana* (thale cress), 200

Arab-Islamic tradition. *See* Avicenna

Arber, Werner, 188, 189

archaebacteria, genome sequences of, 200

aristocratic families, and hereditary disease, 56, 57–58

Aristotle: on causation, 23; Galileo's break with tradition of, 39; logic of genealogy and, 1

Aristotle on generation, 6; Dino del Garbo and, 54; epigenesis and, 30; Harvey and, 23, 24–25; inherited bodily characters and, 19, 42, 43; sex roles and, 24–25, 93

Aromatari, Giuseppe degli, 29

*Ascaris* (giant roundworm), 81, 88

Asilomar conference, 195, 197, 198

assemblage, 183–84

Astbury, William, 164, 167

atavism: Bernard on, 85; Darwin's concern with, 78; genealogy and, 122; psychiatric meaning of, 120. *See also* regression

Aub, Joseph, 164

autocatalysis, 152, 153, 156, 176

automated DNA sequencing, 193

Avery, Oswald, 164, 178–79

Avicenna, 53, 54, 57

Bachelard, Gaston, 39, 163, 187

Bachofen, Johann Jacob, 119

bacteria: in gene technology, 188–89, 190, 195–96, 197, 198, 204; genome sequences of, 200; as model organisms, 173, 174, 174, 178–80; patenting of, 195–96

bacteriophages: carrying pieces of bacterial genome, 180; electron microscopy of, 167; genome sequences of, 190, 200; lambda phage, 189; as model organisms, 153, 174, 176–78, 179, 180; T4 genome fine structure, 150

Bakewell, Robert, 66

Baltimore, David, 188

bastardization, 138. *See also* hybridization

Bastian, Adolf, 110

Bateson, William, 114, 128, 138, 157

Baur, Erwin, 150, 151, 156–57, 158

Bawden, Frederick, 175

Bayh-Dole Act, 195

Beadle, George, 154, 159, 175–76, 179

Becker, Erich, 153

Beijerinck, Martinus Willem, 137, 174

Benzer, Seymour, 150

Berg, Paul, 175, 189, 194–95

Bernal, John Desmond, 167, 175

Bernard, Claude, 84–85, 89, 127, 170

big science, 201

Binet, Alfred, 107

biochemistry: new specialty of, 163; radioactive tracers in, 170–72

bioinformatics, 201, 213

biology: central problem of, around 1800, 35; divisions of, and "general biology," 127, 157; limitations of genocentrism in, 217; medieval literature on, 48; molecularization of, 162; named as a science around 1800, 34. *See also* molecular biology

biomedical platforms, 164

biometrical school, 108, 109; Davenport's involvement with, 177; gradualism of, 155; Mendelians versus, 114–15

biophors, 87, 142, 143
biophysics, 166, 184
biopolitical context: of classical ge-
    netics, 185; of molecular genetics,
    217
biopolitical dispositive, 12
biotechnology industry, 187, 193,
    195–98. *See also* gene technology
birth control, 99
Blakeslee, Albert, 149, 151, 177
blending versus non-blending inheri-
    tance, 90–91, 93
blood, 53, 204, 206, 208; breeding
    and, 64; circulation of, 21; gene-
    alogy and, 122; generation and,
    25, 48–49; kinship and, 48–49,
    119–20, 121; nobility and, 48–
    49, 102; race and, 103. *See also*
    menstrual blood; noble blood;
    sickle-cell anemia
blood relationships, 44, 45, 48–49,
    51, 80, 121
blood transfusion experiments, 78
Blumenbach, Johann Friedrich, 35,
    66, 92, 106
Boas, Franz, 110, 112–13, 124, 125
Bodmer, Walter, 211
Bohr, Niels, 176
Boivin, André, 179
Bonnet, Charles, 35
botanical gardens, 8, 59–60, 61, 63,
    129, 130, 140–41
Botstein, David, 192, 198
Boveri, Marcella, 151
Boveri, Theodor, 81, 151, 157, 169
Boyer, Herbert, 189, 195
Bracton, Henry, 48
Bragg, Henry, 167
Bragg, Lawrence, 167
Braun, Alexander, 116
bread mold. See *Neurospora crassa*
    (red bread mold)
breeders: as bourgeois role models, 73;
    Darwin's reliance on knowledge
    of, 74, 75, 77; de Vries's references
    to, 86; heredity as a force and, 92,

134; Maupertuis's use of observa-
    tions by, 72; Mendel's connections
    with, 133; naturalists and, 64–66,
    69, 86; Russian tradition of, 159
breeding, animal and plant: agricul-
    tural stations and, 133, 135, 158;
    early Mendelian genetics and, 94,
    133–36, 158; new experimental
    genetics and, 128; racial concepts
    and, 102; smart, 204. *See also*
    hybridization; Mendel, Gregor
Brenner, Sydney, 198, 214
Bridges, Calvin, 146
Brinster, Ralph, 205
Broca, Paul, 106
Brown, Louise, 216
Buchner, Eduard, 170
Buffon, Comte de (Georges-Louis
    Leclerc): human races and, 66, 68,
    106; hybridization experiments
    of, 63; meaning of "reproduction"
    introduced by, 15; notion of or-
    ganic molecules, 35–36, 39; on
    species, 34–35, 60–61, 75
Burian, Richard, 165
Burton, Robert, 42
Butenandt, Adolf, 163

*Caenorhabditis elegans* (roundworm),
    199, 200, 214
Cambrosio, Alberto, 164
Campbell, Keith, 215
Camper, Petrus, 106
cancer: caused by gene therapy, 209;
    tests for hereditary disposition to,
    210
cancer research: fractionation of cyto-
    plasm in, 169, 180; molecular
    genetics in context of, 164–65; rat
    in, 173
Canguilhem, Georges, x, 159
Cantor, Charles, 201
Capecchi, Mario, 205
capitalist economies: biotechnology
    industry in, 187; knowledge re-
    gime of heredity in, 8

Carlsberg Laboratory, 136, 140
Caskey, Thomas, 207
Caspari, Ernst, 153
Caspersson, Torbjörn, 164, 167
*castas* system, 66–68, 67, 102, 103, 104
causation: Aristotelian view of, 23; Kant on organisms and, 35. *See also* action at a distance
Cavalli-Sforza, Luca, 203
cDNA libraries, 190
Celera Genomics, 200
cell theory, 80–82; Mendel's exposure to, 130, 133. *See also* cytology; nucleus of cell
cellular fractionation techniques, 166–67, 169–70, 172
central dogma of molecular biology, 183, 188
cephalic index, 106, 112, 124
Chakrabarty, Ananda, 195–96
Chargaff, Erwin, 179, 181
chimeric organisms, 206, 216
chromosomes: annotated maps of, 218, 218; anomalies of, 151, 152; asymmetry of sex roles and, 93; Correns's theory and, 143; of *Drosophila*, 146, 152; early observations and naming of, 81; mapping by sequence analysis, 191–92; position effect and, 152; radium-induced alterations in, 149; reification of gene and, 151; Weismann's theory and, 87, 87. *See also* plasmids
chromosome theory of heredity, 151
Churchill, Frederick, 81, 82, 85, 140
*cis*-genetic plants, 204
cistron, 150
classical genetics: autocatalysis and heterocatalysis in, 152, 153–54, 154, 156, 176; biopolitical context of, 185; cell theory and, 82; conceptual problems of, 139–46; epistemological significance of, 156, 217; establishment as a discipline, 129, 156–60; Galton's pat-

ent/latent distinction and, 80; phage research and, 176; as phenomenological science, 156; population genetics and, 155–56; shift to molecular discourse from, 161–62; theoretical diversity culminating in, 90; time span of, 161. *See also* gene mapping; Mendelian genetics; model organisms; transmission genetics
Claude, Albert, 164, 167, 169
clones: bacterial colonies as, 180; botanical concept of, 135; of whole genes, 189, 195
cloning: human, 216; patented expression vector for, 196; reproductive behavior and, 196; of vertebrates, 215–16
code. *See* genetic code
Cohen, Daniel, 199
Cohen, Stanley, 189
cohort, 116
Cold Spring Harbor, 149, 150, 177–78, 179, 181, 198, 199; Eugenics Record Office at, 121, 121, 122, 123; symposia, 177, 181, 198
Collins, Francis, 199, 200
colloidal protoplasm, 166
colonial rule: extermination of native tribes in, 103–4; polygenism/monogenism debate and, 125
computer technology, 185. *See also* databases of genomic information
Comte, Auguste, 116
connate diseases, 53–54, 56
constant hybrids, Mendel's interest in, 132
constant varieties, 4, 130
continued generation, Linnaean concept of, 31–32
corn. *See* maize (corn) genetics
Cornell University, 158, 177
correlations, statistical, 108, 109–10, 112, 113, 114
Correns, Carl: conceptual problems of experimental genetics and, 139, 140–45, 144; institutional affilia-

tions of, 156, 157, 158; on Mendel's rule, 138
cousin marriages, 50–51, 119–20, 123; animal breeders' strategy resembling, 135
creatianism, 22–23, 52
creationism, 22; of Linnaeus, 32–33, 75
Crick, Francis, 162, 181, 183, 200
cross-fertilization, 63, 138. *See also* crossing experiments; hybridization
crossing experiments, 131, 134, 139, 145, 148, 152, 162, 176; new forms of, 174, 180
Crystal, Ron, 208
cybernetics, 185
cystic fibrosis, 208, 209
cytogenetics, experimental, 151, 152, 157, 177
cytology, 81, 82, 127, 151. *See also* cell theory
cytoplasmic forms of heredity, 149

Dahlgren, Eva F., 2
Dahlgren, Karl Vilhelm Ossian, 1–2
Darlington, Cyrill, 11
Darwin, Charles: cousin marriages in family of, 50–51; *Descent of Man*, 104, 120; on "force of inheritance," 92; Galton's theory of heredity and, 78–80; heritable elements postulated by, 91; horizontal dimension of life and, 37, 37–39; Mendel's experiments and, 130; naturalistic concept of heredity and, 2; *On the Origin of Species*, 37, 37–38, 74–75, 76–77; pangenesis hypothesis of, 76, 77–78, 90; population genetics and, 155–56; Prosper Lucas and, 74, 75, 76; retrospective methods of, 125; *The Variation of Animals and Plants under Domestication*, v, 77–78; William James on, v; xenia and, 140
Darwin, Erasmus, 56, 75
databases of genomic information, 201, 203, 213

data delivery and access, 201–2
*Datura* (crimson weed), 151, 177
Davenport, Charles B., 121, 122, 177
deCODE Genetics, 202–3
degeneration: eighteenth-century naturalists' meaning of, 61; eugenicists' talk of, 122; nineteenth-century discourse on, 103, 104; psychiatric concern for, 120–21. *See also* eugenics
degenerative diseases, 57
Delbrück, Max, 149, 150, 153, 176–77, 179, 184
Deleuze, Gilles, 183
del Garbo, Dino, 52, 53–54, 57–58
DeLisi, Charles, 198
Demerec, Milislav, 177
democracy, Tocqueville on, 50
demographic transition, 96, 97
deoxyribonucleic acid. *See* DNA
Descartes, René, 6, 20, 23
determinism, biological, 185–86, 210–11
development: Correns's dilemma about dispositions and, 142–43; Darwin's pangenesis and, 77–78; distinction between heredity and, 16, 80–81, 146, 159; as function of cytoplasm, not nucleus, 86; Galton on, 80; Haeckel on, 92; Kühn's work on gene action in, 153–55, 154; of model organisms, 148; molecular genetics and, 155, 157; organization of genetic material and, 152; postgenomic era and, 211, 212, 213–15; Weismann on, 87–89. *See also* generation
developmental maps, 218
de Vries, Hugo: Galton's discontinuous evolution and, 108; *Intracellular Pangenesis*, 85–86; Mendelian genetics embraced by, 137–38, 141; *Mutation Theory*, 148, 155; Weismann compared to, 87, 89, 91
diachronic relations of descent, 21, 38–39, 94, 116

diathesis, 56, 120

diseases, hereditary, 42, 43–44, 52–58, 75; new experimental genetics and, 158; sickle-cell anemia, 186, 206–8, 209. *See also* medicine

diseases, infectious, detection of pathogens in, 210

dispositions: as *Anlagen*, 58–59, 69; breeders' reference to, 65; to cancer, 210; Correns on, 141–43, 145; in generalized nineteenth-century concept of heredity, 72; to hereditary diseases, 56; synchronic dimension of, 21, 39

DNA: amplification of, by PCR, 192–93, 204; complementary (cDNA), 190; elucidation of structure of, 162, 168, 181, *182*, 183, 200; forensic testing of, 203–4; identified as hereditary material, 178–79; patent for product consisting of, 196; recombinant, 188, 189, 195, 196, 206; restriction fragment length polymorphisms of, 192, 210; separation of fragments of, 193; sequencing of, 190–92, *191–92*, 193–94, 199–200; ultraviolet microscopy of, 167; X-ray studies of, 164, 167, 181, *182*. *See also* gene; genome research

DNA chips, 200, 212, 215

DNA ligases, 188

DNA polymerases, 188, 190, 193

Dobzhansky, Theodosius, 126

dog breeding, 48, 63, 64, 72, 73

Dolly, 215

domesticated plants and animals, 61, 64; Darwin on, v, 77–78

dominant-recessive character pairs, 131, 132

*Drosophila melanogaster*, 128, 146, 147, 148; Beadle and Ephrussi's work on gene action in, 154, 175; chromosome banding patterns of, 152; Demerec's work with, 177, 178; genome sequence of, 200;

German research of 1920s on, 158; molecular developmental biology of, 213–14; Muller's work with, 149, 150, 151; mutants of, 149, 152; replaced as model by bacteriophage, 153

Duchesneau, François, 81, 82

*Echinus. See* sea urchin

*E. coli. See Escherichia coli*

economy of nature: Darwin on, 75; Linnaeus on, 33–34

*ego*, 46

electron microscope, 163–64, 167, 175

electrophoresis, 166, 190, 193, 201

Ellis, Emory, 176

Ellis Island, 124

emboîtement, 29, 31

embryonic stem cell research, 196, 216

embryos, screening of, 210

endogamy, 119–20, 124, 125. *See also* inbreeding

Engels, Friedrich, 103, 120

engineering, 185. *See also* genetic engineering; molecular engineering

English Civil War, racial interpretation of, 102

Enlightenment, 66, 72, 102

enzymes, 162–63; in biotechnological production processes, 197–98; cell fractionation and, 169; "one-gene–one-enzyme" hypothesis, 176; protein synthesis and, 183; as tools for genetic manipulation, 188–89; in vitro research with, 170, 180. *See also* ferments

*Ephestia kühniella* (flour moth), 154, *154–55, 175*

Ephrussi, Boris, 154, 159, 175

epigenesis: cell theory and, 82; hereditary disease and, 55; inheritance as a force and, 92; Kant's anthropology and, 59; Kühn's sense of, 153, 154; premodern view of, 22, 24, 28, 30; Wolff on, 31, 36

epigenetics, 202, 211, 213
epigenomics, 213
epistemic objects of heredity, xi–xii; arising with experimental genetics, 128, 139, 218; gene as, 126, 146, 156; model organisms and, 173–74; peculiarities of, in biological research, 7; reified, xi, 149, 217; as spatial structures in twenty-first century, 218
epistemic space of heredity, xi, 218; anthropological sources and, 66; of Darwin's natural selection, 38; Darwin's view of characters in, 77; defined by hybridization literature, 64; of Galton's theory, 7; nineteenth-century manifestations of, 72, 75, 129, 138; reconfiguration of social life and, 8; statistical and genealogical analyses in, 126; temporal dimension of, 75
equivocal generation, 18, 26, 34
*Erbgut*, 194
erythropoietin, 204
*Escherichia coli*: bacteriophages of, 167; genome sequencing of, 199, 200; as model organism, 173, 174, 179–80; patented expression vector for use in, 196
eugenics: analytic procedures serving, 126; animal and plant breeding as model for, 135; broad movement of, 8–9, 13; classical genetics and, 185; Dahlgren's support for, 2; Davenport's involvement with, 177; demographic changes and, 95–96; Galton's program of, 8, 11; hard versus soft heredity and, 99–100; ideological conditions of, 100–101; Lorenz's skepticism about, 122; national initiatives for, 97–98; National Socialists in Germany and, 9, 97, 99, 101, 102, 158; new experimental genetics and, 158; organizations promoting, 9, 98; popular literature on, 100; rare

before nineteenth century, 52; Social Democratic proponents of, 98
Eugenics Record Office at Cold Spring Harbor, 121, 121, 122, 123
European expansion, 3. *See also* acclimatization; colonial rule
Evans, Martin, 205
evolution: as change in gene frequencies, 155; conservation of developmental genes in, 214–15; developmental biology and, 157, 159; discontinuous, Galton on, 108; experimental genetics of twentieth century and, 157, 177; human, control of, 186; Jacob on, 206; Lamarck's theory of, 62; Mendel's exposure to theory of, 133; Nägeli's alternative to Darwin, 82, 83–84; natural history transformed into study of, 127; nineteenth-century debates about heredity and, 74, 76, 80; Pearson's mathematical contributions on, 108; postgenomic era and, 212; synthesis of 1930s, 155–56, 157; transmission genetics and, 146, 159. *See also* Darwin, Charles; natural selection; species formation
evolutionary maps, 218
evolutionary psychology, 114
exogamy, 119–20
exons, 190
experimental systems, 139, 169–73, 180, 184
expression of characters: as heterocatalysis, 152, 153–54, 154, 156; Weismann on environmental influences on, 89, 90. *See also* gene expression
eye: color of, 124, 152, 153; development of, 214

Fåhraeus, Robin, 166
falcons, 48, 64
family lines, Boas's research on, 124
female seed, 22–23, 25. *See also* seed

feminists, 99
ferments, 153, 154, 162. *See also*
    enzymes
Fernel, Jean, 42
fingerprinting, genetic, 203–4
Fischer, Emil, 162
Fischer, Eugen, 124, 125, 158
Fisher, Ronald, 155
FlavrSavr tomato, 204
Flemming, Walther, 81
flour moth. See *Ephestia kühniella*
    (flour moth)
foods, genetically altered, 196–97,
    204–5
forces, hereditary, 91–92; breeders'
    understanding of, 92, 134. *See
    also* vital forces
forensic medicine, 203–4
formal genetics, 145
Fortun, Mike, 203
Foucault, Michel: biopolitical disposi-
    tive and, 12; discourse of heredity
    and, 69; epistemological breaks
    and, 7–8; on Mendel, 132–33;
    on organization as key biological
    concept, 34
Fraenkel-Conrat, Heinz, 175
French Revolution, 50, 56, 102–3
freshwater polyp (*Hydra*), 35, 36, 61
fruit fly. See *Drosophila melanogaster*

Gaddesden, John of, 54, 57
Galenic tradition, 25, 42, 53, 54
Galton, Francis, 4, 6–13; blended in-
    heritance and, 91; genealogical
    structures of, 117, 117–18; *Heredi-
    tary Genius*, 11–13, 41, 69, 107–
    8, 117, 117; inheritance as a force
    and, 92; Lorenz's reservations
    about, 122; naturalistic concept
    of heredity and, 2–3; on nature and
    nurture, 113–14; new science of
    genetics and, 139; political anal-
    ogy of, 9–11, 12; political views
    of, 10; post office analogy of, 6–7,
    12; statistics and, 89, 107–10, 109;
    theory of heredity, 78–80; "A
    Theory of Heredity," 6–7, 9–10;
    use of term "heredity," 41
Galton Laboratory for National
    Eugenics, 109
Gamow, George, 185
Gans, Eduard, 49–50
García-Bellido, Antonio, 213–14
Gärtner, Carl Friedrich, 63–64, 130,
    131
Gaudillière, Jean-Paul, 158
Gausemeier, Bernd, 123
Gayon, Jean, 75–76, 79, 92, 155
Gearhart, John, 216
Gehring, Walter, 214
Gelsinger, Jesse, 208
gemmules, 77–78, 80, 83, 84, 142
gender, 5, 110, 118. *See also* sex
gene: Benzer's criticism of the term,
    150; century of, 161; concept be-
    yond straightforward coding, 202;
    conserved in evolution, 214–15;
    Correns's reservations about, 145;
    as epistemic object, 126, 146, 156;
    Johannsen's reservations about,
    140, 145; jumping, 150, 173;
    materialization of, 184; multiple
    operational definitions of, 156;
    as new term coined by Johannsen,
    128, 145; "one-gene–one-enzyme"
    hypothesis, 176; ownership of,
    197; protein as assumed material
    of, 175, 176, 178; sequencing of,
    190–91; twentieth-century images
    of, 217; unknown nature of, in
    1950, 151–52. *See also* DNA
gene action: knockout mice and,
    205–6; Kühn's work on, 153–55,
    154. *See also* gene expression
genealogy: ancient logic of, 1; applied
    breeding research and, 135–36;
    Darwin's reference to, 37; Lorenz's
    scientific enterprise of, 122–23;
    medical and psychiatric uses of,
    57, 120–21, 121, 123; medieval
    explosion of activities in, 46,

47, 48; in Nägeli's theory, 84;
nineteenth-century bourgeois interest in, 51, 122; persistent figures of thought using, 39; political
control served by, 126; safeguarding family property and, 46, 48,
49; traditional representations of,
121–22. See also kinship
gene chips. See DNA chips
gene expression, 202; in development,
214, 215; DNA chips and, 212,
213. See also expression of characters; gene action
gene mapping, 146, 150–51, 152, 153,
180. See also maps of genetic and
postgenetic data
Genentech, 195, 196
gene pool, 206
generation: equivocal, 18, 26, 34; Hippocrates on, 19, 25; Linnaeus's
agnosticism about, 31; Lucas on
inheritance versus, 76; male and
female roles in, 22, 24–25, 28, 29,
93; Maupertuis's concern with,
73; metaphors of blood and, 48–
49, 51; paternity disputes and, 51;
premodern view of, 16–21; present-
day discourse of, 39; secondary
connotation beginning around
1800, 35, 115–17; shift to discourse of heredity from, xi, 8, 21,
39, 42–43; spontaneous, 18, 26,
34. See also Aristotle on generation; development
gene technology, 129, 162, 188–94;
commercialization of, 187, 193,
195–98; debates about risks of,
194–95, 196–97; genetically modified organisms, 194, 204–6; government regulation of, 195–96;
green, 204. See also genome
research
gene therapy, 196, 208–10
Généthon, 199, 200
genetically modified organisms. See
gene technology

genetic code: cryptologic efforts to
decipher, 185; discourse of, 184;
elucidation of, 168, 181, 183. See
also genome research
genetic diagnosis, 210, 212
genetic engineering, 189–90, 194,
205–6. See also gene technology;
molecular engineering
genetic fingerprinting, 203–4
genetics: emergence of, 128–29. See
also classical genetics; formal genetics; Mendelian genetics; molecular genetics; population genetics; transmission genetics
genetic testing companies, 203
genocentrism in twentieth-century
biology, 217
genome research, 198–204; data management in, 201–2; individual and
ethnic identity in, 202–4; organizations involved in, 200–201. See
also gene technology
genomics, 8, 129, 162, 187, 193
genotype: in classical genetics, 156;
Johannsen's concept of, 140, 141,
142, 145, 149
genus, in ancient logic, 1
GenVec, 208
German genetics community of 1900–
1933, 156–58
germs: alternative conceptions emerging in eighteenth century, 35–37;
Darwinian theory and, 38, 39; of
Galton's stirp, 6, 7, 9, 10, 11, 79–
80; invisibility of, 39; preexistence
in, 28, 29–30; in preformationist
theories, 28–29, 36
Gesner, Conrad, 17
Gilbert, Walter, 190
giraffe, 17
global capitalism, and biotechnology
industry, 187
globalization, and concept of
heredity, 3, 8
Goldschmidt, Richard, 152, 156, 157,
158

Goltz, Dietlinde, 25
Goody, Jack, 45
Gottweis, Herbert, 197
gout, 56
Gregorian reforms, 45
Griesemer, James, 159
Griffith, Frederick, 164, 178
Gros, François, 202
guinea pigs, 173
Gurdon, John, 215

Habsburg lip, 122
Hadorn, Ernst, 214
Haeckel, Ernst, 92
*Haemophilus influenzae*, 199
Haldane, J. B. S., 155
Haller, Albrecht von, 36
Hansen, Emil Christian, 136, 140
hard versus soft heredity, 55, 56, 57, 89–91, 99–100
Hardy, Godfrey Harold, 123–24, 155
Hardy-Weinberg law, 123–24, 155
Hartmann, Max, 139, 157
Harvey, William, 6, 21–22, 23–28, 30; Linnaeus's theory compared to, 34; vertical relations of descent and, 37, 38–39
Harwood, Jonathan, 156–58
Hegel, Georg Wilhelm Friedrich, 49–50
Heinrich, Klaus, 1
Herbert, William, 63
Herder, Johann Gottfried, 104, 113
hérédité, 41, 71
heredity: at beginning of twenty-first century, 15, 211, 212; Darwin's use of term, 39; deviant characters in early modern thinking about, 43–44; distinction between development and, 16, 80–81, 146, 159; emergence of generalized concept of, 72, 73–74; emergence of naturalistic concept of, 2–4; forces associated with, 91–92, 134 (*see also* vital forces); as information, 183, 184–85, 193, 217, 218; Johannsen's genotype definition

of, 145; juridical meaning of, 5–6, 41–42, 44 (*see also* inheritance: of property); late nineteenth-century speculation about, 89–94; as object of technical manipulation, 187–88; organized knowledge production and, 39; Pearson's definition of, 114; in separate domains of discourse, 8, 42–43, 44–45, 69, 71, 72, 73; shift from discourse of generation to, xi, 8, 21, 39, 42–43; terminological history of, 41–42, 71. *See also* epistemic objects of heredity; epistemic space of heredity; generation; genetics; knowledge regime of heredity; political dimensions of heredity
Hérelle, Félix d', 176
heritability: plant and animal breeders' concern with, 134; statistical concept of, 114, 115
heritage, 3, 4, 94
Hertwig, Oscar, 81, 82
Hertwig, Richard, 81
heterocatalysis, 152, 153–54, 154, 156
*Hieracium* (hawkweed), 132
Hippocrates, 19, 25
Hoagland, Mahlon, 170
Hobbes, Thomas, 28
Hogeboom, George, 167
holism, 211–12
Holley, Robert, 190
Holocaust, 185
Hood, Leroy, 193
horizontalization of life: in Darwin's theory, 39, 94; end of eighteenth century, 37; radicalized by genetic engineering, 206. *See also* synchronic dimension of heredity
horticulture, 64, 130. *See also* botanical gardens
Hotchkiss, Rollin, 167
human cloning, 216
human genetics: diversity of populations in, 203; genome research on, 198–204, 209; Kaiser Wilhelm In-

stitute and, 158; laboratory manipulation of, 197; number of genes in, 202; used to control evolution, 186. *See also* anthropology; anthropometry; eugenics; race
Human Genome Diversity Project, 203
Human Genome Organization (HUGO), 199
Human Genome Project, 199–202, 211
human origin in Africa, 185
human population genetics, 123–24, 126, 158
human reproductive technologies, 216
humoral pathology, 53, 54, 55
Hwang Woo Suk, 216
hybridization, 63–64, 65; Correns and, 140, 141; by genetic engineering, 206; Johannsen and, 140, 145; Mendel in tradition of, 130, 137–38; Naudin's work on, 130–31; premodern view of, 16–19, 17; of pure lines, 128, 129; in sheep breeding, 134, 165. *See also* breeding, animal and plant; crossing experiments; variation
*Hydra* (freshwater polyp), 35, 36, 61

Icelandic genome data, 202–3
imagination of pregnant woman. *See* maternal imagination
inbreeding, 134. *See also* endogamy; pure lines
incest prohibitions, medieval, 45, 48. *See also* cousin marriages
industrialization, 8, 96, 97; purity in, 136–37; second industrial revolution, 129, 133
industry: basic research and, 163–64, 187; biotechnology, 187, 193, 195–98; instrument development by, 165; radioisotopes produced by, 171. *See also* gene technology
information: databases of, 201, 203, 213; heredity conceived as, 183, 184–85, 193, 217, 218; lock-and-

key principle replaced by, 178; personal genetic information, 197
informational thinking, 177, 184–85
inheritance: blending versus non-blending, 90–91, 93; laws of, 107, 109–10; of property, 5, 44, 45–46, 48–50, 51. *See also* acquired characters, inheritance of
Institute for Genomic Research, The (TIGR), 199
insulin, bacterial production of, 195, 204
intelligence tests, 107
interchanges, Harvey's metaphor of, 27, 33
intermittent hereditary characters, 76–77
introns, 190
in vitro experimental systems, 169–71, 173, 180, 194
in vitro fertilization, 210, 216
in vitro genetics, 180
*ipse*, 46

Jackson, David, 193
Jacob, François, ix–x; on evolution as tinkering, 206; informational view of genetics and, 184, 185; on organization as key biological concept, 34; regulatory genes postulated by, 214; on shift from generation to heredity, 8, 16, 42
Jaenisch, Rudolf, 194
Jasanoff, Sheila, 190
Jefferson, Thomas, 116
Jeffreys, Alec, 203
Jennings, Herbert Spencer, 149
Jews: eugenics and, 100; studies of, 124, 125
Joerges, Bernward, 165
Johannsen, Wilhelm: conceptual problems of experimental genetics and, 139–40, 143, 145, 146; Galton's discontinuous evolution and, 108; "gene" coined by, 128; genotype concept of, 140, 141, 142, 145, 149; organization of genetic material

Johannsen, Wilhelm (*cont.*)
  and, 152; pure lines and, 128, 134,
    135, 136, 139, 140, 145, 148
John of Gaddesden, 54, 57
Jollos, Victor, 149–50
journals of genetics, 128, 156, 157
jumping genes, 150, 173
juridical meaning of heredity, 5–6, 41–
    42, 44. *See also* inheritance: of
    property

Kaiser Wilhelm Institutes, 145, 157,
    158, 163, 167, 175
Kaiser Wilhelm Society, 123, 157–58
Kant, Immanuel, 35, 42, 58–59, 68, 71
Kausche, Gustav, 167
Kautsky, Karl, 98
Kay, Lily, 165, 168, 170, 178, 218
Keating, Peter, 164
Kendrew, John, 167
Kevles, Daniel, 205
Khorana, Gobind, 181
Kielmeyer, Carl Friedrich, 34, 69
kinship: ancient logic of, 1; at dawn
    of modernity, 51; Foucault on
    power and, 12; Germanic and
    Roman systems of, 45, 46; rigor-
    ous nineteenth-century analysis
    of, 101, 115–26, *117, 118*. *See
    also* cousin marriages; genealogy;
    inheritance: of property
Kjeldahl, Johan, 140
knockout mice, 205–6
knockout technology, 215
knowledge regime of heredity, x; be-
    ginning in multiple domains, 44;
    as biopolitical dispositive, 12–13;
    dichotomies in, 4; epistemic ob-
    jects and, xii; global roots of, 3;
    historical connections forming, 8;
    shift in, around 1900, 138–39
Koch, Robert, 136
Kölliker, Albert, 82
Kölreuter, Joseph Gottlieb, 63, 131
Kornberg, Arthur, 188
Koyré, Alexandre, 39
Kraepelin, Emil, 123

Kroeber, Alfred L., 113, 120
Kühn, Alfred, 153–55, *154,* 159, 175

Lamarck, Jean-Baptiste de, 62, 75. *See
    also* acquired characters, inheri-
    tance of
Lander, Eric, 209
landraces, 134, 137
Laurembergius, Petrus, 64
law of ancestral inheritance, 108–9,
    *109,* 114–15
laws of generation, Linnaeus's belief
    in, 33–34
laws of inheritance, 107, 109–10. *See
    also* Hardy-Weinberg law; law of
    ancestral inheritance; Mendel's
    rules
Leder, Philip, 196, 205
Lederberg, Joshua, 179, 186
Leeuwenhoek, Antony van, 29
Lépine, Pierre, 164
leprosy, 54
Lesky, Erna, 21
Lévi-Strauss, Claude, 19
Lewis, Ed, 213–14
Liceti, Fortunio, 29
like begets like, 16, 19, 26–27, 100, 101
linkage: Correns's observation of,
    143; gene mapping and, 146
Linnaeus, Carl, 30–34; Ahlströmer
    and, 65; on constant varieties,
    130; creationism of, 32–33, 75; on
    human varieties, 69, 104–5, *105,*
    106; *Peloria* variety of toadflax
    and, 61, 62
Lipmann, Fritz, 164
Lipphardt, Veronika, 125
Little, Clarence C., 149
"lock and key" principle, 162–63
Locke, John, 17–18, 19
logic, ancient, 1
López Beltrán, Carlos, 41, 71
Lorenz, Ottokar, 122
Louis, Antoine, 57
Löwy, Ilana, 158
Lucas, Prosper, 41, 74, 75, 76, 92
Lugt, Maaike van der, 53

Luria, Salvador, 167, 176, 179
Lyell, Charles, 103–4
Lysenko, Trofim, 159, 183, 186

MacKenzie, Donald, 98
MacLeod, Colin, 178
macromolecules, 166–67, 170, 172.
    *See also* nucleic acids; proteins
Maine, Henry James Sumner, 119
maize (corn) genetics, 148, 149, 150,
    152, 173, 177
majorat, 46, 48, 49, 50, 52
Malebranche, Nicolas, 29
Mannheim, Karl, 116
maps of genetic and postgenetic data,
    218, 218. *See also* gene mapping
Mark, Hermann, 167
Marx, Karl, 49, 103, 116
Massachusetts General Hospital, 170,
    173
maternal imagination, 18, 51, 93
Matthaei, Heinrich, 181
Maupertuis, Pierre-Louis Moreau de,
    72–73, 75
Maxam, Allan, 190
Mazzolini, Renato, 91, 105
McCarty, Maclyn, 178
McClintock, Barbara, 150, 152, 173,
    177
McKusick, Victor, 199
McLaughlin, Harry, 122
McLennan, John F., 119
medicine: forensic, 203–4; genealog-
    ical diagrams in, 121, 121; gene
    therapy in, 196, 208–10; molecu-
    lar, 186, 187, 196, 206–11; mouse
    model in, 148; new experimental
    genetics and, 158; population
    databases and, 203; socialization
    and politicization of, 96–97. *See
    also* cancer; diseases, hereditary;
    psychiatry
Melchers, Georg, 175
menageries, 59, 72
Mendel, Gregor: abbot at monastery
    of, 102; assumptions made by,
    131–32, 137; breeders' methods

and, 66; cell theory and, 82; ex-
    periments of, 63, 128, 129–32,
    143, 148; formal object ("factor")
    of, 138; Foucault on, 132–33; gen-
    eration concept and, 116; meta-
    phors of inheritance used by, 64;
    rehabilitated in Brno, 183; triple
    achievement of, 138; unnoticed by
    contemporaries, 130, 132
Mendelian genetics: applied breeding
    research and, 135–36; biometrics
    versus, 114–15; blending versus
    non-blending inheritance and, 91;
    of eye color in Fischer's study,
    124; law of ancestral inheritance
    and, 115; limited early applica-
    tions of, 158–59; mutations in,
    150; rediscovery of, 137, 143, 152;
    reductionism of, 211; Rüdin's
    schizophrenia study and, 123–24;
    of sickle-cell anemia, 186; social
    technologies and, 98; Soviet re-
    jection of, 159. *See also* classical
    genetics
Mendel's laws, rediscovery of, 137–
    38. *See also* Mendel's rules
Mendel's rules, 128, 138, 151
menstrual blood, 22, 25, 48
menstruation, 54
Mercado, Luis, 52, 54–56, 57
Mercuriale, Girolamo, 19
merino sheep, 65, 134
messenger RNA, 181, 188, 190, 212
metabolic maps, 218
metabolism: DNA chip technology
    and, 212–13; enzymes (ferments)
    in, 153, 154, 163; postgenomic
    era and, 211
mice. *See* mouse
micelles, of Nägeli, 83, 142
microbiology, 127, 129, 133, 158, 164
microscope: in cytology, 151; elec-
    tron, 163–64, 167, 175; Galton's
    references to, 6, 7; late nineteenth-
    century techniques with, 81–82;
    in McClintock's chromosomal
    studies, 152; ultraviolet, 167

microsomes, 167, 169
Mill, John Stuart, 116
Miramon, Charles de, 49
miscegenations: of Latin American
    colonial castes, 66–68; between
    species, 18, 19
Mitchurin, Ivan Vladimirovitch, 159
mitochondria, 166, 169
mixed races, 112, 124, 125
model organisms, 127–29, 139, 145–
    46, 148; bacteria as, 173, 174,
    174, 178–80; bacteriophages as,
    153, 174, 176–78, 179, 180;
    *Caenorhabditis elegans*, 199, 200,
    214; in developmental biology,
    214; flour moth *Ephestia kühniella*,
    154, 154–55; genome sequencing
    of, 199, 200; mutations in, 150;
    *Neurospora crassa*, 174, 175–76,
    178, 179; new in molecular biol-
    ogy, 173–80; suppliers of, 149.
    See also *Drosophila melanogaster*;
    *Escherichia coli*; maize (corn)
    genetics; mouse
molecular biology: automation of
    techniques in, 193–94; basic ver-
    sus applied research in, 195, 197,
    201; central dogma of, 183, 188;
    emergence of, 162; experimental
    systems in, 162, 169–73; inter-
    nationality of, 173, 185; late
    twentieth-century transformation
    of, 187; new model organisms
    in, 173–80; Nobel Prizes in, 168;
    origins of term, 166, 168; public
    funding of research in, 169; tex-
    tual metaphors of, 193–94; trans-
    disciplinary research in, 168, 172–
    73, 181, 185
molecular developmental biology,
    213–15
molecular engineering, 188
molecular genetics: cooperation of
    basic research and industry in,
    163–64; developmental perspec-
    tive in, 155, 157; emergence of,
    161–65; epistemological signifi-

cance of, 217; experimental sys-
    tems in, 169–73; genetic engineer-
    ing options in, 194–98; genome
    research in, 198–204; informa-
    tional thinking in, 177; as new as-
    semblage, 183–84; new research
    technologies in, 165–69, 180–81,
    190–94, 201, 212–13; political
    salience of, 185–86; present status
    of, 217–18; in vitro systems for,
    173. *See also* molecular biology
molecular medicine, 186, 187, 196,
    206–11. *See also* medicine
Monod, Jacques, 184, 185, 214
monogenism, 125
Montaigne, Michel de, 43, 56
Morange, Michel, 214–15, 218
Morgan, Lewis H., *118*, 118–19
Morgan, Thomas Hunt: Boveri's re-
    port translated by, 81; Delbrück's
    visit to, 153, 176; *Drosophila* re-
    search of, 146, *147*; on formal
    character of his program, 155;
    gene maps of, 146, 152; Lewis as
    student of, 213; Lysenko's oppo-
    sition to, 183; Muller as student
    of, 146, 151; on nonhereditary
    modifications, 149; transmission
    genetics in style of, 159
Moss, Lenny, 155
mouse, 148, 149, 173; annotated chro-
    mosome 4 map of, *218*; cloning
    of, 215–16; eye development in,
    214; growth hormone gene in-
    serted into, 205; integrating for-
    eign DNA into, 194; knockout,
    205–6; patented oncomouse, 196,
    205
Muller, Hermann Joseph, 146, 149,
    150, 152, 186
Mullis, Kary, 192
muscular dystrophy, gene-therapeutic
    trials for, 208
mutation: in alternative theory of evo-
    lution, 155; Baur's factorial/small
    distinction, 150; Beijerinck's en-
    couragement of de Vries to study,

137; in *Drosophila*, 146, 152, 214; in *E. coli*, 179–80; gene mapping by use of, 150–51; induction of, 149, 151, 176; in Kühn's epigenetic program, 154; in mathematical population genetics, 155; in phages, 176, 179; in pure lines of model organisms, 148; techniques for producing point mutations, 193; unit of, 156
muton, 150
*Mycoplasma capricolum*, 199

Nägeli, Carl Wilhelm von, 82–84, 89, 91; Correns and, 140, 141, 142; de Vries's ideas and, 85, 86–87; generation concept and, 116; Mendel's correspondence with, 130, 132, 133; new science of genetics and, 139, 140; on nonhereditary modifications, 149; Weismann's ideas and, 86–87, 88
Napp, Cyrill Franz, 102, 134
Nathans, Daniel, 188, 189
National Socialist eugenics, 9, 97, 99, 101, 102, 158
natural history, 59–63; Darwin's evidence for variation from, 77, 89; of mankind, 66; transformed into evolutionary biology, 127
naturalists: practical breeders and, 64–66, 69. *See also* degeneration; species; varieties
natural selection: eugenics as substitute for, 98; experimental science and establishment of, 75–76; extermination of native peoples and, 104; population genetics and, 155–56. *See also* selection
nature versus nurture, 69, 113–14, 115
Naudin, Charles, 63, 130–31, 132
Nazi regime. *See* Holocaust; National Socialist eugenics
neo-Darwinism, 155–56
Neufeld, Fred, 178
*Neurospora crassa* (red bread mold), 174, 175–76, 178, 179

Nirenberg, Marshall, 181, 183
nobility, 49, 58, 64, 102. *See also* noble blood; noble maladies
noble blood, 48–49, 102
noble maladies, 56, 57–58
non-naturals, 2, 53
nucleic acids: complementary base pairing properties of, 212; paradigm of, 178, 179; synthesis of sequences of, 193. *See also* DNA; RNA
nucleotides, radiolabeled, 171
nucleus of cell, 81, 82, 93; asymmetry of sex roles and, 93; in cloning techniques, 215–16; in de Vries's theory, 86; in Nägeli's theory, 84; reification of gene and, 151; in Weismann's theory, 88–89; Wilson on, 85
Nussbaum, Moritz, 82
Nüsslein-Volhard, Christiane, 214

Ochoa, Severo, 181
oncomouse, 196, 205
"One-gene–one-enzyme" hypothesis, 176
ontogeny recapitulating phylogeny, 92
Orel, Vít zslav, 133
organic molecules, Buffon's notion of, 35–36, 39
organization, as key biological concept, 34
original sin, 11, 90
origin myths, 1
ovism, 29, 31, 93

Painter, Theophilus, 152
Palmiter, Richard, 205
Pálsson, Gísli, 203
pangenesis, 23; Darwin's hypothesis of, 76, 77–78, 90; de Vries's speculations on, 85–86, 142; Galton's rejection of, 78, 79
paper chromatography, 179
paradigm shifts, 8, 129
*Paramecium*, 149
Parisano, Emilio, 29

Parnes, Ohad, 115–16
Pasteur, Louis, 136
patents: on genetically altered organisms, 195–96, 205; on oncomouse, 196, 205; on PCR technology, 193; purity promises and, 137; on recombinant DNA, 196; on transgenic plants, 204; Venter's conflict with NIH about, 199
paternity disputes, early modern, 51
paternity testing, 203–4
patriarchal powers: Harvey's concept of generation and, 24–26, 28; inheritance rights and, 50; preexistence theories and, 30, 52
Paul, Diane, 100
Pauling, Linus, 167, 179, 186, 206–7
*pax* gene family, 214
PCR (polymerase chain reaction), 192–93, 204
pea (*Pisum sativum*), 128, 131, 141
Pearson, Karl, 108–9, 111, 112, 113–14, 115
pedigree breeding, 134–36
pedigrees, 121–22
*Peloria*, 61, 62
Pemberton, Stephen, 207, 209
Perutz, Max, 167
Pfankuch, Edgar, 167
phages. *See* bacteriophages
pharmacogenetics, 210
*Phaseolus vulgaris* (princess bean), 139
phenomenotechnical gestalt switch, 187–88
phenotype: in classical genetics, 156; Johannsen's concept of, 141
Pirie, Norman, 175
*Pisum sativum* (pea), 128, 131, 141
Plagge, Ernst, 153
plant seed, 60, 108; production of, 64. *See also* breeding, animal and plant
plasmids, 188, 189, 190, 191, 204
platforms, biomedical, 164
Plato's *Republic*, 52
Pliny, 17

pneumococcus (*Streptococcus pneumoniae*), 174, 178–79
Polanyi, Michael, 167
political dimensions of heredity, 3–4, 12–13; anthropologists in service of, 125–26; Galton's analogies and, 9–10, 11–12; generations as carriers of sovereignty, 116; long-term historical processes and, 8; molecular genetics and, 185–86; public health systems and, 96–97; Virchow's concept of organism and, 91. *See also* eugenics; inheritance: of property; patriarchal powers; race
polygenism, 125
polymerase chain reaction (PCR), 192–93, 204
*Polyommatus phlaeas*, 89
polyploidy, 151
Pomata, Gianna, 56
population genetics: databases for, 203; in evolutionary biology, 157, 158; horizontal relations in, 206; human, 123–24, 126, 158; mathematical models in, 155; theoretical orientation of, 159
Porphyry, 1
Porter, Theodore, 91
position effect, 152
postgenomics, 4, 187, 202, 211–18
preexistence theories, 28–30, 31, 52
preformationist theories, 23, 26, 28–29, 36; classical genetics as, 155; hereditary disease and, 55; Kühn's orientation away from, 153, 154
pregnancy, 16, 39, 54, 55. *See also* maternal imagination
prenatal diagnosis, 210
prepotency, 64–65
Prichard, James Cowles, 103, 104
primogeniture (majorat), 46, 48, 49, 50, 52
princess bean (*Phaseolus vulgaris*), 139
procreation: generation as synonym for, 115; like begets like in, 19. *See also* generation

prostitution, and eugenics, 99
proteins: as assumed hereditary material, 175, 176, 178; radiolabeled amino acids in, 171; separation techniques for, 166; X-ray structure analysis of, 167. *See also* enzymes
protein synthesis: cell fractionation research and, 169, 173; gene technology for, 190, 195, 198; genetic code and, 183; role of nucleic acids in, 181, 183, 188
proteomics, 211, 213
protoplasm, 166
protozoa: conjugation in, 180; cytoplasmic heredity and, 149
protozoology, 158
psychiatry, 120–21, 121, 123–24, 158
public health systems, 96–97
Punnett, Reginald, 128
pure lines, 128, 129, 136, 145–46, 148; of Johannsen, 128, 134, 135, 136, 139, 140, 145, 148
purity: in industrial production, 136–37; of race, 124, 125, 126

Quetelet, Adolphe, 107

Rabinow, Paul, 183–84
race: anthropology and, 66–69, 101, 102–7, 110, 112–13, 124, 125–26; breeding and, 65, 136; *castas* system and, 66–68, 67, 102, 103, 104; claim of nonexistence for, 126; classical genetics and, 185; eugenic thinking and, 2, 13, 101; Kant on, 58–59; Linnaeus on, 69; mixed, 112, 124, 125; postwar molecular genetics and, 185; pure, 124, 125, 126. *See also* landraces
racial body (*Volkskörper*), 185
racial hygiene, 95, 97, 98, 158
racism: genomic data and danger of, 203; of scientists like Fischer and Rüdin, 125; social Darwinism and, 90

radiation genetics, 149, 151, 152
radioactive labeling, 170–72
rat liver system, 173
Ray, John, 17–18, 60
Réaumur, René-Antoine Ferchault de, 17
recombinant DNA, 188, 189, 195, 196, 206
recombinant microorganisms, 204
recombination, 146, 150, 156
recon, 150
reductionism, 211–12
regeneration of body parts, 35
regression, 65, 78, 108, 109–10. *See also* atavism; reversion
Rehobother Bastards, 124
reification of genetic objects, xi, 149, 151, 217. *See also* epistemic objects of heredity
reproduction: of difference, 69; generation and, 15–16; of higher organisms, 138, 215–16; of species, Linnaeus's concept of, 32–33. *See also* generation
restriction enzymes, 188, 189–90, 191–92
restriction fragment length polymorphism (RFLP), 192, 210
Retzius, Anders Adolf, 106
reverse transcriptase, 188, 190
reversion, 78, 80, 138. *See also* regression
ribosomal RNA, 181
rice genome project, 199
Rimpau, Wilhelm, 86
RNA: alternative splicing of, 213; cell activity patterns and, 212; in protein synthesis, 181, 188, 190; three classes of, 181, 183; of tobacco mosaic virus, 175; ultraviolet microscopy of, 167
RNA polymerases, 188
Roberts, Richard, 190
Rockefeller Foundation, 123, 168
Rockefeller Institute, 164, 167, 169, 173, 174, 178
Roger, Jacques, 21

roundworm. See *Caenorhabditis elegans* (roundworm)
Rous, Peyton, 164, 169
Rüdin, Ernst, 123–24, 125, 158
Ruska, Ernst, 163
Ruska, Helmut, 167, 175
Ruzicka, Leopold, 163

Sabean, David, 48, 51
*Saccharomyces cerevisiae*, 199, 200. *See also* yeast
Sachs, Julius, 137
Sanger, Frederick, 190
Sanger, Margaret, 99
Sanger Centre, 199, 200
Schell, Jeff, 204
schizophrenia, 123–24
Schmidgen, Henning, x
Schneider, Walter, 167
Schramm, Gerhard, 175
Schrödinger, Erwin, 176–77
Schultz, Jack, 167
science, special character of, x
scientific breeding, 135
scientific ideology, 159
scientific revolutions, 128, 129
scintillation counters, 172
Scythian lamb, 17, *17*
sea urchin, 81, 151
Sebright, John Saunders, 65
second industrial revolution, 129, 133
seed, 48, 55, 64; Descartes on, 20; Dino del Garbo and Mercado on, 54; Harvey on, 24, 26; Hippocrates on, 19; Montaigne on, 43. *See also* female seed; plant seed
selection: of bacteria in cultures, 180; in mathematical population genetics, 155; of phages in cultures, 176; prenatal diagnosis and, 210. *See also* natural selection
self-fertilization, 134–36
self-made man: successful breeder as, 73; Victorian ideology of, 69
severe combined immunodeficiency (SCID), 208–9

sex, 22, 38, 50, 60; determination of, 93, 143; role in generation, 22, 24–25, 28, 29, 93. *See also* gender
sex-linked inheritance, 147
sexual hormones, 163
sexuality: classical genetics and, 157, 185; Foucault on power and, 12
sexual life, rationalization of, 126
sexual reproduction, 36; classical genetics linked to, 206; Darwin on, 77–78; as experimental tool, 138; Linnaeus on, 30, 32, 63
Sharp, Philip Allen, 190
sheep breeders, 65, 133–34
Shinn, Terry, 165
sickle-cell anemia, 186, 206–8, 209
similarity between parents and offspring, 18–19, 51; Darwin on, 74; Harvey on, 22; heritability and, 115
Singer, Maxine, 175
Sinsheimer, Robert, 198
skin color, 66–68, 104–6, *105*
skull measurements, 106–7. *See also* cephalic index
slavery, 125
Sloan, Philip, 68
smart breeding, 204
Smith, David, 198
Smith, Hamilton, 188, 189
Smith, Lloyd, 193
Smithies, Oliver, 205
social anthropology, 119–20
social control, through biological control, 168
social Darwinism, 90, 101
social reformers: anthropologists as, 125–26; eugenic arguments of, 13, 99, 100
social status: eugenics and, 13, 101; hereditary diseases and, 56–58; race and, 13, 69
Société Royale de Médecine, Paris, 52, 56, 57
sociobiology, 114
somatostatin, 195

Sommer, Robert, 123
soul, 22–23, 52
Soviet Union, 159
species: in ancient logic, 1; Buffon on, 34–35, 60–61, 75; genetic engineering across boundaries of, 206; human races seen as, 125; Linnaeus on, 32–34, 60, 75; naturalists' view of, 66; new view of, around 1800, 34–35, 129; nineteenth-century questions about, 75; practical breeders and, 66; Ray's 1683 definition of, 18. *See also* varieties
species formation: acquired characters and, 61–62; de Vries on mutation and, 148; by hybridization, 63, 130
specificity, biological, 163, 184
Spencer, Herbert, 41, 90
spontaneous generation, 18, 26, 34
Stadler, Lewis, 149
Stalin, 159
Stanley, Wendell M., 174, 175
statistical correlations, 108, 109–10, 112, 113, 114
statistical methods, 101, 107–15; Boas's use of, 110, 112, 124; Galton's interest in, 89, 107–10, *109*; Lorenz's reservations about, 122; mathematical development of, 107, 109–10; Mendel's use of, 131, 133, 138; multivariate, 109; Pearson's development of, 108–9, *111*, 112, 113–14, 115; social goals of researchers using, 126
Staudinger, Hermann, 166
Stefánsson, Kári, 203
Stein, Emmy, 151
stem cell research, 196, 216
Stent, Gunther, 177
sterilization laws, 9, 99
Sterne, Laurence, 20
Stewart, Timothy, 196
*stirp*, Galton's theory of, 6–7, 9–12, 79, 90, 109
*stirps*, in genealogy, 46
Stocking, George, 103

Strasburger, Eduard, 81, 82
*Streptococcus pneumoniae* (pneumococcus), 174, 178–79
Strohmeyer, Wilhelm, 123
Stubbe, Hans, 162
Sturtevant, Alfred, 146, 152
Sulston, John, 200
Sunder Rajan, Kaushik, 193
Sutton, Walter, 151
SV40 virus, 189
Svalöf experimental station, 86, 135, 158
Svedberg, Theodor, 164, 166
Swanson, Robert, 195
sweet pea (*Lathyrus odoratus*), 108
synchronic dimension of heredity, 21, 33, 39, 94, 116. *See also* horizontalization of life
synthetic biology, 190, 194
systems biology, 213, 217
Szybalski, Waclaw, 189

target theory (*Treffertheorie*), 149, 150
Tatum, Edward, 175–76, 179–80
Tay-Sachs disease, 209
technologies, for molecular genetics research, 165–69, 180–81, 190–94, 201, 212–13
Temin, Howard, 188
Terman, Lewis M., 107
textual metaphors, 193–94
The Institute for Genomic Research (TIGR), 199
Thierry, Augustin, 102–3
Thompson, Warren S., 96
Timofeeff-Ressovsky, Nikolai, 149, 150, 158
Tiselius, Arne, 166
tobacco mosaic virus (TMV), 167, 174, 175, 178, 180
Tocqueville, Alexis de, 50
traducianism, 23, 51–52
*Tragopogon pratensis*, 63
transcription factors, 214
transfer RNA, 181, 190
transgenic animals, 196, 205–6

transgenic microorganisms, 204
transgenic plants, 196, 204–5
transmission, unit of, 156
transmission genetics, 139, 146, 148, 154, 155, 159; reductionism of, 211. *See also* classical genetics
Trembley, Abraham, 35, 36
*Tristram Shandy* (Sterne), 20
*truncus*, 46
Tschermak-Seysenegg, Erich von, 137, 141
twin research, 114
Twort, Frederick, 176
Tylor, Edward, 119–20, 125–26

ultracentrifugation, 164, 166–67, 169, 172, 175, 180
UNESCO Statement on Race of 1951, 112
Unger, Franz, 130
urbanization, 8, 96, 97

variation: Boas's studies of race and, 113; Darwin on, 74–75; de Vries's experimental approach and, 86; eighteenth-century explorations of, 61–64; Haeckel on, 92; hybridization and, 63; inheritance of acquired characters and, 61–63; Kant's anthropology and, 58–59; Mendel's findings and, 138; nineteenth-century questions about, 75–76. *See also* hybridization; mutation
varieties: breeders and, 66, 75; constant, 4, 130; inheritance of acquired characters and, 62; Kant on races as, 58; Linnaeus's definition of, 33; Maupertuis on, 73; naturalists' observations of, 60, 61, 64
Venter, Craig, 199, 200
*Vererbung*, 41–42, 44; Kant's concept of, 42, 58, 71
Vettel, Eric, 187
Vilmorin, Louis, 86
Vilmorin family business, 134, 135, 136

Virchow, Rudolf, 81, 82, 86, 91, 110
viruses: genome sequences of, 200; as model organisms, 174–75, 180; Rous chicken sarcoma virus, 169; SV40, 189. *See also* bacteriophages; tobacco mosaic virus (TMV)
vital forces: banished by molecular biology, 184; Kant's aversion to, 73; Maupertuis's belief in, 73; superfluous to Galton's theory, 80; Virchow on monarchic principle of, 91; in Wolff's epigenetic position, 36. *See also* forces, hereditary

Wailoo, Keith, 207, 209
Waldeyer, Wilhelm, 81
Warming, Eugen, 136
Waters, Ken, 217
Watson, James, 162, 181, 183, 193, 199, 200
weaning, 16
Weaver, Warren, 168
Webber, Herbert J., 135
Wedgewood and Darwin families, 50–51
Weinberg, Wilhelm, 123–24, 155
Weindling, Paul, 97
Weismann, August, 81, 82, 86–89, 87, 88, 91; Correns's avoidance of speculations of, 141, 142, 143; on environmental influences, 89, 90; Johannsen's criticism of speculations of, 140; Kroeber's use of theory of, 113; legacy of, in molecular genetics, 183; social uses of theory of, 100
Wettstein, Fritz von, 145
Wiener, Norbert, 185
Wieschaus, Eric, 214
Wilmut, Ian, 215
Wilson, Edmund Beecher, 85
Wilson, James, 208
Windaus, Adolf, 163
Winnacker, Ernst Ludwig, 211
Wittgenstein, Ludwig, v

wolf children, 68, 69
Wolff, Caspar Friedrich, 31, 36
women's rights movement, eugenic
    arguments in, 99
Wright, Sewall, 155
Wundt, Wilhelm, 107

xenia, 140, 141
*Xenopus laevis*, 215
X-ray–induced mutants, 149; in
    *Neurospora crassa*, 176
X-ray structure analysis, 167; of DNA,
    164, 167, 181, 182; of fibers, 164,
    167; of tobacco mosaic virus, 175

yeast: DNA chip with genes of, 212;
    sequencing of genome of, 199, 200
yeast artificial chromosomes (YACs),
    193
Yule, Udny, 114–15

Zacchia, Paolo, 51
Zamecnik, Paul, 164, 170
*Zea mays. See* maize (corn) genetics
zebrafish (*Danio rerio*), 214
Zimmer, Carl, 149, 150
Zinder, Norton, 180
zoological gardens, 59
Zworykin, Vladimir, 164